Results and Problems in Cell Differentiation

Series Editors:
W. Hennig, L. Nover, U. Scheer

25

Springer-Verlag
Berlin Heidelberg GmbH

Rolf Ohlsson (Ed.)

Genomic Imprinting

An Interdisciplinary Approach

With 46 Figures

 Springer

Dr. R. Ohlsson
Dept. of Animal Development
and Genetics
University of Uppsala
Norbyvägen 18 A
75236 Uppsala
Sweden

ISSN 0080-1844
ISBN 978-3-662-21956-0

Library of Congress Cataloging-in-Publication Data
Genomic imprinting: an interdisciplinary approach / Rolf Ohlsson (ed.). p. cm. – (Results and problems in cell differentiation: 25) Includes bibliographical references and index.
ISBN 978-3-662-21956-0   ISBN 978-3-540-69111-2 (eBook)
DOI 10.1007/978-3-540-69111-2
1. Genomic imprinting.
I. Ohlsson, R. (Rolf)   II. Series. QH607.R4 vol. 25 [QH450] 571.8'35 s–dc21 [572.8'65]

Library of Congress Catologing-in-Publication Data

© Springer-Verlag Berlin Heidelberg 1999
Originally published by Springer-Verlag Berlin Heidelberg New York in 1999
Softcover reprint of the hardcover 1st edition 1999

Production: PRO EDIT GmbH, D-69126 Heidelberg
Cover Concept: Meta Design, Berlin
Cover Production: design & Production, D-69121 Heidelberg
Typesetting: Best-set Typesetter Ltd., Hong Kong
SPIN 10647723   39/3137 – 5 4 3 2 1 0 – Printed on acid-free paper

Preface

The form and function of every living organism on the earth depends on the complex regulation of gene expression. This is carried out by controlling and interdigitating spatial and temporal patterns of gene activity during the lifetime of eukaryotic organisms. This is most dramatically apparent during early stages of development, when new types of cells and organs are being formed, often during very short time spans. To achieve this, it is vital that developmentally important genes can be kept in inactive or active states which are stably inherited in the soma. Indeed, it is now wellknown that the propensity for a gene to be transcribed or silenced is stably propagated through many cell generations, even from parent to progeny. This phenomenon constitutes a type of extragenetic or epigenetic memory of cell identity and developmental potential which has been fundamental to the evolution of complex lifeforms, such as the reader of this book.

This monograph focuses on a particular aspect of the epigenetic control of gene function: genomic imprinting. This defines a phenomenon where some genes or whole chromosomes can be silenced, activated, or even deleted depending on their parental origin. The impact of genomic imprinting is most clearly seen in the areas of cancer, clinical genetics, and development. Many of the processes associated with genomic imprinting can be observed in plants, yeast and man, for example, and may constitute, therefore, principles which are very conserved on an evolutionary scale. This perception underlies the organization of the section *Players of Imprinting*, which is aimed at providing an in-depth service to the reader who wants an introduction to or to learn more of principles that may be fundamental to epigenetic control of gene function in general and genomic imprinting in particular.

The fact that the phenomenon of genomic imprinting does not fully obey the laws of Mendelian inheritance has delayed its acceptance in the scientific community. From its obscurity during the 1980s, genomic imprinting has in more recent times received increased attention from scientists from a wide variety of disciplines. I would like to dedicate this book, therefore, to the pioneers of genomic imprinting who, by displaying scientific excellence and rigourous thinking, managed to break the tradition of old dogmas and establish a new type of scientific discipline. I am happy to say that some of these pioneers have also contributed to this book. Hopefully, the buyer of this book will enjoy reading these and other contributions, as much as I have.

Finally, I would like to express my gratitude to all of the colleagues, in particular Dr. David Haig, who have made this book possible.

Uppsala January 1999   *Rolf Ohlsson*

Contents

Imprinting and Paternal Genome Elimination in Insects
Glenn Herrick and Jon Seger

Imprinting and X-Chromosome Inactivation
Mary F. Lyon

The Mechanisms of Genomic Imprinting
Bernhard Horsthemke, Azim Surani, Tharapel James, and Rolf Ohlsson

Human Diseases and Genomic Imprinting
Judith G. Hall

Genomic Imprinting and Cancer
Benjamin Tycko

Players of Imprinting

Mechanisms of Transcriptional Regulation
Gary Franklin

Epigenetic Control of Gene Expression
Aharon Razin and Ruth Shemer

Polycomb Silencing and the Maintenance of Stable Chromatin States
Vincenzo Pirrotta

Domains and Boundaries in Chromosomes
Tatiana I. Gerasimova and Victor G. Corces

A Role for Modifier Genes in Genome Imprinting
C. Cristofre Martin and Carmen Sapienza

Allelic *Trans*-Sensing and Imprinting
Andràs Pàldi and Yann Jouvenot

Nuclear Architecture
Wallace F. Marshall and John W. Sedat

Appendix: Imprinted Genes and Regions in Mouse and Human
Colin V. Beechey

Kinship and Genomic Imprinting

Robert Trivers[1] and Austin Burt[2]

1
Introduction

Genomic imprinting refers to parent-specific gene expression, that is, to a difference in gene expression depending on which parent contributed the gene. In the usual case, an allele is silent when inherited from one parent while the identical stretch of DNA would be active were it inherited from the other, but sometimes this effect is seen in some tissues and not others, or there is only a quantitative difference in gene expression, depending on parent of origin (see Horsthemke et al., Chap. 5, this vol.).

This ability to be expressed according to parent-of-origin has striking implications for a gene's degree of relatedness to its parents – and, thus, to all other relatives differentially related through them, where degree of relatedness to another is the chance for any given gene in one individual that there is an identical copy in another individual by direct descent from a common ancestor. Consider an autosomal gene and its relatedness to its half-sibling related only through the mother. For the conventional nonimprinted gene, we argue as follows. There is a 1/2 chance the gene came from mother and, if so, a 1/2 chance it was passed on to the sibling; $r = (1/2) \times (1/2) = 1/4$ (Fig. 1). But imprinted genes carry knowledge of where they come from. A maternally active gene is definitely found in mother and there is a 1/2 chance she passed it on to any sibling, so r for the maternally active gene is 1/2. A paternally active gene came from father not mother so there is a 0 chance she passed that gene on to a half-sibling, hence r for a paternally active gene is 0 (Fig. 1).

For nonimprinted genes, conflict is not expected between maternal and paternal autosomes over appropriate action toward others. Being equally ignorant of where they come from, so to speak, they compute the same probabilistic or average degree of relatedness (in this case, 1/4). But with knowledge come divergent r's of 0 and of 1/2. This creates two kinds of conflict. One occurs over evolutionary time in which the spread of selfish paternally active

[1] Anthropology, Rutgers, New Brunswick, New Jersey 08903-0270, USA
[2] Dept. of Biology, Imperial College, Silwood Park, Ascot, Berks. SL5 7PY, UK

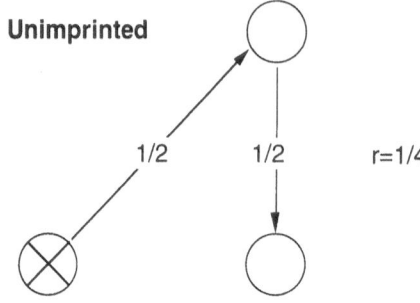

Unimprinted

1/2   1/2   r=1/4

Fig. 1. Degrees of relatedness of unimprinted and imprinted genes found in one individual (*circle with cross*) to its half-sibling, related through a common mother. Genes are assumed to be autosomal. Imprinted genes are either maternally active (*ma*) or paternally active (*pa*)

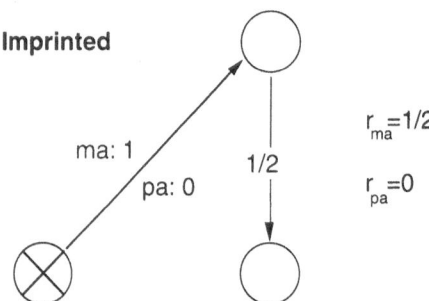

Imprinted

ma: 1
pa: 0   1/2

$r_{ma}=1/2$

$r_{pa}=0$

genes naturally selects for opposing maternally active genes (and vice versa) while both select (albeit less strongly) for opposing nonimprinted genes. The second kind of conflict is imagined to occur within an individual whenever opposing genes interact against each other.

This general approach to the meaning of genomic imprinting may be called the kinship theory of imprinting and is due primarily to the work of David Haig, who first introduced the approach (Haig and Westoby 1989; Haig and Graham 1991; Moore and Haig 1991) and who has been most active in developing it further (Haig 1992a,b, 1994, 1996a,b, 1997a,b). Other theories have been advanced to explain the adaptive significance of imprinting (rev. in Hurst 1997), but those to date seem to lack both supporting logic and evidence (Haig and Trivers 1995), while Haig's theory explains a broad pattern of evidence, with no facts clearly in contradiction. One need assume no more than Mendelian genetics and the action of natural selection. Once parent-specific gene expression appears, it must by logic evolve the biases Haig describes, paternally active genes acting to further paternal relatives, i.e., patrilines, and maternally active genes, matrilines. If there are additional selection pressures we need to take into account to understand imprinting, we do not yet know what they are.

Here we review the kinship theory of imprinting and in the process attempt to explain:

 - the taxonomic distribution of imprinting,
 - the early effects of imprinted genes,
 - the expected behavioral effects of imprinted genes,
 - the genetic characteristics of imprinted genes, and
 - the bodily distribution of paternally active vs. maternally active tissue.

2
Imprinting and Conflict over Parental Investment

The importance of measuring exact degrees of relatedness to each parent was first appreciated in the context of mother-offspring and sibling relations, especially in mice. To the degree that a female's lifetime reproduction is not entirely fathered by a single male, an individual's paternal genes will be selected to value mother and maternal siblings less than will its maternal genes. (In fact, even when a female's offspring are all fathered by the same male, there may still be selection for selfish paternal alleles insofar as the father can quickly replace the mother with another female.) In the extreme case, where all siblings, present and future, are fathered by different males, paternal genes will value neither mother nor maternal half-siblings, while maternal genes will value siblings by 1/2 and mother's future reproductive success similarly. In the more usual case, at least some of one's siblings will be maternal half-siblings, so paternally active genes are expected to be associated with greater demands on maternal investment, at a presumed cost to both mother and present or future siblings. Maternally active genes are expected to have opposite effects. (Selection on successive offspring of a single female differs somewhat from selection acting on members of a litter simultaneously competing over maternal investment, but this is a distinction we shall not pursue in this chapter; see single versus multiple straw models in Haig 1992a.)

A series of facts regarding imprinted genes in mice, humans, and two species of plants are consistent with these simple expectations.

1. *Taxonomic Distribution.* The kinship theory correctly predicts that imprinting will occur in mammals and flowering plants (see Messing and Grossniklaus Chap. 2, this Vol.), both of which show extensive post-zygotic investment by the mother or by maternal tissues, and its absence from *Drosophila* and *Caenorhabditis*, which lack any form of postzygotic investment (or known kin interactions). More detailed predictions await further evidence (see below).

2. *Individual Genes.* Among imprinted genes that have early growth effects (i.e., while the offspring is still deriving resources from the mother), we expect paternally active genes to increase growth and maternally active genes to decrease growth. This prediction can be tested by examining knockout mutants, in which both copies of a gene have been inactivated: we expect that knocking-out paternally active genes will reduce growth, and knocking-out maternally

Table 1. Gene knockouts (KO) and uniparental disomies (UPD) in mice (M) and humans (H) categorized according to effect on growth. Growth enhancement is predicted to be more common in maternal KOs and paternal UPDs than in paternal KOs and maternal UPDs. Entries refer to individual genes (KOs), chromosomal regions (mouse UPDs), or whole chromosomes (human UPDs). Many other knockouts and uniparental disomies have no obvious growth effect. Compiled from Beechey (Appendix, this Vol.), Hurst & McVean (1997), Itier et al. (1998), and references therein; several ambiguous cases have been left out (e.g. mouse paternal UPD 2 Distal and 7 Prox, human maternal UPD 15)

	Growth enhanced	Growth retarded
Maternal KOs and Paternal UPDs	KO: *Igf2r, H19* UPD(M): 11 Prox UPD(H): 11	KO: *Mash2, p57KIP2*(?) UPD(M): 7Central, 17Distal, 18(?) UPD(H): 6, 14
Paternal KOs and Maternal UPDs	KO: – UPD(M): – UPD(H): –	KO: *Igf2, Peg3, Grf1* UPD(M): 7Distal, 11Prox, 18(?) UPD(H): 7

active genes will increase growth. A complicating factor is that any genetic mutation or abnormality, including a knock-out, is more likely to decrease growth than increase growth. Consistent with these expectations, knockouts of maternally active genes sometimes enhance growth (2 of 4 cases), while knockouts of paternally active genes do not (0 of 3 cases) (Table 1). Unfortunately, sample sizes are too small to justify statistical tests of significance.

3. Uniparental Disomies. A uniparental disomy is a pair of homologous chromosomes or chromosome arms in an otherwise normal genotype which come from a single parent and hence are imprinted in the same direction (see Hall, Chapter 6, this Vol.). When uniparental disomies have growth effects, paternal ones are sometimes growth-enhancing (2 out of 7 cases) while maternal ones never are (0 out of 4) (Table 1). While this difference is not statistically significant, it is consistent with the kinship theory of imprinting, especially if we assume that uniparental disomies usually disrupt development and hence retard growth.

4. Chimeras. Mice chimeras are useful in studying imprinting when these chimeras consist of a mixture of wild-type cells and either androgenetic cells (both sets of chromosomes from a male, hence paternally active) or parthenogenetic/gynogenetic (both sets from a female, hence maternally active). Mice chimeras consisting of a mixture of androgenetic cells (paternal) and wild type are, as expected, usually larger than individuals with wild-type cells alone, which in turn are larger than chimeras consisting of parthenogenetic cells (maternal) mixed with wild-type (Fundele et al. 1997). These differences tend to disappear after weaning, again as expected since there is little scope for internal genetic conflict over growth when the resources to support growth are supplied from outside the family.

5. Data from Corn and Arabidopsis. The evidence from plants is broadly consistent with Haig's theory (Messing and Grossniklaus, Chap. 2, this Vol.).

A maternally active gene in *Arabidopsis* strongly retards embryo growth (Grossniklaus et al. 1998), while a paternal uniparental disomy in maize increases kernel size and the corresponding maternal disomy has the opposite effect (Lin 1982). Some imprinted genes, such as those affecting color of the kernel, are difficult to interpret, while others, such as those affecting zein accumulation presumably affect a protein resource of value to the embryo and mother alike (maternally active genes decrease zein in the endosperm).

3
Other Situations Where Genomic Imprinting Might Be Expected

3.1
General Introduction to Kinship Theory; Warning Calls and Sociality

Of course, kinship interactions extend beyond the parent-offspring relationship in many species. In these, imprinting is expected whenever effects on patrilines differ from those on matrilines. For example, in many mammals females live their lives in close proximity to relatives and have many important interactions with them. These may include warning others of danger, shared use of space (and defense of it from conspecifics), social grooming, shared aspects of reproduction, cooperative food-gathering (in some species, hunting) etc. Insofar as these actions affect matrilines and patrilines differently – as often they must – there is potential for imprinting. The chief difference between these effects and conventional parent-offspring effects is that conventional effects are expected to act early in development and may have profound ramifying consequences for juvenile growth and survival. By contrast, later kinship interactions may select for effects which have little to do with size but a lot to do with behavior, brain physiology, interactions with others and internal psychological conflict. In our own species, the latter may interact in a complex and important way with systems of deceit and self-deception, which also promote mental fragmentation (see below).

Consider alarm calls: in many ground squirrels adult females warn other closely related females of the approach of predators. In Belding's these females are more related to each other on their maternal side than on their paternal, and they adjust frequency of calling to the number of closely related females living nearby (Sherman 1977, 1980). For nonimprinted genes, alarm-calling is a unitary event with a single goal: warning genetically related females of danger. For imprinted genes there may be psychological bifurcation and conflict. One can easily imagine maternally active genes taking the lead in generating a sentry-like mentality, with the approach of a predator eliciting a loud cry warning that "a predator is coming, a predator is coming!" while a fearful, inner paternal self tells you to keep silent, to run and to hide. Degree of sociality between neighbors also correlates positively with kinship and Holmes and Sherman (1982) have made the striking discovery that 1-year-old females in nature prefer their litter-mate full sisters over their litter-mate half sisters. Nothing is known of imprinting in this group, but paternally active genes would gain the most from discriminating correctly between siblings based on their paternity.

3.2
The Taxonomic Distribution of Imprinting

Viviparous Vertebrates. In mammals and flowering plants offspring are in intimate contact with their mothers (or maternal tissue) during critical stages of early development. They are in a position to actively extract additional resources. The various mechanisms that the human fetus appears to employ to gain additional resources from mother has been beautifully described by Haig (1993). All of these mechanisms are subject to selection for imprinting. It is at present unknown whether imprinting has evolved outside of mammals and flowering plants but it is certainly expected in viviparous species, such as some lizards, snakes, frogs, fish and arthropods, all of which have evolved intimate parent-offspring association during early development. In pipefish and seahorses, intimate connections are established between the offspring and its father – and broods may consist of offspring contributed by more than one mother, so that maternally active genes would be expected to play the same selfish role normally played by paternally active genes in species such as ourselves.

Haplodipoid Social Insects. Social insects provide abundant opportunities throughout the life cycle for imprinting to evolve, but haplodiploid species provide some special opportunities (Haig 1992b; see also Herrick and Seger, Chap. 3, this Vol.). In this regard, they are exactly like X chromosomes (see below). For example, under complete monogamy (single insemination of the queen) a worker ant's paternal genes are identically related to those of other females but unrelated to males. As far as ratio of investment is concerned, paternal genes prefer all-female, while maternal prefer 1:1 (Haig 1992b). In hostile interactions between two sister workers (or incipient queens) paternal genes would be expected to value the other female as much as self while maternal genes would see the other as related by only 1/2. Since the mother is equally related to her daughters, a selection pressure exists for her to silence her own genes affecting interactions among her daughters the better to express her mate's since the latter will express her own interests (D. Haig, pers. comm.). (Of course, any tendency toward multiple mating by females reduces these effects.) Note the potential for conflict over whether a gene affecting transfer of investment ought to proceed. A mother might wish to shut down such genes whenever they cause transfer between sisters in which $B < C$ (where

B and C are benefit and cost to the two sisters) while a maternal gene in each daughter would wish to be active as long as B < 2C. (Conflict between imprintor and imprinted genes is taken up in greater detail for diploid species below.) Suggestive evidence for imprinting in a haplodiploid species has recently been uncovered for the hymenopteran parasitoid *Nasonia vitripennis* (Dobson & Tanouye 1998).

Diploid on Haploids. Postzygotic investment and the potential for maternal-paternal conflicts of interest also occur in those species where a diploid phase grows on a haploid. For example, in *Neurospora* and other hyphal ascomycetes, a single fertilization event produces a dikaryon and eventually leads to the production of many haploid spores, and paternally derived genes will be selected to increase the number of spores, more or less regardless of the consequences for the supporting maternal mycelium. Will the ascomycetes be the next group in which genomic imprinting is discovered? They do have methylation (e.g., Foss et al. 1993), and, interestingly, apomixis derived from sex is rare among fungi (Carlile 1987).

Similar diploid-on-haploid life cycles occur in bryophytes (e.g., mosses) and red algae. The traditional explanation for this postfertilization amplification of the diploid phase is that it represents a way for the mother to increase the number of progeny from a single rare fertilization event. Alternatively, perhaps it is a way for the offspring to parasitize the mother, and if so we might expect paternally imprinted genes to have taken the lead in this. Indeed, much of the evolution of these diploid tissues/individuals may have been pushed along by paternally active genes attempting to better exploit the haploid mother.

3.3
Species with Biparental Care (e.g., Birds and Humans)

The expected case for birds is by no means straightforward. Substantial early investment occurs in the form of an egg prior to the arrival of the paternal genome, so imprinting is unlikely to affect very early development, as it does in mammals. On the other hand, post-natal parental care is nearly universal. But it is also usually provided by both parents. Insofar as this leads to monogamy – with the frequent production of full siblings – there is selection away from imprinting, but even "monogamous" bird species sometimes display surprising levels of extrapair paternity (e.g., 30–50%; reviewed in Birkhead and Møller 1992; Westneat and Sherman 1997; see also Dixon et al. 1994 and Stuchbury et al. 1994) and this will select for the usual bias toward paternal genes exploiting maternal investment. Across species of birds, levels of offspring begging increase as a function of increasing percentage of extrapair parentage (Briskie et al. 1994), but this is expected for selection acting on unimprinted genes as well as paternally active ones.

Even without extrapair parentage in monogamous species it seems unlikely that male and female interests in reproduction are ever identical. One parent may outsurvive the other and breed with someone new and become effectively polygamous, measured over a lifetime. This permits selection to act on imprinted genes to extract investment from the opposite-sex parent, i.e., paternally active genes exploiting maternal investment, and maternally active genes exploiting paternal investment.

These possibilities are especially important in our own species, where a long period of parental investment is often shared by the two sexes (and usually unequally). In turn, degrees of relatedness among siblings will only approach 1/2, while availability of other relatives may often show a bias by sex of parent, e.g., patrilocality leading to greater paternal than maternal relatedness to neighbors.

The long period of parental investment also selects for an interaction between parental indoctrination and genomic imprinting. A parent may be selected to mold its offspring's behavior so at to be appropriately biased toward that parent's relatives and such parental indoctrination is expected to fall on more fertile genomic ground (= full genetic cooperation) when the offspring genes involved in the interaction are imprinted in the same direction. Thus, father's molding offspring behavior to be paternally biased will succeed best when working on paternally active genes in the progeny.

3.4
Inbreeding and Dispersal

When an individual inbreeds, he or she will be related to the partner through the mother or the father or both. When related through mother and father (e.g., toward a full sib) maternal and paternal genes will usually calculate the same degree of relatedness to the partner (in this case, 1/2). In the more usual cases, r will only be measured through one parent (e.g., half-sib or cousin). Consider a female about to engage in an extrapair copulation with her maternal half-brother (that is, they will not raise the child together). Maternal genes in her see the half-sib as related by 1/2, increasing her relatedness to her own offspring by 1/4. Against this must be set the costs of inbreeding, e.g., inbreeding depression. If the gain in relatedness outweighs the costs of inbreeding, maternally active genes will be selected to promote inbreeding. By contrast, paternally active genes are expected to look askance at this kind of inbreeding. They enjoy no gain in relatedness but surely suffer the costs of inbreeding, so paternally active genes will be selected which discourage inbreeding on the maternal side. One can easily imagine maternally active genes in females which delight in other maternally active genes (and think kissing cousins are cute), while paternally active genes look on in a sullen and moralistic way, emphasizing the biological defects so generated! Exactly the opposite attitudes to inbreeding are expected in the male about to mate with his maternal half-sister:

his maternal genes will worry about saddling the half-sister with low fitness offspring, while his paternal genes will take a Devil-may-care attitude and urge him on – Go for it!

Dispersal is also a social act whose evolution is influenced by kinship, insofar as it reduces the probability of mating and competing with relatives. If we imagine an ancestral state with no (sex bias in) dispersal, the average relatedness of an individual to the entire social group (as opposed to just littermates) will usually be greater through paternal genes than through maternal genes, due to higher variance in male reproductive success. This will be particularly true if we consider to individuals of similar age, who are the more likely mates and competitors (for genetic evidence, see Altmann et al. 1996). Thus, dispersal away from the social group will benefit paternal genes more than maternal ones, and we might expect paternally imprinted genes to have taken the lead in its evolution. In mammals, males usually disperse more than females (though humans are a notable exception), which will tend to decrease relatedness through paternal genes (except when there is only one dominant male or when related males disperse together), and which effect is stronger will depend on the exact details of the social system. This is a situation in which we might expect even the *direction* of imprinting to vary from species to species (for a possible example of such a gene, see Williamson et al. 1996).

3.5
Imprinting and the Sex Chromosomes

It has long been appreciated that degrees of relatedness measured across the sex chromosomes may differ from those measured across autosomes. For example, a Y is related to the Y in its son or paternal half-sib by 1 and to a daughter or sister by 0. The chief difference between the sex chromosomes and the autosomes is that selection for imprinting occurs more frequently, and with greater force, on the sex chromosomes.

Consider a paternally active gene on an X chromosome: under extreme polyandry (i.e., each of a female's offspring fathered by a different male) it will be related to maternal siblings by 0, just as for autosomes. Under the opposite condition – complete monogamy – selection for paternally active genes on autosomes dies away, but a new opportunity emerges for X-linked genes. Each paternal X is fully related to every sister and not at all related to any brother. In short, among a current sibship, a paternally active gene which consumed brothers to benefit self or sisters would spread. A maternally active X gene would value brothers and sisters equally and half as much as self (thus more nearly approximating mother's best wishes). The same argument applies to the Y chromosome: under polyandry it values neither kind of sibling, but under monogamy it values self and brothers to the exclusion of sisters. (The one difference is that the Y can be regarded as permanently imprinted since it need never reverse imprints between generations – Hurst 1994.)

The kind of effect that might be produced is suggested by evidence from gerbils and mice that sex of one's intrauterine neighbor has a strong effect on development (Clark and Galef 1995). A female developing between two males is masculinized to her own disadvantage, development is slowed and – compared to a female raised between two females – lifetime reproductive success (in the lab) cut by 1/2 (Clark et al. 1986; Clark and Galef 1988). The same pattern is found for males: benefits from male neighbors, costs from female (Clark et al. 1992; Clark and Galef 1995). Is this an example of imprinted sex chromosomes favoring their own kind *in utero*? Paternal Xs are expected to try to create a female-benefiting uterine environment and Ys a male-benefiting one, regardless of cost to the opposite sex. Unfortunately, little is known of the degree of polyandry in nature for either gerbils or mice, or whether degree of monogamy across species correlates with degree of intrauterine effects. The few observations from gerbils suggest that a female in the wild easily mates with more than one male per season (Ågren et al. 1989).

Note that a mammal species switching from polyandry to monogamy would simultaneously reduce selection for imprinting on autosomes while selecting for sex-specific benefits from the paternal X and Y. An interaction between autosomes and the paternal X might be expected in the monogamous species. Paternal Xs could activate paternal autosomal genes otherwise newly silenced or might oppose movement to biallelic expression of maternal genes. A Y in males might have similar effects. There is evidence from a monogamous *Peromyscus polionotus* recently derived from the polyandrous *P. maniculatus*, that offspring size strongly depends on which species contributes the paternal genes. When the polyandrous species contributes them, offspring are born giants; when the monogamous species contributes them, offspring are born midgets (Dawson 1965). This could result from unimprinted genes or from imprinted ones; the latter possibility is being actively investigated by Shirley Tilghman and her colleagues (as recounted in Mestel 1998).

Skuse et al. (1997) have made a striking discovery concerning the human paternal X chromosome. It appears to have an imprinted gene or genes which affect positively degree of sociability and social intelligence. The evidence comes from women with only one X chromosome (Turner's syndrome): their single X may be maternal (less sociable and socially aware) or paternal (more so). By inference, these paternal X genes improve a female's ability to detect the feelings and emotions of others, to interpret their body language, to understand her effects on them, to reason with them when upset, and so on.

Men, of course, lack a paternal X chromosome, and this may be the function of linking the sociability gene to paternal X expression, i.e., to permit the trait to be sex-limited. It would be interesting to model the spread of such a gene in competition with a sex-limiting gene which, in effect, counts X chromosome number, as in *Drosophila*, or "responds" to absence of the Y, as in mammals.

Such a paternally active sociality gene may also spread because of the high relatedness ($r = 1$) on the paternal X between females with the same father. Species of mammals differ considerably in the degree to which a single male may breed with most of the females in a given area or group, within a breeding season or between seasons. In some areas of India, for example, a single male langur monkey may breed all or almost all of a group of as many as 15 adult females for an average of about 5 years (Hrdy 1977). This may generate an enormous 5-year cohort of females in the group who are clonally related for their paternal X chromosome. Of course, in humans local variation in male reproductive success will be far smaller, only polygamy creating extended groups of paternal sisters outside the nuclear family.

None of this is possible in marsupials, in which the paternal X is inactivated in (or, in some species, eliminated from) the soma of all females, so that it would be interesting if an eventual knowledge of the genetic structure of placental and marsupial X chromosomes reveals a systematic difference for genes affecting female sociality. Preferential inactivation of the paternal X in marsupials – and the trophoblast of mice – has been interpreted as a device to shut down selfishly imprinted genes on the paternal X (Moore and Haig 1991) and we have seen above that these are more likely to occur there than on comparable autosomes. However, the evidence strongly suggests that a paternally active gene on the X causes the rest of the X to become inactive (Lee and Jaenisch 1997) so that some additional argument is needed to solve this puzzle.

In placental animals there is random inactivation in the soma, but clones of cells with the same active X can be seen in morphology (e.g., coat color patterns in cats) and Tan et al. (1995) have now shown that in mice this is also true of the brain: alternating radials bands of cells in the cortex express either predominantly the paternal or the maternal X chromosome. Are these bands ever in conflict, especially in monogamous species – and, if so, what kinds of mental states are thereby generated?

4
Evolution of the Imprinting Apparatus

4.1
The Evolution of the On-Off System

Haig (1997a) has recently shown that for genes whose effects are additive (no dominance), alleles which are appropriately imprinted are adopting an evolutionarily stable strategy (ESS) in competition with nonimprinted alleles whenever inclusive fitness payoffs are positive for one parental allele and negative for the other. An interesting implication of his analysis is that small differences in gene expression due to parent of origin are unstable and are soon replaced by alleles with greater differences until expression becomes monoallelic (Mochizuki et al. 1996; Haig 1997a).

Consider, for example, an unimprinted gene whose expression increases fetal growth and thereby offspring fitness. Imagine that it has reached an ESS level of expression. Now introduce an allele which, when paternally inherited, has slightly greater expression, while maternal alleles produce the usual effect. Assume this change increases inclusive fitness so that the allele spreads. Now a new allele with the same paternal effect but whose maternal allele *reduces* expression so that total expression for a homozygote is identical to the unimprinted homozygote will itself spread. A new allele with yet higher paternal expression spreads, but is superseded by one with even lower maternal expression, and so on, until maternal expression has been reduced to zero, at which point selection will favor the level of paternal expression which maximizes paternal inclusive fitness. The same argument applies at a locus at which increased maternal expression increases inclusive fitness: paternal expression is soon reduced to zero. Haig calls this the loudest voice prevails principle. It provides a ready explanation for the fact that most imprinted genes are entirely inactive in one sex or at most biallelic in some tissues only. Of course, such all-or-nothing expression is much easier to detect than are small differences in gene expression, which nevertheless are known for some imprinted genes.

4.2
Rates of Evolution of Imprinted Genes

McVean and Hurst (1997) predicted that intragenomic conflict would cause imprinted genes to be relatively fast-evolving, since improvements on each side would seem continually to be favored. A test of the prediction is the rate of nonsynonymous DNA substitutions (those that change protein structure, Ka) compared to synonymous ones (those that do not, Ks). Where Ka/Ks is high, directional selection is indicated. Their own analysis of data from imprinted genes in mice and rats, compared to nonimprinted genes not involved in mother-fetal interactions, showed a non-significant tendency for Ka/Ks to be higher for imprinted genes ($p = 0.12$). Using a more sensitive parametric analysis of covariance, with regression lines forced through the origin (intercepts are not significantly different from each other nor zero), we compute $p = 0.04$ for the difference in regression coefficients of Ka on Ks for imprinted vs. nonimprinted genes. That is, nonsynonymous substitutions are relatively more frequent in imprinted genes, as expected. A Mann-Whitney test for differences in Ka/Ks ratios is likewise significant ($p = 0.04$). Curiously, the difference seems to be at least as much a reduced Ks as an increased Ka. However, this same analysis on an expanded dataset shows no significant difference (L. D. Hurst, pers. comm.) Clearly, the issue is not yet settled.

4.3
Imprinted Genes and Introns

Hurst et al. (1996) suggested that imprinted genes might have fewer and smaller introns than nonimprinted genes, on the assumption that this would increase rates of transcription, themselves favored because of intragenomic conflict. Alternatively, fewer and smaller introns might afford a smaller target for antagonistic genes to attack. Certainly, there is a highly significant effect in the predicted direction, so strong as to be impervious to minor decisions regarding which data to include and which to exclude (Hurst et al. 1996; Haig 1996c; McVean et al. 1996). But we would like to see evidence on the alleged benefits of few and small introns. For example, among unimprinted genes, are transcription rates correlated with intronic content?

4.4
Can Parents Facultatively Adjust the Degree of Imprinting in Their Progeny?

In the above discussion we have assumed that a gene is always imprinted in a given direction, e.g., paternally active, maternally silent, but in theory an advantage could be gained if the parent could – during gametogenesis – adjust the level of imprinting on its genes to factors that correlate with kinship interactions among progeny (Haig 1994, 1996a). For example, a male bird engaging in an extrapair copulation might gain in inclusive fitness if he passed on relatively more selfish-acting genes to progeny than when he copulated with his own mate. So far as we know, there is no direct evidence which bears on this for any organism. On the other hand, Haig (1994) has pointed out that the inverse association between length of parental sexual cohabitation and frequency of pregnancy-induced hypertension (Robillard et al. 1994, see also Trupin et al. 1996) is consistent with this possibility, since short associations might lead males to activate more selfish paternal alleles for their progeny, causing maternal hypertension (seen as a device to increase blood flow to the placenta). The bird example may tax a male's adaptive capacities, since the imprint probably needs to be established more than a few days before copulation (unless the male segregates two kinds of sperm) but the human adjustment could be straightforward and easy to achieve.

4.5
Can Genetic Memory Extend Back More Than One Generation?

Haig (1997b) has drawn attention to another interesting possibility. Imprinted genes may benefit by responding to relatedness effects of more than one generation. For example, in a female-philopatric species (e.g. many mammals) a female's maternal genes come from the immediate area while her paternal ones migrated in from elsewhere. Thus, in her daughters, a female's maternal genes will be more related to other females than will her paternal genes and would benefit from acting accordingly. In other words, in a female there are two classes of maternal genes, those which come from grandmother and those from grandfather. Insofar as they act to further grandparental interests appropriately, they will displace imprinted genes that act only to further maternal interests. It is easy to imagine how imprinting might become cumulative when genes continue to be passed down the same-sex lineage, since imprints need not be reversed, but there are also opposing selection pressures acting on unlinked genes that are either unimprinted or oppositely imprinted.

4.6
Conflict over Imprinting

There must exist some machinery that establishes imprints and which does so differently in each sex (see Horsthemke et al., Chap. 5, this Vol.). This implies the existence of genetic modifiers which associate a given imprint with a given sex. These can be linked to the imprinted locus or not and could act in *cis* (along the same chromosome), *trans* (between chromosomes) or both (see Paldi and Youvenot, pages 271ff this Vol.). In any case, an analysis of selection pressures suggests that there is scope for conflict between an imprinting gene and an imprinted one, insofar as the imprinted gene will always be found in the offspring in which it is expressed, while the imprinting gene will occur at random among progeny. Also, in the progeny, all three classes of genes, maternally active, paternally active, and nonimprinted, may disagree over the existence of an imprint at another locus (Burt and Trivers *in press*).

The difference of opinion within parents between imprintor genes and imprinted genes is easy to describe. An imprinted gene (e.g., maternally active) comes from one parent with certainty (the mother). Hence it is related to mother's other offspring by 1/2 and values itself twice as much as it values them (all effects on individuals corrected for reproductive value). An imprintor gene in mother can be imagined to imprint both chromosomes, causing all progeny to act the same. Such a gene is selected to maximize maternal interests, i.e., total surviving progeny, half of which carry at random a copy of the imprintor gene. There must exist circumstances under which the imprinted gene is selected to be imprinted [$B > (1/2)C$], but an imprintor gene is not selected to apply the imprint [$B < C$] – indeed, is selected to remove it should it appear. The target gene may be selected to mimic other imprinted genes, by acquiring their recognition sequences, which in turn selects for the imprinting apparatus to make ever finer discriminations. This kind of conflict might easily generate a tendency for imprinted genes to appear in clusters, as imprinted genes disfavored by imprintors would make use of mechanisms operating at other nearby loci where imprintors and imprinted genes are both being positively selected. Some imprinted genes in mice and humans are, indeed, found in clusters (see Appendix and Horsthemke et al., Chap. 5, this Vol.).

Alternative explanations for clustering posit a general difficulty in becoming imprinted (Haig 1997a) or an imprinting process which is costly (Mochizuki et al. 1996). It is noteworthy that molecular biologists have described complex interconnected causal pathways acting in both *cis* and *trans* within these imprinting clusters (e.g., Forne et al. 1997; Webber et al. 1998).

Conflict in the opposite direction is easily imagined as well. For example, consider a maternally active growth inhibitor, which benefits mother at a cost to offspring. Since the offspring with the gene always suffers the cost, it will at times prefer not to be active (i.e., imprinted) even though an imprintor gene – valuing all progeny equally – will be selected to apply the imprint.

Similar arguments apply for paternally active genes, so that in both male and female germlines one expects some imprints to be easily and stably acquired (both imprintor and imprintee benefit), some to be imposed and lost, and some perhaps only gradually acquired. Their expected effects in progeny may depend in part on selection acting then at other loci. Even when imprintor and imprinted agree on the imprint, nonimprinted genes in progeny (since they calculate only average coefficients of relatedness to others) may be selected to remove the imprint (e.g., through demethylation, Chaillet 1995) or to ignore it (e.g., by using an alternative promoter, Vu and Hoffman 1994). More severe disagreements may arise between the two kinds of imprinted genes (maternal and paternal). Whether wholesale demethylation occurring early in mouse fetal life (Razin and Shemer 1995) is partly an organism-wide effort to reduce the frequency of imprints, paternal and maternal, is completely unknown. Note that once evolution has established a stable equilibrium in the paternal and maternal tug of war, a given imprinted gene (e.g., *Igf2*) may only contribute positively to inclusive fitness when present with its oppositely imprinted antagonist (e.g., *Igf2r*) (Haig 1997a). This implies that the selective removal of individual imprints may be a dangerous game to play, with wholesale removal the only one to have a chance of being functional.

5
The Functional Interpretation of Tissue Effects Found in Chimeric Mice

Some of the most intriguing facts regarding imprinting are indirect. They come from the study of chimeric mice, that is, mice whose cells are a mixture of types, in this case, wild-type cells mixed with cells whose double set of genes came from a single parent, i.e., the father (androgenetic) or the mother (gynogenetic or parthenogenetic). These chimeras present less radical phenotypes than might be supposed. At all loci, there is some degree of normal gene expression (from the wild-type cells), while the strength of the uniparental effect is expected to reflect relative cell frequency in the chimeras. By contrast, in knockout mice or uniparental disomies, some uniparental effects are imposed on all cells.

Data from numerous chimeras show a general effect in which size of chimera is greater the lower the frequency of maternal cells and (separately) the greater the frequency of paternal ones (Fundele et al. 1997). More strikingly, differently derived cells proliferate and survive in some kinds of tissues and disappear from others. Although the creation of chimeras is time-consuming and sample sizes small, consistent patterns emerge (Table 2). Paternal cells do well in the skeleton (chondrocytes), the hypothalamus and the cell layer that will produce enamel of teeth, while maternal ones do well in the brain itself, especially the neocortex, and the cell layer that will produce the dentin of teeth. In very early development, paternal cells proliferate in extraembryonic tissues and maternal cells in the embryo proper. Paternal cells also show up in brown adipose fat.

The meaning of these facts is mostly obscure. If neocortex is especially oriented toward social interactions involving kin and associated conscious mentation (see below), a maternal dominance may make sense, while the hypothalamus, involved directly in appetites, might represent a more narrowly selfish orientation, hence paternal dominance (Trivers 1997). Psychologically it is easy to imagine one inner voice saying, "I like family, family is important", while the other counters, "I'm hungry!". Enamel is believed to be tissue expensive in rare minerals, while brown adipose fat is rich in energy for immediate consumption, hence both may show a selfish (paternal) bias.

It is interesting that adult male chimeric mice with a large complement of parthenogenetic cells in the brain are quick to aggress against other males (Allen et al. 1995). The interpretation of this finding is not obvious but we prefer the following. If dispersal is sufficiently male-biased that maternally active genes are selected to take the lead in causing that dispersal (see above), they should also cause the dispersing males to avoid each other. Quick aggression as maternal relatedness is detected would seem a natural device to space out maternally related males. The intention is not to injure the opponent, only

Table 2. Mouse tissues showing a bias in chimeras toward paternal (androgenetic) or maternal (parthenogenetic or gynogenetic) cell contribution

Paternal	Maternal
extraembryonic	embryo
hypothalamus	cortex
chondrocytes	dentin
enamel	
brown adipose fat	

Data from Fundele and Surani (1994), Fundele et al. (1994, 1995a,b), and Allen et al. (1995).

to drive him away quickly. Our argument rests on the assumption that males with parthenogenetic cells are detecting each other as close maternal relatives, perhaps because of their shared parthenogenetic cells (which also show up prominently in the olfactory mucosa and olfactory bulb).

6
Deceit and Selves-Deception

It is generally recognized that natural selection may favor intraspecific deception in a wide range of contexts, including sexual misrepresentation, false warning cries, parent-offspring relations and a host of others (Trivers 1985). It is less widely recognized that selection for deception and avoiding being detected may select for self-deception in the deceiver the better the hide the ongoing deception from others. If conscious knowledge of ongoing deception is accompanied by stress, for example, knowledge may be rendered unconscious so as not to reveal the stress. In short, mental fragmentation – with some internal conflict – may serve the individual by hiding useful information from others (reviewed in Trivers 1985).

Genomic imprinting opens up new vistas in self-deception, which might almost be called "selves deception". While not the only source of internal genetic conflict over actions toward relatives, genomic imprinting pits half of the nuclear genotype against the other half, easily selecting for biased information flow within the organism which may resemble classical self-deception. For example, there is growing evidence that different parts of the body may express maternal and paternal factors differently. In chimeras, as we have seen, parthenogenetic cells survive and proliferate in the neocortex, while androgenetic cells do well in the hypothalamus. If these differences reflect effects of imprinting, then the neocortex may be relatively maternal in imprinted gene expression and the hypothalamus paternal. The neocortex may be more maternal because it is extensively involved in interactions with various relatives, while the hypothalamus is more intimately connected with such egoistic impulses as hunger and growth. If the neocortex helps allocate resources to others by giving signals from the hypothalamus a given weight in its calculations, then selection for paternal effects in the hypothalamus may favor a heightened signal so as to compensate for maternal efforts toward down-regulation from the neocortex, and vice versa.

Haig (1996b) has modeled the analogous situation of mother-fetal conflict in which fetal hormones have replaced maternal hormones as control agents for various maternal states pertaining to blood flow, blood sugar levels, etc. In mother-fetal conflict the two are separate organisms related by 1/2 in which the fetus has preferential access to the mother's bloodstream, where it can augment her hormone levels for desired downstream effects. In conflict due to genomic imprinting, the two are unrelated genome halves of the same individual which may be differentially expressed in different parts of the body, favoring biased forms of communication between these parts. These could also involve hormones and be subject to the same escalation of hormone production by some tissue, countered by increasing inertness to the hormones in other tissue. This, of course, is on top of any *within*-tissue conflict between maternal and paternal genes (over levels of hormone production, for example, or sensitivity to its effects).

Insofar as a paternal orientation may often be the more selfish one and we wish to suppress evidence of such selfishness to others, there may be a tendency for the paternal viewpoint to more often be unconscious. Whether stable alternative personalities could be associated with paternal and maternal genes is, of course, completely unknown. Recent evidence strongly suggests that the left hemisphere of the human brain is, under certain circumstances, much more likely than the right to deny reality and to rationalize the denial (Ramachandran 1997, Ramachandran and Rogers-Ramachandran 1996). It would be most interesting if imprinted genes showed a bias by hemisphere in the relative expression of paternal and maternal genes!

Acknowledgments. We thank David Haig for sharing his ideas and his expertise, Gordon Getty for his interest and support and Laurence Hurst, Rolf Ohlsson, Paul Sherman, and Margo Wilson for their helpful comments.

References

Ågren G, Zhou Q, Zhong W (1989) Ecology and behavior of Mongolian gerbils, *Meriones unguiculatus*, at Xilinhot, Inner Mongolia, China. Anim Behav 37:11-27

Allen N, Logan K, Lally G, Drage D, Norris M, Keverne B (1995) Distribtuion of parthenogenetic cells in the mouse brain and their influence on brain development and behavior. Proc Natl Acad Sci USA 92:10782-10786

Altmann J, Alberts S, Haines S, Dubach J, Muruthi P, Coote T, Geffen E, Cheesman D, Mututua R, Saiyalel S, Wayre R, Lacy R, Bruford M (1996) Behavior predicts genetic structure in a wild primate group, Proc Natl Acad Sci USA 93:5797-5801

Bender R, Surani A, Kothary R, Li L-L, Furst D, Christ B, Fundele R (1995) Tissue-specific loss of proliferative capacity of parthenogenetic cells in fetal mouse chimeras. Roux's Arch Dev Biol 204:436-443

Birkhead T, Møller A (1992) Sperm competition in birds. Academic Press, London

Briskie J, Naugles C, Leech S (1994) Begging intensity of nestling birds varies with sibling relatedness. Proc R Soc Lond B 258:73-78

Burt, A, Trivers R (*in press*). Genetic conflicts in genomic imprinting. Proc R Soc Lond B

Carlile MJ (1987) Genetic exchange and gene flow: their promotion and prevention. In: Rayner ADM, Brasier CM, Moore D (eds) Evolutionary biology of the fungi. Cambridge University Press, Cambridge, pp 203-214

Chaillet JR, Bader DS, Leder P (1995) Regulation of genomic imprinting by gametic and embryonic processes. Genes Dev 9:1177-1187

Clark M, Galef B (1988) Effect of uterine position on rate of sexual development in female Mongolian gerbils. Physiol Behav 42:15-18

Clark M, Galef B (1995) Prenatal influences on reproductive life history strategies. Trends Ecol Evol 10:151–153

Clark M, Spencer C, Galef B (1986) Reproductive life history correlates of early and late sexual maturation in female Mongolian gerbils (*Meriones unguiculatus*). Anim Behav 34:551–560

Clark M, Tucker L, Galef B (1992) Stud males and dud males: intra-uterine position effects on the reproductive success of male gerbils. Anim Behav 43:215–221

Dawson W (1965) Fertility and size inheritance in a *Peromyscus* species cross. Evolution 19:44–55

DeChiara T, Efstratiades A, Robertson E (1990) A growth-deficiency phenotype in heterozygous mice carring an insulin-like growth factor II gene disrupted by targetting. Nature 345:78–80

Dixon A, Ross D, O'Malley S, Burke T (1994) Paternal investment inversely related to degree of extra-pair paternity in the reed bunting. Nature 371:698–700

Dobson SL, Tanouye MA (1998) Evidence for a genomic imprinting sex determination mechanism in *Nasonia vitripennis* (Hymenoptera; Chalcidoidea). Genetics 149:233–242

Forne T, Oswald J, Dean W, Saam J, Bailleul B, Dandolo L, Tilghman S, Walter J, Reik W (1997) Loss of the maternal *H19* gene induces changes in *Igf2* methylation in both *cis* and *trans*. Proc Natl Acad Sci USA 94:10245–10248

Foss HM, Roberts CJ, Claeys KM, Selker EU (1993) Abnormal chromosome behavior in *Neurospora* mutants defective in DNA methylation. Science 262:1737–1741

Fundele R, Surani A (1994) Experimental embryological analysis of genetic imprinting in mouse development. Dev Genet 15:515–522

Fundele R, Bober E, Arnold H, Grim M, Bender R, Wilting J, Christ B (1994) Early skeletal muscle development proceeds normally in parthenogenetic mouse embryos. Dev Biol 161:30–36

Fundele R, Li L-L, Herxfeld A, Barton S, Surani A (1995a) Proliferation and differentiation of androgenetic cells in fetal mouse chimeras. Roux's Arch Dev Biol 204:494–501

Fundele R, Barton S, Christ B, Krause R, Surani A (1995b) Distribution of androgenetic cells in fetal mouse chimeras. Roux's Arch Dev Biol 204:484–493

Fundele R, Surani A, Allen N (1997) Consequences of genomic imprinting for fetal development. In: Reik W, Surani A (eds) Genomic imprinting. Oxford University Press, Oxford, pp 98–117

Grossniklaus U, Vielle-Calzada J-P, Hoeppner MA, Gagliano W (1998) Maternal control of embyogenesis by MEDEA a *Polycomb* group gene in *Arabidopsis*. Science 280:446–450

Haig D (1992a) Genomic imprinting and the theory of parent-offspring conflict. Semin Dev Biol 3:153–160

Haig D (1992b) Intragenomic conflict and the evolution of eusociality. J Theor Biol 156:401–403

Haig D (1993) Genetic conflicts in human pregnancy. Q Rev Biol 68:495–532

Haig D (1994) Cohabitation and pregnancy-induced hypertension. Lancet 344:1633–1634

Haig D (1996a) The altercation of generations; genetic conflicts of pregnancy. Am J Reprod Immunol 35:226–232

Haig D (1996b) Placental hormones, genomic imprinting, and maternal-fetal communication. J Evol Biol 9:357–380

Haig D (1996c) Do imprinted genes have few and small introns? BioEssays 18:351–353

Haig D (1997a) Parental antagonism, relatedness asymmetries, and genomic imprinting. Proc R Soc Lond B 264:1657–1662

Haig D (1997b) The social gene. In: Krebs J, Davies N (eds) Behavioural ecology, 4th edn. Blackwell, Oxford, pp 284–306

Haig D, Graham C (1991) Genomic imprinting and the strange case of the insulin-like growth factor II receptor. Cell 64:1045–1046

Haig D, Trivers R (1995) The evolution of parental imprinting: a review of hypotheses. In: Ohlsson R, Hall K, Ritzen M (eds) Genomic imprinting. Cambridge University Press, Cambridge, pp 17–28

Haig D, Westoby M (1989) Parent-specific gene expression and the triploid endosperm. Am Nat 134:147–155

Holmes W, Sherman P (1982) The ontogeny of kin recognition in two species of ground squirrels. Am Zool 22:491–517

Hrdy S (1977) The Langurs of Abu. Harvard University Press, Cambridge

Hurst LD (1994) Embryonic growth and the evolution of the mammalian Y chromosome. 1. The Y as an attractor for selfish growth factors. Heredity 73:223–232

Hurst LD (1997) Evolutionary theories of genomic imprinting. In: Reik W, Surani A (eds) Genomic imprinting. Oxford University Press, Oxford, pp 17–28

Hurst LD, McVean GT (1997) Growth effects of uniparental disomies and the conflict theory of genomic imprinting. Trends Genet 13:436–443

Hurst LD, McVean G, Moore T (1996) Imprinted genes have few and small introns. Nat Genet 12:234–237

Itier JM, Tremp GL, Léonard JF, Multon MC, Ret G, Schweighoffer F, Tocque B, Bluet-Pajot MT, Cormier V, Dautry F (1988) Imprinted gene in postnatal growth. Nature 393:125–126

Lau M, Stewart C, Lie Z, Bhatt H, Rotwein P, Stewart C (1994) Loss of the imprinted *Igf2*/cation-independent mannose-6-phosphate receptor results in fetal overgrowth and perinatal lethality. Gene Dev 8:2953–2963

Lee J, Jaenisch R (1997) The (epi)genetic control of mammalian X-chromosome inactivation. Curr Opin Genet Dev 7:274–289

Leighton P, Ingram R, Eggenschwiler J, Efstratiades A, Tilghman S (1995) Disruption of imprinting caused by deletion of the *H19* gene region in mice. Nature 375:34–39

Lin B-Y (1982) Association of endosperm reduction with parental imprinting in maize. Genetics 100:475–486

McVean GT, Hurst LD (1997) Molecular evolution of imprinted genes: no evidence for antagonistic co-evoluion. Proc R Soc Lond B 264:739–746

McVean GT, Hurst LD, Moore T (1996) Genomic evolution in mice and men: imprinted genes have little intronic content. BioEssays 18:773–775

Mestel R (1998) The genetic battle of the sexes. Nat Hist 107(2):44–49

Miyoshi N, Kuroiwa Y, Kohda T, Shitara H, Yonekawa H, Kawabe T, Hasegawa H, Barton S, Surani M, Kaneko-Ishino T, Ishino F (1998) Identification of the *Meg 1/Grb10* imprinted gene on mouse proximale chromosome 11, a candidate for the Silver-Russell syndrome gene. Proc Natl Acad Sci USA 95:1102–1107

Mochizuki A, Takeda Y, Iwasa Y (1996) The evolution of genomic imprinting. Genetics 144:1283–1295

Moore T, Haig D (1991) Genomic imprinting in mammalian development: a parental tug-of-war. Trends Genet 7:45–49

Ramachandran V (1997) The evolutionary biology of self-deception, laughter, dreaming and depression: some clues from anosognosia. Medical Hypotheses 47:347–362

Ramachandran V, Rogers-Ramachandran D (1996) Denial of disabilities in anosognosia. Nature 382:501

Razin A, Shemer R (1995) DNA methylation in early development. Hum Mol Genet 4:1751–1755

Robillard P-Y, Hulsey T, Perianin J, Janky E, Miri E, Papiernik E (1994) Association of pregnancy-induced hypertension with duration of sexual cohabitation before conception. Lancet 344:973–975

Sherman P (1977) Nepotism and the evolution of alarm calls. Science 197:1246–1253

Sherman PW (1980) The limits of ground squirrel nepotism. In: Barlow GB, Silverberg J (eds) Sociobiology, beyond nature/nurture? AAAS Selected Symposium 35, Westview Press, Boulder, Colorado, pp 505–544

Skuse D, James R, Bishop D, Coppini B, Dalton P, Aamodt-Lieper G, Bacarese-Hamilton M, Creswell C, McGurk R, Jacobs P (1997) Evidence from Turner's syndrome of an imprinted X-linked locus affecting cognitive function. Nature 387:705–708

Stutchbury B, Rhymer J, Morton E (1994) Extrapair paternity in hooded warblers. Behav Ecol 5:384–392

Tan S-S, Faulkner-Jones B, Breen S, Walsh M, Bertram J, Reese B (1995) Cell dispersion patterns in different cortical regions studied with an X-inactivated transgenic marker. Development 121:1029–1039

Trivers R (1985) Social evolution. Benjamin/Cumming, Menlo Park, California

Trivers R (1997) Genetic basis of intrapsychic conflict. In: Segal N, Weisfeld G, Weisfeld C (eds) Uniting psychology and biology. American Psychological Assoc, Washington, DC pp 385–395

Trupin L, Simon L, Eskenazi B (1996) Change in paternity: a risk factor for preelampsia in multiparas. Epidemiology 7:240–244

Vu TH, Hoffman AR (1994) Promoter-specific imprinting of the human insulin-like growth factor-II gene. Nature 371.714–717

Webber A, Ingram R, Levorse J, Tilghman S (1998) Location of enhancers is essential for the imprinting of H19 and Igf2 genes. Nature 391:711–715

Westneat D, Sherman P (1997) Density and extra-pair fertilizations in birds: a comparative analysis. Behav Ecol Sociobiol 41:205–215

Williamson CM, Schofield J, Dutton ER, Seymour A, Beechey CV, Edwards YH, Peters J (1996) Glomerular-specific imprinting of the mouse gs-alpha gene – how does this relate to hormone resistance in Albright hereditary osteodystrophy? Genomics 36:280–287

"Whether stable alternative personalities could be associated with paternal and maternal genes is, of course, completely unknown. Recent evidence strongly suggests that the left hemisphere of the human brain is, under certain circumstances, much more likely than the right to deny reality and to rationalize the denial (Ramachandran 1997, Ramachandran and Rogers-Ramachandran 1996). It would be most interesting if imprinted genes showed a bias by hemisphere in the relative expression of paternal and maternal genes!

Genomic Imprinting in Plants

Joachim Messing[1] and Ueli Grossniklaus[2]

1
Introduction

Epigenetic programming is most likely the least understood part of the control of gene expression and too broad a subject to consider in a single chapter. The difficulty in studying its role in gene expression is that very few Mendelian mutations cause arrest in epigenetic programming and that chromatin changes occurring many cell divisions before transcription starts are difficult to monitor by biochemical means (Lund et al. 1995a). Given this complexity, we shall focus here on one example of epigenetic programming that has already been genetically exploited: parental genomic imprinting. One of the major advantages of studying plant versus animal development is based on the generation of alleles of genes affected in epigenetic programming so that biochemical methods can be applied by comparing tissues of different genetic origin. Here, we consider allelic variations and mutations that specifically focus on epigenetic programming at the gamete level.

Genomic imprinting is a specific example of epigenetic programming that occurs prior to fertilization during gametogenesis. Imprinted genes show differential expression depending on the sex of the parent from which they are inherited (John and Surani 1996; Neumann and Barlow 1996; Reik 1996). Due to the presence of imprinted loci, parental genomes are not functionally equivalent. Thus, normal postfertilization development requires the contribution of both a paternal and maternal genome in addition to the specialized cytoplasm of the female gamete. A disturbance of normal imprinting can lead to severe developmental aberrations or cancer (Feinberg 1993; Reik and Maher 1997). Genomic imprinting is often referred to as uniquely mammalian. However, imprinting also plays a crucial role for seed development in flowering plants (angiosperms) (Lin 1982, 1984; Birchler and Hart 1987). Whereas mammalian embryogenesis strictly requires the genetic contribution of both parents (Barton et al. 1984; McGrath and Solter 1984; Surani et al. 1984), such a

[1] Waksman Institute, 190 Frelinghuysen Road, Rutgers, The State University of New Jersey, Piscataway, New Jersey 08854-8020, USA
[2] Cold Spring Harbor Laboratory, 1 Bungtown Road, PO Box 100, Cold Spring Harbor, New York 11724, USA

requirement does not exist in the embryos of all plants. In angiosperms, double fertilization occurs to yield two products, the diploid zygote and the usually triploid primary endosperm (Fig. 1). Formation of gynogenetic and androgenetic haploids in many plants (Kimber and Riley 1963; Sarkar and Coe 1966; Kermicle 1969) and of asexually derived embryos in apomictic genera (Nogler 1984) suggests that imprinting does not play a crucial role for embryogenesis in these species, though it may exist in others. In contrast, correct genomic imprinting is required for successful endosperm development in maize (*Zea mays*), where it has convincingly been demonstrated at the level of genomes, chromosomes and individual genes (Kermicle and Alleman 1990; Walbot 1996). Interspecific and interploidy crosses suggest that imprinting is also important for endosperm development in other species (Nishiyama and Yabuno 1978; Johnston et al. 1980; Haig and Westoby 1991).

2
Different Roles of Genomic Imprinting in Animals and Plants

Although phenomena of epigenetic gene regulation related to imprinting have been observed in many organisms, genomic imprinting as described here has

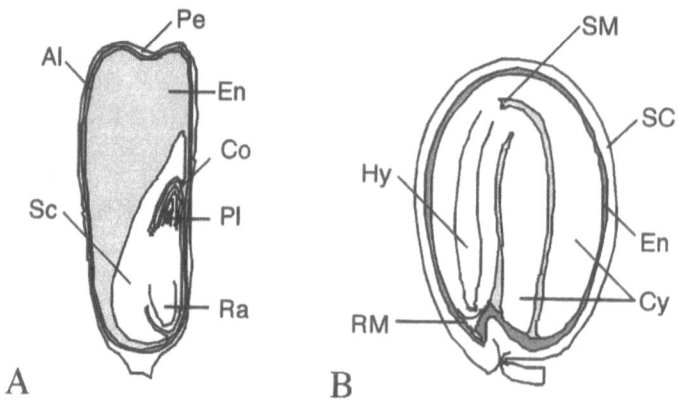

Fig. 1A, B. Schematic diagram of seeds of maize and *Arabidopsis.* **A** Maize kernel with persistent endosperm. The endosperm (*En*) makes up about 80% of the dry weight of the seed and the embryo occupies a lateral position within the kernel. The outermost layer of the endosperm tissue forms the aleurone (*Al*), a pigmented monolayer of cells, which is covered by the maternal pericparp (*Pe*) tissue, the equivalent of the seed coat in the maize kernel. The embryonic tissue in the maize kernel comprises the scutellum (*Sc*), the radicle (*Ra*), the coleoptile (*Co*), and the plumule (*Pl*), which bears five to six immature leaves in the dry seed. **B** *Arabidopsis* seed. The embryo occupies the entire seed and only a single cell layer of endosperm tissue is left at maturity. The endosperm is formed transiently and becomes absorbed during seed development. The embryo comprises the two cotyledons (*Cy*), the hypocotyl (*Hy*), and the root (*RM*) and shoot meristems (*SM*), and is protected by a maternally derived multilayered seed coat (*SC*). The seeds are not drawn to scale. A maize kernel is 8 to 9 mm long, an *Arabidopsis* seed about 0.5 mm. (After Kiesselbach 1949, Meinke 1994)

been reported only in mammals and flowering plants. In both, reproduction is characterized by a placental habit, that is the embryo or seed receives all of its postfertilization nutrients from the female parent. In mammals, extraembryonic (trophectoderm and primitive endoderm) and maternal tissues (the placenta is a more or less tight association of maternal and extraembryonic tissues) play a crucial role in the acquisition of nutrients for the embryo. In angiosperms, complex interactions between embryo, endosperm, and maternal sporophytic tissues characterize seed development. It is thought that the endosperm is important for providing nutrients to the developing embryo. However, a wide range of developmental patterns have been described for endosperm development (Murray 1988; Johri et al. 1992; Lopes and Larkins 1993). For instance, in cereals such as maize, the endosperm is persistent and required for postgermination development of the seedling (Fig. 1). In contrast, many dicotyledonous species, including *Arabidopsis thaliana*, produce nonpersistent endosperm which is degraded by the time the embryo matures (Fig. 1), or the endosperm forms only rudimentary or not at all, as in the Orchideacaea. Thus, the relative importance of endosperm, embryo, and maternal tissues in nutrient acquisition is expected to vary among genera. In species with nonpersistent or poorly developed endosperm, the embryo proper or extraembryonic tissues such as the suspensor (Yeung and Meinke 1993) may partially or fully perform the nurturing function of the endosperm.

Haig and Westoby (1989, 1991) proposed that imprinting evolved in organisms with a placental habit as a consequence of an intragenomic conflict over the allocation of nutrients from the mother to her offspring. Provided that the mother carries offspring from more than one father over the time of her life span, the interest of paternal and maternal genes within an embryo or seed differ: the father directly benefits from an increase in size (and consequently fitness) of his offspring but does not experience any cost, whereas the mother benefits most if her resources are allocated equally to all her offspring including potential future offspring from a different father. The theory predicts that (1) some parentally controlled genes (e.g., imprinted loci) should influence the growth rate of the embryo or seed, and (2) that paternal control is expected to promote growth whereas maternal expression should tend to reduce it. This conflict leads to a parental tug-of-war where the sperm genome counteracts factors involved in nutrient acquisition that are controlled by the egg (Moore and Haig 1991). Evidence supporting this adaptive explanation has come from studies on androgenetic and gynogenetic mouse embryos and from functional analyses of imprinted genes in mice and humans (Haig and Graham 1991; Cassidy 1995; Jänisch 1997; Reik and Maher 1997). Until recently, support for the parental conflict theory in plants has been restricted to evidence derived from interspecific and interploidy crosses (Haig and Westoby 1989, 1991). One also could speculate that the function of the endosperm factors, Ef1–4 on chromosome arm 10L, are consistent with the theory: endosperms lacking a paternal copy of 10L produce small kernels. However, additional maternally derived copies of 10L cannot restore kernel size (Lin, 1982). This argues that Ef1–4 are preferentially expressed when inherited from the father, but Birchler and Hart (1987) have shown that seed size in the absence of Ef1–4 can be further reduced by factors on other chromosomes (maternal enhancement) in a dosage dependent manner. Since we do not know whether Ef1–4 are epigenetically changed during gametogenesis (independence of grand-parentage), much more has to be learnt about these factors before their role with regard to Haig and Westoby's hypothesis can be firmly established. Furthermore, the few imprinted loci in maize studied at the single gene level do not appear to influence the growth rate of the seed. This may be related to the fact that genomic imprinting in angiosperms may not only play a role in postfertilization development of the seed but also in cellular diffe-rentiation processes unique to plants which are not subject to a parental conflict.

3
Gametophytic Development, a Unique Process in Plants

Because of the unique features of sexual reproduction in flowering plants, gametes have a different developmental history than in animals where meiotic products differentiate directly into sperm and egg cells. In contrast, meiotically derived spores of plants undergo several mitotic division cycles to form multicellular haploid organisms, the male (microgametophyte, pollen) and female gametophytes (megagametophyte, embryo sac) (Fig. 2). The gametes differentiate later in the development of the gametophytes. Thus, cell lineages from gametophytic cells may be derived from the mitotic amplification of epigenetic states that can be established during the early stages of gametogenesis. In the majority of species the megagametophyte consists of seven cells belonging to four cell types: three antipodals at one pole, the egg cell and two synergids at the other, and a large binucleated central cell in the middle (Grossniklaus and Schneitz 1998). The male gametophyte usually con-sist of three cells, with a vegetative cell that harbors two sperm cells. Double fertilization involves the fusion of two pairs of cells: the egg cell is fertilized to become the zygotic embryo, while the two polar nuclei in the central cell fuse with a sperm nucleus to form the primary endosperm nucleus. The latter contains a triploid genome in most species, with two maternal contributions.

In maize and *Arabidopsis*, as in most species, female and male gametophytes are derived from a single meiotic product (Maheswari 1950). Thus, except for gene dosage, the two fertilization products are genetically twins. However, if epigenetic differentiation occurs between meiosis and ferti-lization, the twins could differ epigenetically. One reason to suspect such a developmental differentiation in the female gametophyte is because it is the only phase in plant development where the migration of nuclei may be in-

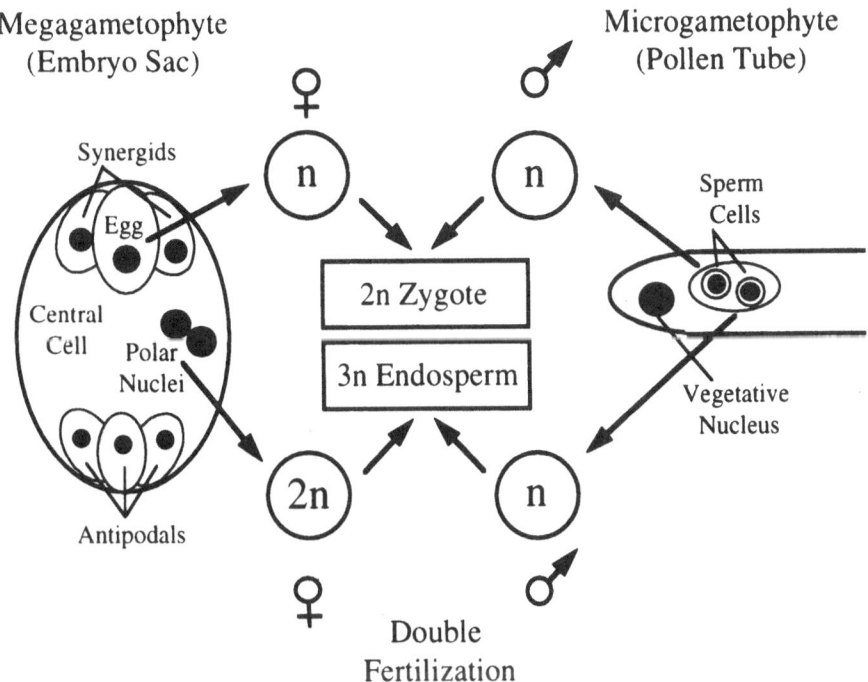

Fig. 2. Mega- and microgametophyte and their roles in plant reproduction. The female gametophyte (*left*) contains the egg and central cell both of which participate in double fertilization. The male gametophyte (*right*) releases two sperm cells into one of the synergids. From there, they migrate to their target cells and one fuses with the egg to form the diploid zygote, while the other unites with the central cell to produce the triploid endosperm. The primary endosperm nucleus carries two maternal genomes originating from the polar nuclei and one paternal genome from the sperm.

volved in cell specification and differentiation. Positional information specified by particular regions of the syncytial female gametophyte may then be translated into epigenetic differences between its cell types. If this is correct, genomic imprinting in plants may play a role both prior to fertilization in cellular differentiation as well as postfertilization during embryo and endosperm morphogenesis. These two aspects of genomic imprinting in plants occur in parallel and may influence each other.

A major difference of genomic imprinting in these two cases might be related to the role of the sperm. Epigenetic changes that specifically differentiate between egg and central cell and which are propagated mitotically during embryo and endosperm development would not be subject to a parental tug-of-war. This is because in most plant species, the two sperm cells are functionally equivalent and fertilize egg and central cell randomly (Dumas and Mogensen 1993; Russell 1993). Thus, epigenetic differences between embryo and endosperm can only be established in the female gametophyte prior to fertilization. Because the male contributes equally to both fertilization products, no counteracting epigenetic changes specific to embryo or endosperm are expected. In contrast, epigenetic changes that affect seed development could be subject to a parental conflict but paternal expression would affect the endosperm and embryo equally. This initial male equivalence could be modified by effects caused by the specialized cytoplasms of egg and central cell, cell type-specific epigenetic changes as outlined above, or gene dosage. Indeed, most examples of genomic imprinting in plants specifically affect the endosperm, although few imprinted genes have been studied in detail.

4
Plant Development Versus Defense Mechanisms

Still, plants are a very rich source of phenotypes of aberrant epigenetic programming. For instance, transposable elements and transgenes are frequently recognized as aberrant genetic information that should be avoided by the plant expression and recombination machinery. Suppressing these sequences by DNA modification systems has been referred to as silencing or defense response (Matzke and Matzke 1998). Usually, the silencing process is gradual and might occur somatically or increase over several generations. Gametic imprinting requires the opposite process. It involves an activating step or presetting step, a term proposed by McClintock (Schwartz 1982). This exemplifies how epigenetic programming is involved in both plant development and defense. Illegitimate recombination (e.g., viruses, transgenes, or transposons) or unequal crossingover might trigger cellular defense mechanisms that lead to the modification of transgenes, transposons and/or repetitive DNA sequences. However, these modifications could be reversed if they come under the control of a developmental program. Interesting examples are muted transposable elements like *Ac/Ds* and *Spm* that "cycle" between an active and inactive state. In both cases, the tagged genes are also expressed in the endosperm. Activation of the transposable element (cycling) can lead to gene expression of the tagged gene in endosperm tissue. However, activation is associated with transmission of the element through the female gametophyte (Chomet 1988; Fedoroff and Banks 1988).

Gene silencing has been regarded as a complication for genetic engineering of plants rather than a new field of basic research. Nevertheless, DNA insertions by transformation have provided a new avenue to define *cis*-acting features of epigenetic programming (Que and Jorgensen 1998). However, it is unclear at this time whether they are models for recombination or repair mechanisms (Kermicle 1996), which are certainly significant areas of epigenetic programming and could shine light for instance on certain cancers which arise somatically. In our discussion here, gametic epigenetic switches are restricted to the study of tissue-specific gene expression in the developing seed. Therefore, we are not exhausting the work related to plant development but concentrate on the phenotypes of allelic variants affecting seed gene products whereas other aspects such as paramutation are reviewed elsewhere (Kermicle 1996; Hollick et al. 1997).

5
Activation of Certain *R* Alleles During Female Gametogenesis

Most allelic variants described so far relate to genes that are specifically expressed during endosperm development. It is quite striking that in case of the *r* (red color) locus in maize, only *R* alleles specifying gene expression in endosperm (*R* class) exhibit gametic imprinting. Those that specify pigment synthesis in other plant organs (*r* class) do not show any difference regardless of the gametic origin of the allele (Brink et al. 1970). Furthermore, *R* alleles that are sensitive to gametic imprinting always exhibit an increase in gene expression in endosperm tissue, but not in other plant organs. For instance, when *R* is inherited from the female it confers full color to the aleurone layer of the endosperm whereas the aleurone is only partly pigmented or mottled if *R* originates from the sperm (Fig. 3A, B; Kermicle 1970). Because the male contribution is not discriminating between embryo and endosperm, plants derived from mottled kernels are expected to carry *R* in a state conferring a mottled phenotype. The state of *R*-mottled is reversed or preset to full color if transmitted through the polar nuclei, but remains mottled if inherited by the male. Since either of the sperm nuclei can fertilize the egg nucleus of the embryo sac, any change in epigenetic state of *R* should then be visible in both the endosperm and other plant organs. This is not the case. Changes of phenotype are limited to endosperm tissue, again pointing to a difference between the egg and central cell carrying the two polar nuclei. Since the epigenetic change marks the *r* locus during megagametophyte development for increased gene expression, the change appears to go from an inactive form to an active form of gene expression.

There are several other interesting aspects of imprinting at the *r* locus in maize. Because the two polar nuclei contribute two maternal genomes, any allele transmitted through the female gametophyte creates a balance of two maternal to one paternal allele in the endosperm. An increase in dose would also be consistent with increased gene expression. However, the elegant studies of Kermicle (1970) have shown that the unequal phenotype from reciprocal crosses of certain *R* alleles in the endosperm is dependent not on dosage, but rather on gametic origin. By taking advantage of B-chromosome-associated nondisjunction in the second mitotic division of the generative nucleus in the pollen, an endosperm nucleus can be formed that is tetrasomic (Birchler 1993). This permits a balancing of the *r* alleles from both female and male gametes in the endosperm. Another approach involves the production of a somatically induced, segmental disomic endosperm. The *r* locus is located on chromosome 10. If a composite *Ds* element is placed between the centromere and *R*, chro-

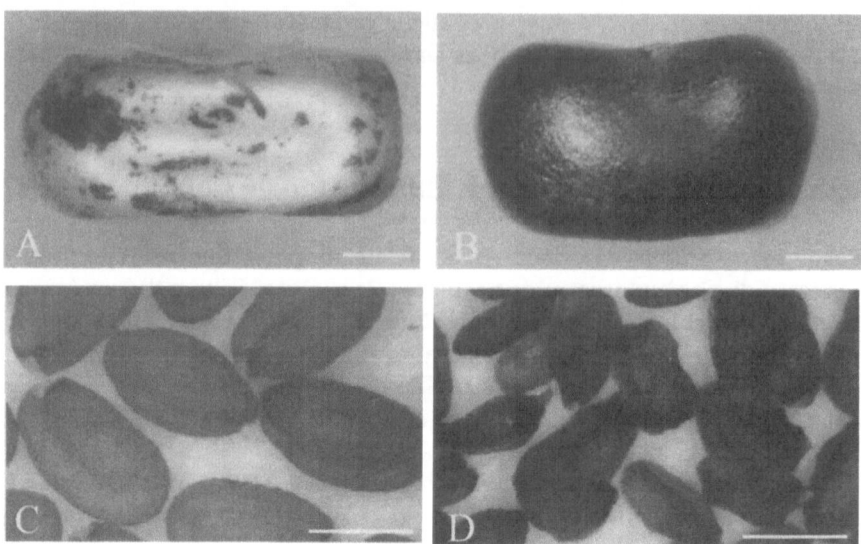

Fig. 3A–D. Phenotypes of seeds illustrating the effect of imprinted loci. **A, B** Maize kernels carrying *R-r:standard* (*R-r*) alleles displaying mottled **A** and full color **B** aleurone pigmentation. The phenotypes shown reflect paternally and maternally inherited *R-r* alleles which had prior exposure to *R-stippled* (*R-r/R-st*) and were crossed to colorless tester plants (*r-g/r-g*). If inherited paternally, such *R-r* alleles display a mottled phenotype **A**, if transmitted through the female they produce a fully colored aleurone **B**. **C, D** *Arabidopsis* seeds that inherited a mutant *medea* (*mea*) allele through the pollen **C** or the female gamtophyte **D**. If *mea* is inherited paternally the seeds develop normally **C**, but they abort and collapse when the mutant allele is inherited maternally **D**. *Bar* **A, B** = 1500 μm; **C, D** = 300 μm

mosome breakage can be induced somatically by introducing *Ac* through the pollen, yielding a mosaic endosperm tissue with respect to *R* dosage. This permits the visualization of phenotypes next to each other that would depend on gene dosage (Kermicle and Alleman 1990). In either case, when *R* is quadraplex, triplex, or duplex, the phenotype is dependent only on the gametic origin of the *r* allele. This again confirms that gene expression is determined by an epigenetic change established in the female gametophyte prior to fertilization.

The *R*-mottled phenotype does not reflect a complete silencing, but rather a mosaic of various levels of gene expression (Fig. 3A). R is a transcription factor, but its own regulation is not well understood yet. However, because of its ability to specify pigment accumulation in so many different plant tissues, it would not be surprising if a number of *trans*-acting factors bind to the upstream region of *R*, contributing to its transcriptional activation. In this case, one would expect to find a number of independent enhancers in the promoter of the *r* gene, some of them affected by genomic imprinting and others not. Interestingly, maternal derepression of *r* (*mdr*), located on chromosome 4, is a *trans*-acting factor that is required for the maternal activation of *R* (Kermicle and Alleman 1990). When *mdr* is inherited maternally together with *R*, the aleurone is no longer fully colored (Kermicle, 1978). Since *mdr* is a loss-of-function mutation, *Mdr* wild-type function is required for full color expression of *R*, indicating that the maternal state is an activated one (see also legend to Fig. 3; Kermicle, 1995). However, *mdr* is unlikely to be a general imprinting factor (imprintor) because maternal derepression is a more global process, whereas *mdr* is specific for *R* as far as is known. On the other hand, it is clear that *mdr* is sensitive to a change in the epigenetic state of *R*. It can only function if *R* is epigenetically changed. Therefore, *mdr* is likely to be one of the DNA-binding proteins that can control *R* gene expression, but is blocked from binding to its target site if *R* is turned off epigenetically (e.g., transmitted through pollen).

6
Imprinted Alleles of the Delta Zein Regulator, *dzr1*

Similar to *R*, *dzr1*, a regulator of storage protein synthesis, exists in allelic variants that are mute when pollen-transmitted, but activated in the reciprocal cross (Chaudhuri and Messing 1994). One complication is the fact that so far no deletions or nulls of *dzr1* have been identified and that all crosses are based on heteroallelic interactions. Furthermore, *dzr1* has not been cloned yet and its function, the differential mRNA stability of a high-methionine storage protein (10-kDa zein) gene, is only a biochemical phenotype. Therefore, any allelic screens based on transposable elements (e.g., site-directed insertional mutagenesis) have to wait until cloned sequences of *dzr1* are available. Nevertheless, the differential accumulation of the 10-kDa zein is easily monitored in Western blots. Moreover, differential accumulation of the 10-kDa zein can serve as a critical threshold for seed methionine in animal feed that has dramatic impact on animal growth (Fig. 4). There are basically two types of heteroallelic interactions of *dzr* alleles. One results in an additive (gene dosage), the other in a dominant effect. The 10-kDa zein accumulates at a high level in BSSS53 and at a low level in W64A and Mo17. Combination of BSSS53 and W64A exhibits an additive dosage effect, with two doses of *dzr1*+BSSS53 showing more expression of *dzs10*, the gene encoding the 10-kDa zein protein, than one dose. However, *dzr1*+Mo17 can eliminate high-level expression of the 10-kDa zein protein in crosses with BSSS53, producing a dominant negative effect in the heteroalleleic interactions of *dzr1*+Mo17 and *dzr1*+BSSS53. This effect is dependent on transmitting the *dzr1*+Mo17 allele through the female gametophyte. Cognizant of the uniparental activation, a (*dzr1*+BSSS53 × *dzr1*+Mo17) × *dzr1*+BSSS53 backcross can show the segregation of the dominant negative effect as an allele of *dzr1*. Furthermore, it can show that activation occurs only in the immediate parentage and is independent of grandparentage; note that *dzr1*+Mo17 was pollen-transmitted in the previous generation. Thus, as was observed for *R*, a paternal state of *dzr1* (no reduction

Fig 4. Feed efficiency can be controlled by genomic imprinting. The picture shows two examples of 16-day-old Ann Arbor chicks grown on a corn-based diet. The two different groups of animals were fed with a diet without supplemented methionine that differs only by the genetic background of the corn meal. The two meals were derived from the hybrid BSSS53×Mo17 and the hybrid Mo17×BSSS53. Due to genomic imprinting of *dzr1*+Mo17 allele, the first hybrid had high levels of methionine and the second one low levels. This difference in methionine amounted to about a 50% difference in weight gained over a 2 week feeding trial of 2 day-old animals (Messing and Fisher 1991).

in 10-kDa zein expression) can be reverted or preset to the maternal state (reduction in 10-kDa zein expression) if transmitted through the megagametophyte.

7
The Role of the Female Gametophyte in Genomic Imprinting in Plants

In this respect, it is interesting that endosperm is a terminal differentiated tissue and does not participate in the next generation. For both, the *R* and *dzr1* locus, the maternal state is not recovered if transmitted through the male gametophyte. This is either because (1) imprinting affects only the central but not the egg cell and the paternal state represents the ground state, or (2) because the maternal state becomes reset to the paternal state during male gametogenesis. Based on the interactions between *R* and *mdr* it appears as if activation (maternal derepression) of *R* involves an active process whereas the paternal mottled phenotype represents the ground state. Therefore, it is likely that R is derepressed in the central cell only and no active resetting is required during male gametogenesis because the embryo carries the allele in its ground state. If this is the case, maternal imprinting at R and *dzr1* reflects epigenetic differences between the egg and central cell which are established prior to fertilization and are then mitotically inherited during endosperm and embryo development. This aspect of imprinting is specific to plants and distinct from parental imprinting in animals which requires a resetting step prior to germline formation. Resetting is an integral part of Haig and Westoby's theory. If the active state were transmitted through both gametes, there would be no uniparental activation of gene expression. Since the sperm does not discriminate between egg and central cell, this plant-specific role of imprinting in cellular differentiation is not subject to a parental conflict.

Genomic imprinting in plants seems to have the unique feature of invoking the activation of genes that are expressed during endosperm development. Since the two gene loci described above exist also in the form of alleles that are active when transmitted through the male, these genes can all be pollen-transmitted in an active form. There is no indication that pollen transmission is associated with an epigenetic change (e.g., gene silencing or reversal of gene silencing). Therefore, gametic imprinting in plants depends on the configuration of the gene in question and the capability of changing such a configuration specifically during the mitotic divisions of the functional megaspore nucleus. During the three mitotic divisions in the embryo sac, no cellularization occurs, but all nuclei assume specific positions within the embryo sac. One is always destined to be the egg nucleus, which is distinct from the two polar nuclei. The change in configuration must therefore involve an allele that is a muted form or ground state of the gene. The configuration must be one that can be changed through mitosis and revert to the active form. In this respect, it is interesting that a number of years ago, Spena et al. (1983) noted that DNA sequences associated with storage protein genes were differentially methylated in leaf and endosperm tissues. DNA methylation is a clonally inherited epigenetic mark that represents an attractive mechanism to distinguish the alleles of imprinted genes (Jänisch 1997).

The major storage protein genes in maize, also called zeins, are exclusively expressed in the endosperm. Although subgroups of zeins are encoded by multigene families and a direct functional analysis of zein genes in a methylated or nonmethylated state has not been carried out yet, it is reasonable to speculate that some methylated sites could be located in regulatory regions of these genes. Similar to the situation with transposable elements, one would expect that the methylated form is the one that is not expressed, while the nonmethylated one is available for gene expression. Lund et al. (1995b) have recently used differences in zeins of different inbreds to correlate a change in methylation with transmission of DNA sequences through the female gametophyte in reciprocal crosses; certain zeins appear to be expressed when their gene sequences are derived from their female parent. A similar situation has been described for tubulins. Tubulins are also encoded by a multigene family; however, in contrast to the zeins, their function is required in all tissues. By taking advantage of polymorphism between different inbred lines, Lund et al. (1995c) could show in hybrid plants that specific tubulins of one parent are specifically expressed in endosperm and that their expression depends on the parent to be the female in the cross. Furthermore, female transmission also correlates with a change in methylation status of the corresponding tubulin gene. The change involves demethylation of tubulin sequences that stay methylated in the embryo or when pollen-transmitted.

When transmitted through the polar nuclei, tubulin and zein sequences always appear to be hypomethylated. Thus, consistent with other observations, removal of methylated sites leads to increased gene expression; for example, DNA from the promoter region of the S (seed color) gene of the r locus shows reduced methylation when inherited from the female, correlating with increased R expression. In aleurones where R was inherited together with mdr through the megagametophyte, methylation at the S gene is increased correlating with reduced R expression in the mottled phenotype (Kermicle 1996). It is clear that methylation changes per se are not sufficient to cause a change in gene expression. It will be necessary to show that these changes affect specific DNA promoter sites, where transcriptional activation is sensitive to methylated DNA sequences (Schläppi et al. 1994). For instance, Mdr function could be required for both demethylation of R sequences and the presetting of R. In either case, there is a good indication that female gametophyte development involves a stage where mitosis leads to differentially methylated DNA sequences, including those that are linked to gene expression during endosperm development.

8
Control of *Arabidopsis thaliana* Seed Development by Genomic Imprinting

Whereas imprinting of R and $dzr1$ represent examples where imprinting may play a role in the differentiation of the two fertilization products, a recent example in *Arabidopsis* provides support for a role of imprinting in seed development that closely resembles the animal situation. In a screen for mutations affecting the gametophytic phase of the life cycle, a mutant was isolated that displayed a gametophytic control of seed development (Grossniklaus et al. 1998). Interestingly, seed abortion results only when the mutant allele is female transmitted, indicating that the gene is turned on during female gametogenesis, but not when transmitted through the pollen (Fig. 3C, D). Because seed abortion is solely controlled by the genotype of the female gametophyte, this mutant was termed *medea* (*mea*) (Euripides, 431 B.C.). Successful seed development requires a maternally inherited wild-type *MEA* allele but is independent of the paternal contribution. This is even the case if *MEA* is transmitted in two copies through the pollen because an additional paternal wild type allele cannot rescue the seed abortion phenotype. It is important to note that, unlike in maize, crosses between diploid and tetraploid plants in *Arabidopsis* produce viable seeds, although their size and morphology is usually less uniform than in crosses involving plants of the same ploidy (Grossniklaus et al. 1998). Thus, the *mea* phenotype is not caused by a dosage effect in the endosperm and tetraploidy does not affect the uniparental activation or lack of activation of its function. Genetic analysis suggested that *mea* either affects a cytoplasmic factor or disrupts an imprinted gene expressed only from the maternal copy. Whether *mea* displays a maternal effect of cytoplasmic or chromosomal nature could not be distinguished by genetic analysis. However, the transposon tagged mutation facilitated the cloning of the *MEA* gene which encodes a SET domain protein, similar to Enhancer of zeste, a member of the Polycomb group (*Pc-G*) (Grossniklaus et al. 1998). The *MEA* mRNA accumulates at very low levels, but can be detected prior to fertilization as well as throughout seed development indicating both maternal and zygotic expression. Recent expression studies by in situ hybridization allowed us to directly visualize transcription at the *MEA* locus (J-P. Vielle-Calzada, M. Hoeppner, and U. Grossniklaus, unpubl.). During the early nuclear divisions of the endosperm, only two out of three *MEA* copies are actively transcribed. This is consistent with the genetic observation that the *MEA* gene is activated during mitosis in the female gametophyte, since the polar nuclei contribute two of the three gene copies to the primary endosperm nucleus. It also shows that the third gene copy, most likely the contribution of the sperm nucleus, remains inactive. Thus, *MEA* is regulated by imprinting and gets activated during megagametogenesis while a paternally inherited allele remains inactive.

Unlike *R* and *dzr1*, *MEA* is required for normal seed development and its absence results in severe developmental defects. In seeds derived from mutant *mea* embryo sacs, both fertilization products are affected (Grossniklaus et al. 1998). Embryo development is severely delayed but *mea* embryos grow to a giant size before they disintegrate during seed desiccation. When *mea* embryos reach the late heart stage they are up to ten times larger in volume than corresponding wild-type embryos. In contrast to the increased proliferation observed in the embryo, there appear to be fewer free nuclear divisions in the endosperm as compared to wild type although morphogenetic progression is normal. Based on the homology to SET domain proteins which are involved in gene regulation by controlling higher order chromatin structure, it is likely that *MEA* controls zygotic target genes involved in cell proliferation. The *mea* phenotype shows striking similarities to phenotypes of mice where maternally expressed genes have been disrupted. Such mice are also larger than wild type consistent with the parental conflict theory. *mea* is the first single gene mutation in plants providing strong support for Haig and Westoby's theory which predicts that maternally controlled genes should restrict embryonic growth.

Due to the complex and ill defined interactions between embryo and endosperm, we do not know whether the *mea* mutation affects primarily the endosperm, the embryo, or both. Since the *MEA* mRNA is expressed in both fertilization products (J-P. Vielle-Calzada, M. Hoeppner, and U. Grossniklaus, unpubl.), it is likely that *MEA* is required for the development of both these tissues. This is in contrast to *dzr1* and imprinted *R* alleles, which are specifically expressed in the endosperm, and is consistent with a role of *MEA* in seed development that is subject to a parental conflict. As outlined earlier, imprinting phenomena may be of importance in either the embryo and/or the endosperm, although imprinting in plants is usually referred to as specific to the endosperm. The situation is likely to vary between species depending on the relative role of embryo and endosperm in the acquisition of nutrients from the mother. Whereas imprinting may be irrelevant to embryogenesis in some species, e.g., in plants where gynogenetic and androgenetic haploids develop normally, it may be crucial for the proper development of the embryo in others as suggested by the *mea* phenotype in *Arabidopsis*. The parental conflict theory predicts parent-of-origin-specific effects only if a mother has offspring from more than one father. Imprinting is, therefore, not expected to evolve in truly self-fertilizing species. *Arabidopsis*, although self-fertilizing, shows a sufficient level of outbreeding in the wild (Koornneef 1994) that each individual is likely to carry offspring from several fathers. The *mea* phenotype represents an example of imprinting in plants that is not differentiating between cell types in the megagametophyte, but rather affects growth during seed development in a way similar to the situation found in mammals.

9
Gametic Imprinting in Plants, a Two-Step Process

All genes described so far in plants that relate to gametic imprinting show without exception that gene activation follows meiosis during the mitotic divisions in the female gametophyte. Although Ef1–4 on maize chromosome 10L are candidates for paternal gene activation, they also could be repressed maternally (Lin 1982). While the MEA gene product is immediately utilized, genes affecting pigment or storage protein synthesis accumulate much later in endosperm development. Therefore, one can assume that gametic imprinting involves two steps. The first one involves an erasure of a *cis*-acting modification of the imprinted gene. The second involves the timing of the expression of a *trans*-acting factor. Since the two steps can be separated by many cell divisions, the first one can be considered as a presetting or pretranscription step. This pretranscription state then represents a new form of genetic memory beyond regular base-pairing that is the basis for normal Mendelian inheritance (Messing 1989; Messing and Fisher 1991). Although the only molecular data available now points to demethylation of local DNA sequences, it is very likely that demethylation simply reflects a change in chromatin structure. For instance, it is known that plants also possess genes that encode proteins that bind to methylated DNA in a nonspecific manner (Zhang et al. 1989). DNA methylation occurs immediately after DNA replication and the newly methylated DNA would attract the nonspecific binding proteins, which in turn would appear as a change in chromatin structure. Such a correlation between a change in methylation and chromatin structure has recently been described for the *P1*-locus in maize (Lund et al. 1995a). The interesting part of such a molecular process is that the change is heritable. Cell lineages would include genomes with the same stable chromatin configuration. Once the methylation imprint is erased, the nonspecific DNA-binding proteins are not attracted to the promoter region anymore, allowing regular nucleosomal organization to occur within gene regions. In this respect, the epigenetic change is a change of chromatin configuration that represents alternate mitotic pathways. It is interesting to note that the genomically imprinted *MEA* locus encodes a protein that is likely to affect chromatin structure, which may itself be related to imprinting. Thus, *MEA* would not only be regulated by imprinting but also play a role in imprinting phenomena at the functional level. Reik et al. (1995) argued that genes interacting with imprinted loci are under selective pressure to become imprinted themselves, as may be the case with *MEA*.

10
Conclusion

In summary, genomic imprinting in plants appears to fulfill two distinct roles. One is specific to plants, where gametogenesis is characterized by postmeiotic nuclear divisions that occur prior to fertilization. Imprinting is established during this phase of development and may be involved in generating epigenetic differences between egg and central cell which are propagated postfertilization during cellular differentiation in the embryo and endosperm. This role of imprinting in differentiation is illustrated by the imprinted *R* and *dzr1* loci, which are endosperm specific, do not affect morphogenesis, and are apparently not subject to a parental conflict. A second class of imprinted loci represented by the recently isolated *MEA* locus share striking similarities to imprinted genes in mammals, control proliferation and growth during seed development, and are consistent with Haig and Westoby's parental conflict theory.

Acknowledgments. Research in the laboratory of J.M. described here has been supported by DOE grant No. DE-FG05-95ER20194 and research in the laboratory of U.G. by the Cold Spring Harbor Laboratory President's Council. U.G. also acknowledges the support of the Janggen-Poehn Foundation and the Demerec-Kaufmann-Hollaender Fellowship in Developmental Genetics. The authors would also like to thank Drs. James Birchler and Hugo Dooner for their critical review of the manuscript and Jim Duffy for drawing Fig. 1. We are particularly grateful to Dr. Jerry Kermicle for helpful suggestions with regard to imprinting at the *r*-locus and for providing the material shown in Fig. 3A and B.

References

Barton SC, Surani MAH, Norris ML (1984) Role of paternal and maternal genomes in mouse development. Nature 311:374–376

Birchler JA (1993) Dosage analysis of maize endosperm development. Annu Rev Genet 27:181–204

Birchler JA, Hart JR (1987) Interaction of endosperm size factors in maize. Genetics 117:309–317

Brink RA, Kermicle JL, Ziebur NK (1970) R expression in maize endosperm, embryos, and seedlings. Genetics 66:87–96

Cassidy SB (1995) Uniparental diosomy and genomic imprinting as causes of human genetic disease. Environ Mol Mutagen 25 Suppl 26:13–20

Chaudhuri S, Messing J (1994) Allele-specific parental imprinting of dzr-1, a posttranscriptional regulator of zein accumulation. Proc Natl Acad Sci USA 91:4867–4871

Chomet P (1988) Characterization of stable and metastable changes of the maize transposable element, Activator. Thesis. State University of New York at Stony Brook.

Dumas C, Mogensen HL (1993) Gametes and fertilizaton: maize as a model system for experimental embryogenesis in flowering plants. Plant Cell 5:1337–1348

Fedoroff NV (1996) Epigenetic regulation of the maize Spm transposable element. In: Russo VEA, Martienssen RA, Riggs AD (eds) Epigenetic mechanisms of gene regulation. Cold Spring Harbor Laboratory Press, Cold Spring Harbor, pp 575–592

Fedoroff NV, Banks JA (1988) Is the suppressor-mutator element controlled by a basic developmental regulatory mechanism? Genetics 120:559–577

Feinberg A (1993) Genomic imprinting and gene activation in cancer. Nat Genet 4:110–113

Grossniklaus U, Schneitz K (1998) The molecular and genetic basis of ovule and megagametophyte development. Semin Cell Dev Biol 9:227–238

Grossniklaus U, Vielle-Calzada J-P, Hoeppner M, Gagliano WB (1998) Maternal control of embryogenesis by MEDEA, a Polycomb-group gene in Arabidopsis. Science 280:446–450

Haig D, Graham C (1991) Genomic imprinting and the strange case of the insulin-like growth factor II receptor. Cell 64:1045–1046

Haig D, Westoby M (1989) Parent specific gene expression and the triploid endosperm. Am Nat 134:147–155

Haig D, Westoby M (1991) Genomic imprinting in endosperm: its effect on seed development in crosses between species, and between different ploidies of the same species, and its implications for the evolution of apomixis. Philos Trans R Soc Lond 333:1–13

Hollick JB, Dorweiler, JE, Chandler VL (1997) Paramutation and related allelic interactions. Trends Genet 13:302–308

Jänisch R (1997) DNA methylation and imprinting: why bother? Trends Genet 13:323–329

John RM, Surani MA (1996) Imprinted genes and regulation of gene expression by epigenetic inheritance. Curr Opin Cell Biol 8:348–353

Johnston SA, den Nijs TPM, Peloquin SJ, Hanneman RE, Jr (1980) The significance of genic balance to endosperm development in interspecific crosses. Theor Appl Genet 57:5–9

Johri BM, Ambegaokar KB, Srivastava PS (1992) Comparative embryology of angiosperms, vols 1, 2. Springer , Berlin Heidelberg New York

Kermicle JL (1969) Androgenesis conditioned by a mutation in maize. Science 166:1422–1424

Kermicle JL (1970) Dependence of the R-mottled aleurone phenotype in maize on mode of sexual transmission. Genetics 66:69–85

Kermicle JL (1978) Imprinting of gene action in maize endosperm. In: Walden DB (ed) Maize breeding and genetics. Wiley, New York, pp 357–371

Kermicle JL (1995) Location, time of action, and dominance relations of an imprintor gene of R-mottled in maize. In: Oono K, Takaiwa F (eds) Modification of Gene Expression and non-Mendelian Inheritance. National Institute of Agrobiological Resources, Tokio, pp 119–134

Kermicle JL (1996) Epigenetic silencing and activation of a maize r gene. In: Russo VEA,

Martienssen RA, Riggs AD (eds) Epigenetic Mechanisms of Gene Regulation. Cold Spring Harbor Laboratory Press, Cold Spring Harbor, pp 267–287

Kermicle JL, Alleman M (1990) Genetic imprinting in maize in relation to the angiosperm life cycle. Development Suppl 1:9–14

Kiesselbach TA (1949) The structure and reproduction of corn. Univ Nebraska Agri Exp Sta Res Bull 161:1–96

Kimber G, Riley R (1963) Haploid angiosperms. Bot Rev 29:480–531

Koornneef M (1994) *Arabidopsis* genetics. In: Meyerowitz EM, Somerville CR (eds) Arabidopsis. Cold Spring Harbor Laboratory Press, Cold Spring Harbor, pp 89–120

Lin B-Y (1982) Association of endosperm reduction with parental imprinting in maize. Genetics 100:475–486

Lin B-Y (1984) Ploidy barrier to endosperm development in maize. Genetics 107:103–115

Lopes MA, Larkins BA (1993) Endosperm origins, development and function. Plant Cell 5:1383–1399

Lund G, Das OP, Messing J (1995a) Tissue-specific DNase I-sensitive sites of the maize *P* gene and their changes upon epimutation. Plant J 7:797–807

Lund G, Ciceri P, Viotti A (1995b) Maternal-specific demethylation and expression of specific alleles of zein genes in the development of *Zea mays* L. Plant J 8:571–581

Lund G, Messing J, Viotti A (1995c) Endosperm-specific demethylation and activation of specific alleles of α-tubulin genes of *Zea mays* L. Mol Gen Genet 246:716–722

Maheswari P (1950) An introduction to the embryology of angiosperms. McGraw-Hill, New York

Matzke MA, Matzke AJM (1998) Epigenetic silencing of plant transgenes as a consequence of diverse cellular defense responses. Cell Mol Life Sci 54:94–103

McGrath J, Solter D (1984) Completion of mouse embryogenesis requires both the maternal and paternal genomes. Cell 37:179–183

Meinke DW (1994) Seed development in *Arabidopsis thaliana*. In: Meyerowitz EM, Somerville CR (eds) Arabidopsis. Cold Spring Harbor Laboratory Press, Cold Spring Harbor, pp 253–295

Messing J (1989) Broadening our understanding of genetic information: beyond base pairing. ASM News 55:255–258

Messing J, Fisher H (1991) Maternal effect on high methionine levels in hybrid corn. J Biotechnol 21:229–238

Moore T, Haig D (1991) Genomic imprinting in mammalian development: a parental tug-of-war. Trends Genet 7:45–49

Murray DM (1988) Nutrition of the angiosperm embryo. Research Studies Press, Somerset, UK

Neumann B, Barlow DP (1996) Multiple roles for DNA methylation in gametic imprinting. Curr Opin Genet Dev 6:159–163

Nishiyama I, Yabuno T (1978) Causal relationships between the polar nuclei in double fertilization and interspecific cross-incompatibility in *Avena*. Cytologia (Tokyo) 43:453–466

Nogler GA (1984) Gametophytic apomixis. In: Johri BM (ed) Embryology of angiosperms. Springer, Berlin Heidelberg New York, pp 475–518

Que Q, Jorgensen RA (1998) Homology-based control of gene expression patterns in transgenic petunia flowers. Dev Genet 22:100–109

Reik W, Feil R, Allen ND, Moore T, Walter J (1995) Imprinted genes, allelic methylation, and imprinted modifier genes of methylation. In: Ohlsen R, Ritzen M (eds) Genomic imprinting – causes and consequences. Cambridge University Press, Cambridge, pp 157–170

Reik, W (1996) The Wellcome prize lecture. Genetic imprinting: the battle of the sexes rages on. Exp Physiol 81:161–172

Reik W, Maher ER (1997) Imprinting in clusters: lessons from Beckwith-Wiedeman syndrome. Trends Genet 13:330–334

Russel SD (1993) The egg cell: Development and role in fertilization and early embryogenesis. Plant Cell 5:1349–1359

Sarkar KR, Coe EH, Jr (1966) A genetic analysis of the origin of maternal haploids in maize. Genetics 54:453–464

Schläppi M, Raina R, Fedoroff N (1994) Epigenetic regulation of the maize *Spm* transposable element: novel activation of a methylated promoter by TnpA. Cell 77:427–437.

Schwartz D (1982) Tissue-specific regulation of gene function: presetting and erasure. Proc Natl Acad Sci USA 79:5991–5992

Spena A, Viotti A, Pirrotta V (1983) Two adjacent genomic zein sequences: structure, organization, and tissue-specific restriction pattern. J Mol Biol 169:799–811

Surani MAH, Barton SC, Norris ML (1984) Development of reconstituted mouse eggs suggests imprinting of the genome during gametogenesis. Nature 308:548–550

Walbot V (1996) Sources and consequences of phenotypic and genotypic plasticity in flowering plants. Trends Plant Sci 1:27–32

Yeung EC, Meinke DW (1993) Embryogenesis in angiosperms: development of the suspensor. Plant Cell 5:1371–1381

Zhang D, Ehrlich KC, Supakar PC, Ehrlich M (1989) A plant DNA-binding protein that recognizes 5-methylcytosine residues. Mol Cell Biol 9:1351–1356

Imprinting and Paternal Genome Elimination in Insects

Glenn Herrick[1] and Jon Seger[2]

1
Introduction

In many insects and other arthropods, males transmit only maternally in-herited chromosomes (White 1973; Brown and Chandra 1977; Nur 1980, 1990a,b,c; Bell 1982; Bull 1983; Lyon and Rastan 1984; Lyon 1993; Wrensch and Ebbert 1993; Brun et al. 1995; Borsa and Kjellberg 1996). This remarkable genetic asymmetry can result from any of three principal systems of paternal genome exclusion, each of which has evolved several times. The most familiar and widespread exclusion system is *arrhenotoky*, in which fatherless males develop from unfertilized eggs and therefore lack paternal chromosomes at all stages of development. Most arrhenotokous systems are genetically haplodiploid, but a few are based on other modes of inheritance (see Nur 1980, 1990c; Bell 1982; Suomalainen et al. 1987). In the two other kinds of exclusion systems, a male's paternally inherited chromosomes are actively *eliminated*: males begin life as seemingly conventional diploid zygotes but then either (1) lose their paternal chromosomes during embryonic development, becoming true maternal haploids (*embryonic* elimination), or (2) exhibit dramatically non-Mendelian patterns of meiosis and spermiogenesis, such that mature sperm carry only maternal chromosomes (*germline* elimination). To denote their formal (transmission-genetic) similarity to haplodiploid arrhenotoky, the embryonic and germline elimination systems are often characterized as *parahaplodiploid* or *pseudoarrhenotokous*.

These three routes to exclusive transmission of maternally inherited chromosomes are connected both mechanistically and historically. For exam-ple, genomic imprints direct both embryonic and germline elimination in many cases, and embryonic systems have evolved from germline systems in some cases (Nur 1980). Arrhenotokous systems do not appear to depend on imprints, but arrhenotoky may have evolved (at least in a few cases) from embryonic elimination, by deletion of the vestigial requirement for fertiliza-tion to initiate male development (Bull 1983; Haig 1993a; Sabelis and

[1] Divison of Molecular Biology and Genetics, Department of Oncological Sciences, University of Utah, Salt Lake City, Utah 84132, USA
[2] Department of Biology, University of Utah, Salt Lake City, Utah 84112, USA

Nagelkerke 1993; Brun et al. 1995; Borsa and Kjellberg 1996). The systems are related dynamically, as well, in that germline elimination, embryonic elimination, and arrhenotoky all seem to represent evolutionary responses to underlying intragenomic "conflicts of interest" between a male's maternally and paternally inherited chromosomes (Brown 1963, 1964; Bull 1983; Haig 1993a,b).

Here we briefly survey paternal genome elimination (PGE) in insects. We emphasize two taxa in which the phenomenology of PGE, and the role of imprinting, are especially well understood. We begin with the dung gnat *Sciara coprophila* (whose study first gave rise to the concept of genomic imprinting), and then turn to the scale insects (a group with thousands of species exhibiting almost every known variation on the theme). Several mechanistic features of scale-insect PGE systems can be interpreted as adaptive evolutionary responses, by maternally inherited genes, to paternal counter adaptations for preventing elimination. We develop a model in which alternating evolutionary "moves" and "countermoves" (fixations of mutations advancing maternal or paternal interests) give rise to the observed diversification of scale-insect PGE systems. The model generates testable predictions about the mechanics and the phylogenetic histories of these systems.

Recent discussions of mammalian imprinting have tended to blur a critical distinction between imprints and their effects. Reference to an "imprinted gene" has come to connote both the *existence of the heritable mark* (the imprint) and the resulting *differential behavior* (the action cued by the imprint). Students of chromosome elimination have usually been careful to observe this distinction (e.g., Nur 1990b), perhaps because its significance is easier to appreciate in these systems than in mammalian systems, where the cued effect is differential gene expression. We argue that mammalian imprinting, too, will be better understood when the imprint itself is clearly distinguished from the actions that may or may not occur in response to it.

Some elimination systems appear not to depend on imprints. For example, intracellular parasitic bacteria of the genus *Wolbachia* manipulate host chromosomes in many arthropod taxa, causing hybrid incompatibility, female parthenogenesis (thelytoky), feminization of otherwise male-destined embryos, and paternal genome elimination (see Breeuwer and Werren 1990; O'Neill et al. 1992, 1997; Rousset et al. 1992; Moran and Baumann 1994; Werren et al. 1995; Lassy and Karr 1996; Schilthuizen and Stouthamer 1997; Werren 1997). *Wolbachia* are sometimes said to imprint host chromosomes destined for destruction, but there is no need to invoke a heritable mark that persists for more than one mitosis in the *Wolbachia*-mediated systems that have been studied cytogenetically. Similarly, the ultraselfish supernumerary *PSR* chromosome of the wasp *Nasonia vitripennis* causes efficient paternal genome elimination (Nur et al. 1988; Werren 1991), but this, too, can be explained by direct effects of *PSR* on the chromosomes that are about to be destroyed. These parasite-mediated chromosome eliminations have much in common with imprint-cued eliminations, but close attention to some important differences (discussed below) should lead to better understanding of the mechanisms and the evolutionary origins of these phenomena. One interesting possibility is that arthropod species infected by *Wolbachia* might become relatively more likely to evolve imprint-mediated elimination systems, if such infections tend to promote the establishment of predisposing conditions that would otherwise be less common (see Varmuza and Mann 1994b).

2
Paternal Chromosome Elimination in *Sciara*

More than 70 years ago, Metz (1925) described meiosis in the male of *Sciara coprophila* (reviewed by Du Bois 1933; Metz 1938; White 1973; Brown and Chandra 1977; Lyon and Rastan 1984; Gerbi 1986). He found that a complete haploid chromosome set is eliminated in meiosis I (Fig. 1d), pinched off from the spermatocyte in a bud (see Fuge 1997). As in many insects, there is no recombination in males. The eliminated chromosomes are paternal.

In addition to this dramatic and complete paternal genome elimination, the chromosome cycle includes several other unusual imprint-cued behaviors, as illustrated in Fig. 1. Male and female zygotes receive one X chromosome and a haploid set of autosomes from the female pronucleus ($X^m A^m$), but they receive two Xs and an autosomal set from the sperm ($X^p X^p A^p$). During the seventh embryonic round of cleavage divisions (Fig. 1b) the *somatic* precursor nuclei lose two X^ps (in males) or one X^p (in females) by "anaphase lag": the X^p chromatids fail to disjoin and are left behind on the metaphase plate (Du Bois 1933; Fuge 1994, 1997; de Saint Phalle and Sullivan 1996). As a consequence, males become somatically $X^m O$ and females become $X^m X^p$. Later, *germline* cells enter an unusual, extended interphase in which the chromosomes are partly condensed and visible ("prochromosomes"); one set is much less condensed than the other, in response to a differential imprint (Fig. 1c; Berry 1941; Rieffel and Crouse 1966). In both males and females one X^p is "extruded" or "dragged" from this interphase nucleus into the cytoplasm, where it later disappears. Thus, the male and female germ lines both become fully diploid ($X^m X^p A^m A^p$).

During meiosis I in the male, the paternal set segregates away from the maternal set and into a cytoplasmic bud that is eliminated from the spermatocyte (Fig. 1d), as first observed by Metz. During meiosis II, the remaining maternal *autosomal* chromatids disjoin and one set is eliminated in a bud, but the two X^m chromatids fail to disjoin and migrate precociously to the sperm-destined pole. The surviving chromosomes ($X^m X^m A^m$) are packaged into a single sperm (Metz 1938). Thus, the male provides double-X sperm to

(a) zygote

$$X^m A^m$$
$$X^p X^p A^p$$

(b) cleavage 7

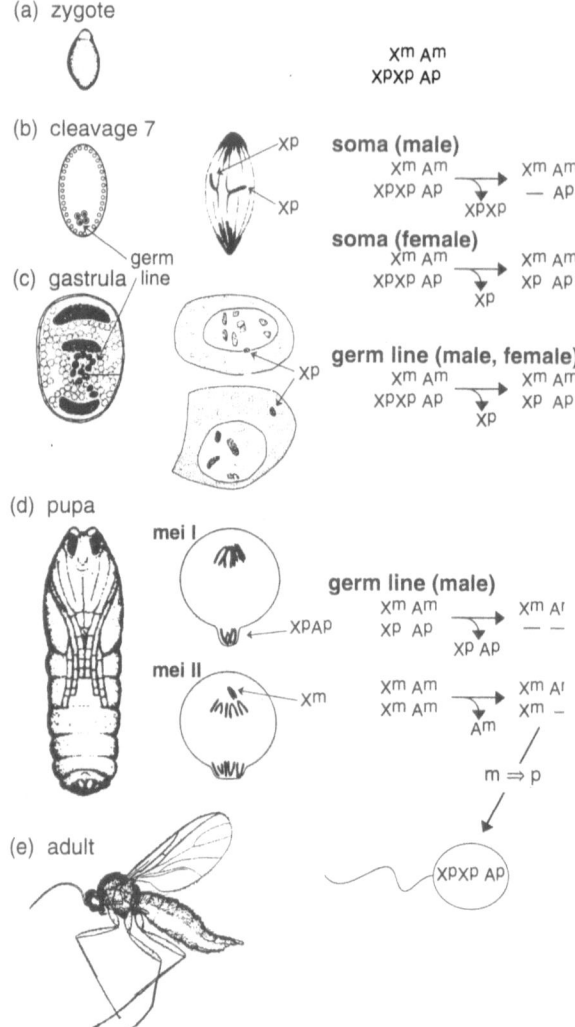

XP

XP

soma (male)
$$X^m A^m \longrightarrow X^m A^m$$
$$X^p X^p A^p \underset{X^p X^p}{\searrow} \quad - A^p$$

soma (female)
$$X^m A^m \longrightarrow X^m A^m$$
$$X^p X^p A^p \underset{X^p}{\searrow} \quad X^p A^p$$

(c) gastrula / germ line

XP

germ line (male, female)
$$X^m A^m \longrightarrow X^m A^m$$
$$X^p X^p A^p \underset{X^p}{\searrow} \quad X^p A^p$$

(d) pupa

mei I

XPAP

mei II

Xm

germ line (male)
$$X^m A^m \longrightarrow X^m A^r$$
$$X^p A^p \underset{X^p A^p}{\searrow} \quad - \; -$$

$$X^m A^m \longrightarrow X^m A^r$$
$$X^m A^m \underset{A^m}{\searrow} \quad X^m \; -$$

$$m \Rightarrow p$$

(e) adult

XPXP AP

Fig. 1a–e. *Sciara* chromosome cycle. Developmental stages are shown at the *left*; corresponding cytogenetic events are shown in the *middle* and then represented diagrammatically at the *right*. See text for a more complete account. Authoritative reviews are provided by Metz (1938) and Gerbi (1986); accessible overviews are provided by White (1973) and by Lyon and Rastan (1984). Here, for simplicity, we ignore the germline limited (L) chromosomes found in most sciarids. *a* The zygotic genotype is diploid, but it includes two paternal X chromosomes ($X^p X^p$). *b* In the seventh cleavage, one cleavage after nuclei migrate to the surface of the embryo and germline cells are set aside internally, all somatic precursors lose two X^ps (in males) or one X^p (in females), during mitosis, by anaphase lag (mitotic figure from Du Bois 1933). *c* In the late gastrula, one X^p is eliminated from each interphase germline cell (drawings from Berry 1941). *d* Meiosis occurs in the pupa. These drawings from Du Bois (1933) show elimination of the paternal set in meiosis I, elimination of a maternal chromatid set in meiosis II, and precocious nondisjunction of X^m chromatids in meiosis II. The surviving maternal chromosomes are packaged into a sperm and will be inherited as paternal. *e* The adult male provides double-X^p sperm to fertilize eggs, completing the cycle. (Insect drawings *a, d, e* from Cole 1969)

fertilize all eggs, completing the cycle. The imprints have been reversed. The male's *maternal* autosomes and X chromosome will behave as *paternal* in his offspring.

2.1
Imprints Defined

In 1938 Metz wrote: ". . . the *difference in behavior* [of chromosomes] is due to a . . . *difference* between [them that] . . . *is impressed* on [them] in the *preceding generation* by the sex of the parent. . . . This . . . *modification persists* for only one generation and is *reversible*." This statement summarizes several original and important insights:

1. Differential chromosome behavior indicates the existence of differential modifications that cue the behavior. Metz used the word "*impressed*." Crouse (1960) later borrowed the term "imprint" from ethology, establishing the modern usage.
2. It is not clear whether the imprints mark the maternal or paternal chromosomes, or both (Lyon and Rastan 1984). Do "paternal imprints" cue elimination, or do "maternal imprints" confer protection? Ultimately, an imprint is a "*difference*" between the maternal and paternal homologs, not an absolute or intrinsic state of either one.
3. The imprints "*persist*" from the parent, through the zygote, at least to spermatogenesis at the end of the male germ line. Thus, the imprints must be replicated along with the chromosome's genetic material for many cell generations, as epigenetic marks.
4. The imprints are "*reversible*". A male transmits his maternal genes as paternal, and a female transmits her paternal genes as maternal.

2.2
When Are Imprints Laid Down?

The chromosome set in the male's sperm has just behaved as a maternal set in meiosis I, but it will behave as a paternal set in the progeny. Thus the imprints can reverse no earlier than meiosis II, or more plausibly, during subsequent spermiogenesis. Alternatively, a paternal identity could be established much later, in the egg, after fertilization but before the first zygotic mitosis (see Latham et al. 1995). In many organisms, the two pronuclei proceed in close apposition, but not fused, through an interphase to the first mitosis, in which the two pronuclear chromosome sets finally are brought together on the metaphase plate. Even at this point, and through several subsequent mitoses, the sets may continue to occupy distinct volumes in the diploid nuclei (*gonomery*: reviewed generally by Wilson 1928; in various insects by Huettner 1924 and Zissler 1992; in humans by Sathananthan 1997;

in interspecific plant hybrids by Leitch et al. 1991; and in fish by Wilson 1928). Diploid males occasionally are produced by parthenogenesis in the scale insect *Pulvinaria hydrangeae* (Coccidae). Both chromosome sets are maternal, yet at the appropriate time in embryogenesis (see below), one set appears to acquire all the properties of a paternally imprinted set (Nur 1963). This suggests that at least in coccids, the imprints may be laid down during early embryogenesis (Chandra and Brown 1975; Brown and Chandra 1977; Nur 1980, 1990b).

Early embryonic imprinting could be common. In the mouse, the two pronuclei also remain separate until the first mitosis; the male pronucleus is more swollen than the female pronucleus (Donahue 1972a,b; Wright and Longo 1988) and may begin S phase earlier (Abramczuk and Sawicki 1975; but see Howlett and Bolton 1985). Pronuclear swapping experiments show that some mouse imprints are laid down before the first mitosis (reviewed by Surani 1986), but additional imprints could be laid down afterward, since the two chromosome sets lie in different parts of the nucleus for at least two more cleavages (Odartchenko and Keneklis 1973; Brandriff et al. 1991). In short, there is reason to suspect that even "paternal imprints" may often be of *maternal* origin, in the sense that they are established in the early embryo by machinery that was present in the egg before fertilization.

2.3
XCE and Action at a Distance

Crouse (1960, 1979) identified a locus on the *Sciara* X chromosome that controls the various differential X behaviors (anaphase lag, interphase extrusion, and meiosis II nondisjunction). When this locus is translocated to an autosome, the autosome exhibits "a program of behavior . . . [that is] indistinguishable from the one enacted normally by the X" (Crouse 1979). This "X Controlling Element" (*XCE*) lies between the centromere and the proximal telomere on the acrocentric *Sciara* X, but when translocated onto an autosome, at a location well away from the centromere, it directs actions in *cis* at great distances. An imprint is clearly involved (Crouse 1979), but where is it? If the imprint marks the centromere, then *XCE* acts on that information over a large distance; or alternatively, if the imprint occurs at *XCE*, then the imprint cues action at a large distance (i.e., at the centromere). Thus an imprint and the locus affected by it may lie far apart.

As Du Bois (1933) observed (Fig. 1b), during the seventh somatic mitosis the X^P centromeres appear to pull apart normally, but their progress is arrested by the chromatids, which remain firmly joined together, causing anaphase lag and elimination of the chromosome. Recent studies confirm this basic observation, add spectacular detail, and suggest that anaphase lag is caused primarily by an imprint-cued failure of chromatid separation rather than by centromere dysfunction (de Saint Phalle and Sullivan 1996). This interpretation implies that *XCE* (and/or the imprint) may affect different cellular structures at different times and places. Although anaphase lag may be caused by a failure of sister-chromatid separation, neither interphase extrusion of an X^P from the germ line, nor meiotic elimination in males, seems likely to be caused (at least primarily) by this mechanism.

The better known but more recently discovered *Xce* locus of the mammalian X chromosome controls the inactivation of its X. Like the *XCE* of *Sciara*, this *Xce* acts in *cis* over great distances (reviewed by Lee and Jaenisch 1997; Heard et al. 1997). The *Xce* region includes the *Xist* gene, which must be expressed to inactivate the X that carries it. *Xce* bears an imprint that influences the expression of its associated *Xist* in extraembryonic tissues, such that the paternal X is preferentially inactivated. In some marsupials, one X (probably paternal) is *eliminated* from the female's soma (reviewed by White 1973; Brown and Chandra 1977; Hayman and Rofe 1977).

Sciara males express both maternal and paternal genes in most tissues, but apparently not in all. Crouse (1966; reviewed by Brown and Chandra 1977) crossed parents carrying balanced X-autosome translocations, and recovered exceptional males with no X^m ($A^m A^P X^P X^P$). These males eliminated one X^P in the germ line, becoming $X^P O$. They were sterile, although males with just one maternal X in the germ line ($X^m O$) were fertile. Crouse concluded tentatively that while a single X^m can support germline function, a single X^P cannot. This implies that the imprint(s) also affect expression of genes needed for germline function.

3
Imprints and Their Effects

A single persistent mark on the *Sciara* X chromosome could cue its various imprint-directed behaviors at different times and places in development, even though the mark caused no visible action in most cells at most times. Similarly, imprints on *Sciara* autosomes appear to cue action only in the spermatocyte; in particular, they do not appear to affect somatic gene expression (Smith-Stocking 1936), as they do in many systems. In any system, we can be aware only of imprints that cue actions we observe. Thus, imprints could exist that cue subtle effects that no one has noticed, and many imprints could have no effects.

If imprints are viewed as distinct from the mechanisms that read them, then it is easier to imagine action at a distance, or several different actions cued by one imprint. For example, methylation of a CpG dinucleotide could affect the functioning of the element within which it sits (e.g., a promoter) in very different ways, depending on aspects of its nuclear environment (e.g., populations of transcription factors and other DNA-binding proteins).

Various mechanisms of imprint-cued action toward specific loci, domains, or chromosomes could evolve independently, with different effects (e.g., elimination or inactivation), and in response to different selective forces.

One important set of selective forces derives from conflicts of interest among genes within a genome (see Haig and Trivers 1995). For example, the maternally and paternally inherited alleles at a locus metaphorically compete for transmission into viable gametes. A statistical bias (meiotic drive), too small to be detected in experiments of practical size, could generate a decisive fitness advantage for the favored allele. The elimination of paternally inherited chromosomes in male *Sciara* dramatically boosts the transmission of maternally inherited genes (see Brown 1963, 1964; Hartl and Brown 1970; Bull 1979, 1983; Haig 1993a,b).

When rare, a maternally inherited allele that caused males to eliminate their paternal chromosomes would enjoy a twofold reproductive advantage because it would appear in all sperms (rather than in half of them). Similarly, a rare allele expressed in mothers that caused their sons to eliminate the father's chromosomes would appear in half (not one quarter) of their sons' sperms. As such an allele increases in frequency, the size of its advantage decreases because it more often causes its own elimination in males, but selection continues to favor it right up to fixation (Brown 1964; Hartl and Brown 1970; Bull 1983). In principle, such a strategy should be feasible in many species with parental imprints, and, in fact, PGE has evolved independently in diverse taxa (White 1973; Brown and Chandra 1977; Nur 1980, 1990a; Bell 1982; Bull 1983; Lyon and Rastan 1984; Nur et al. 1988; Stuart and Hatchett 1988; various chapters in Wrensch and Ebbert 1993; Brun et al. 1995). For example, paternal chromosomes are not transmitted by males of the Hessian fly, *Mayetiola destructor* (Cecidomyiidae), and maternal chromosomes stain especially heavily ("positive heteropycnosis"; White 1973) in both the male and female germ lines, visibly revealing a differential imprint (see White 1973; Brown and Chandra 1977; Stuart and Hatchett 1988; Formusoh et al. 1996). Following a discussion of PGE in scale insects, we briefly describe some less well known but seemingly similar systems in bark beetles and mites.

PGE also occurs in several all-female "hemiclonal" fishes and amphibians with "hybridogenic" reproduction (see Schultz 1977, 1989; Dawley and Bogart 1989; Carmona et al. 1997), and in the more recently characterized hybridogenic stick insects of the *Bacillus rossius-grandii* complex (Mantovani and Scali 1992; Tinti and Scali 1992, 1993, 1995). There may even be hybridogenic plants (Davies 1974; Heslop-Harrison 1990). The hybridogenic fish *Poeciliopsis monacha-lucida* has been especially well studied (see Schultz 1989). Females mate with males of a closely related bisexual species, *P. lucida*. The resulting offspring are all female; these "sperm parasites" (Carmona et al. 1997) express their paternally inherited genes but eliminate the paternal genome in meiosis I of *oogenesis* (Cimino 1972). Thus their eggs carry a clonally propagated haploid maternal genome. By eliminating (and reacquiring) the entire paternal genome each generation, hybridogens maintain populations of diverse F_1 genotypes (see Schultz 1969; White 1978). Like other unisexual vertebrates, hybridogens typically arise as interspecific hybrids, and some hybridogenic "species" have multiple origins (see Dawley and Bogart 1989; Quattro et al. 1991).

Given the very strong selection in favor of PGE, why is it not even more common than it is? It evolves under many different circumstances, and it can remain viable for tens or hundreds of millions of years (unlike fully asexual reproduction; see Bell 1982). Some necessary or predisposing conditions (Whiting 1945; Brown 1977) must be relatively uncommon, but what are they? In our view, the relative rarity of PGE is an interesting and important problem that deserves more attention than it has received.

4
Paternal Chromosome Elimination in Scale Insects

A great diversity of germline and embryonic elimination systems has evolved in the "coccids" (*sensu lato*, now Coccoidea, hence *coccoids*), a suborder within Homoptera that includes mealybugs and other scale insects (Hughes-Schrader 1948; Brown 1977; Brown and Chandra 1977; Nur 1980, 1990a). The principal PGE systems are named after the taxa in which they were first described. The relatively simple *lecanoid* germline system is ancestral to the more complex *Comstockiella* germline system, which in turn gave rise (apparently several times; see below) to *diaspidid* embryonic elimination. PGE may also occur in the testicular tissues of hermaphroditic *Icerya* (Margarodidae) (reviewed by Brown and Chandra 1977; Nur 1980).

4.1
Events in the Early Embryo

In male embryos of species with germline elimination (lecanoid and Comstockiella systems), all chromatids disjoin normally at the sixth mitosis (Nur 1967), but in the next interphase the entire paternal chromosome set ("P-set") fails to decondense; the paternal chromosomes remain transcriptionally inactive, and they stain darkly. This heterochromatic state replicates faithfully in the germ line and in most somatic cell lineages (Fig. 2).

In male embryos of species with embryonic (diaspidid) elimination, paternal chromatids fail to disjoin in an early cleavage division, shortly before the stage at which they would become heterochromatic in species with lecanoid or Comstockiella systems (Brown 1965; Brown and Chandra 1977; Nur 1980). These paternal chromosomes are eliminated by anaphase lag from the nuclei that form on completion of this mitosis, leaving the embryo in a truly haploid state. If a P chromosome escapes elimination at this mitosis, then it remains condensed until the next mitosis, when it is eliminated (Brown 1965). This

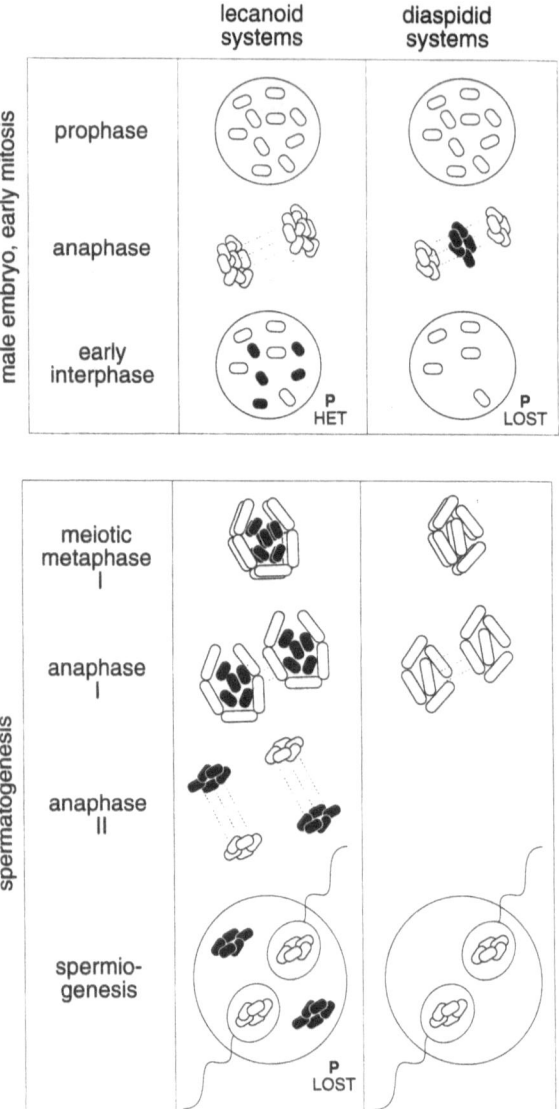

Fig. 2. Paternal genome elimination in male coccoids. Chromosome behaviors in the male embryo and in spermatogenesis are contrasted in species with lecanoid (*left*) and diaspidid (*right*) chromosome systems (reviewed by Brown and Chandra 1977; Nur 1980). Paternal chromosomes are shown as filled (*black*) at stages where they stain or behave differently from maternal chromosomes. *Top panel.* Events at an embryonic mitosis: prophase, anaphase, and the following early interphase. On entering the next interphase, lecanoid paternal chromosomes fail to decondense (heterochromatin). In diaspidid males, the paternal chromosomes are eliminated by anaphase lag. *Bottom panel.* Spermatogenesis. Meiosis in coccoids is inverse, such that the first segregation is equational (sister chromatids disjoin), while the second segregation is reductional (maternal and paternal homologs disjoin). Meiosis I in lecanoid males gives rise to four distinct sets of chromatids: a P-set and an M-set at each pole (first and second images); recombination being absent, meiosis II separates the four sets (third image); in spermiogenesis only the M-sets are packaged into sperm, while the P-sets degenerate (fourth image). In diaspidid spermatogenesis the P-set is absent; meiosis II does not occur; in spermiogenesis the two sets of maternal chromatids are packaged into sperms

suggests that heterochromatization is an essential early step in diaspidid elimination, as it also appears to be in the lecanoid and Comstockiella systems (see below).

Brown and Nelson-Rees (1961) proved that the heterochromatic chromosomes of lecanoid males are paternal by subjecting fathers to high doses of X-irradiation and then observing that their sons carried fragmented heterochromatic chromosomes. Even small fragments can persist and segregate mitotically, because coccoid chromosomes are holocentric (i.e., centromeric function is distributed along the chromosome) (Hughes-Schrader and Ris 1941). Two other conclusions emerged from these experiments. First, the imprint (like centromere function) is *distributed widely* along the length of each chromosome, because fragments behave as paternal in males (Brown and Nelson-Rees 1961; Nur 1990b). Second, the heterochromatic paternal set appears to be *genetically silent*, because sons do not express X-ray-induced dominant lethal mutations when the irradiated chromosomes are inherited from their fathers (in fact, the sons show nearly normal viability), whereas daughters do express such mutations (resulting in poor viability). This inference was corroborated by subsequent studies of nonlethal mutations with visible phenotypes, in which sons expressed the mutations only when inherited from their mothers (Brown and Wiegmann 1969; Nur 1990b).

4.2
Spermatogenesis

As in all Homoptera, meiosis in coccoids is "inverse" (reviewed by Hughes-Schrader 1948; White 1973). In the absence of recombination (e.g., in the males of species with lecanoid or Comstockiella PGE), chromatid-chromatid (equational) disjunction in meiosis I precedes maternal-paternal homologue (reductional) disjunction in meiosis II. The paternal chromosome set enters spermatogenesis in the heterochromatic state that was established at the sixth embryonic mitosis (Fig. 2). At metaphase I the more highly condensed paternal set is "clumped" together at the center of the metaphase plate, nearly surrounded by the maternal set (Schrader 1923; Nur and Brett 1988). Thus the maternal and paternal sets occupy distinct locations at the onset of meiosis. After chromatids disjoin, the spatial arrangement transforms, such that by metaphase II the two sets lie on separate plates. This polarity appears to be the immediate cause of the dramatically non-Mendelian segregation that occurs during anaphase II, when the entire maternal and paternal sets segregate to opposite poles. The heterochromatic P-set does not give rise to sperm (at least not usually, as discussed further below).

The Comstockiella system is a variant of the lecanoid system in which some or all of the paternal chromosomes are eliminated precociously, just prior to meiotic prophase, by an unknown mechanism (they simply disappear!), with the remainder being eliminated later, in meiosis II, by the usual lecanoid mechanism (Kitchin 1970, 1975). This system appears to represent an important transition toward diaspidid-style complete early elimination (Nur 1980). In diaspidid spermatogenesis (Fig. 2), meiosis II is bypassed and the two single-chromatid maternal chromosome sets that segregate in meiosis I are packaged into two sperm.

4.3
Are Gametic or Early Embryonic Imprints Later Converted to Heterochromatin?

Heterochromatin exhibits replicative persistence and reversibility, so the original imprints could be converted to differential chromatin states and then lost in early embryogenesis. More generally, a variety of different kinds of imprints could exist at different loci, in different developmental contexts, and in different taxa.

Studies of supernumerary (B) chromosomes in lecanoid males (Nur 1962; Nur and Brett 1988) suggest that the paternal set is eliminated because it is heterochromatic. Both paternally and maternally inherited B chromosomes exit premeiotic interphase already condensed, but in prophase they take on the less condensed appearance of the maternal A chromosomes, and they remain outside the clump of paternal A chromosomes in metaphase I. In meiosis II they tend to segregate preferentially with the maternal chromosomes, and are thereby included in sperm (Nur 1962; Nur and Brett 1988). This "selfish" accumulation mechanism maintains them within populations, despite their mildly deleterious effects on the individuals that carry them. Certain genetic backgrounds suppress this driving behavior of the B chromosomes, apparently by preventing them from converting to the euchromatic state of the maternal set in prophase I. These B chromosomes (still fully condensed) fail to escape the central paternal chromosome clump at metaphase I and thereby suffer elimination (Nur and Brett 1988). Thus, being heterochromatic would appear to be necessary, and perhaps sufficient, to ensure subsequent elimination in lecanoid systems, and, as discussed above, heterochromatization also appears to play a mechanistic role in diaspidid elimination (see Fig. 2).

4.4
Conditions for the Evolution of PGE in Coccoidea

All coccoids with PGE lack sex chromosomes and have achiasmate male meiosis, but the basal (non-PGE) coccoid lineages retain conventional XX-XO sex determination and male recombination (Fig. 3). Thus, heterogametic sex determination must have been converted to environmental (maternal) control in the line that gave rise to the ancestral PGE system. Conventional male-heterogametic sex determination is incompatible with PGE, because a male who transmitted his maternal chromosomes would thereby transmit an X in every sperm. If every egg also carried an X (from the XX mother), then every offspring of a male exhibiting PGE would be an XX genetic female (see Bull 1979). *Sciara* seems to violate this principle since it has XO males, but offspring sex is determined by the *mother*, not the *father's gamete*. The male germ line is fully diploid ($X^mX^pA^mA^p$), and all sperm carry $X^mX^mA^m$ chromosome complements. The male *soma* becomes *pseudo*-heterogametic ($X^mOA^mA^p$) during embryonic *development*, through differential chromosome *elimination*, as described above (Fig. 1). Thus the "sex chromosomes" of *Sciara* retain only vestiges of a conventional role in gametic sex determination. An analogous transformation of heterogamety occurs in aphids, where parthenogenic females (XX) produce sexual females (XX) and males (XO); in a male-destined egg, one X is eliminated during oogenesis. Males then produce 100% X-bearing sperm (and XX daughters) by aborting spermiogenesis of "O-bearing" meiotic products (reviewed by Blackman 1987).

Why might a system of heterogametic sex determination change to one of maternal control? Haig (1993a,b) points out that a meiotically driving X chromosome would create a female sex-ratio bias, thereby favoring mutations that convert XX zygotes into functional males. Maternal sex determination can also evolve as a way to create female biases, in structured populations where brother-sister matings are common and parents are (as a consequence) selected to reduce their investment in sons (Hamilton 1967, 1979; Bull 1983).

Male recombination also must have been suppressed early in the evolution of coccoid PGE, if not prior to its first appearance. It is incompatible with PGE because it creates chromosomes that are mosaics of maternally and paternally inherited segments. An eliminated chromosome would therefore be both paternal and maternal, in parts, and PGE at one locus would entail "MGE" at others (see Haig and Grafen 1991). Haig (1993a) has proposed a model for the evolution of lecanoid PGE, in which male recombination is suppressed as maternally inherited autosomal alleles "join" an already successfully driving maternal X chromosome. The model thereby derives both maternal sex determination and suppressed male recombination from a single primary event (X-chromosome meiotic drive). This is an attractive feature of the model, although suppressed male recombination (achiasmate meiosis) has evolved in many insect taxa, for reasons that remain obscure. Most of these taxa have conventional heterogametic (XO or XY) sex determination (White 1973; Bell 1982).

4.5
Conflicts of Interest in the Evolution of Coccoid PGE Systems

Coccoid PGE probably arose in the Mesozoic (Fig. 3, transition 2), so modern lecanoid systems may share many subsequently derived features that were not part of the first system. The mutation giving rise to the first system might well

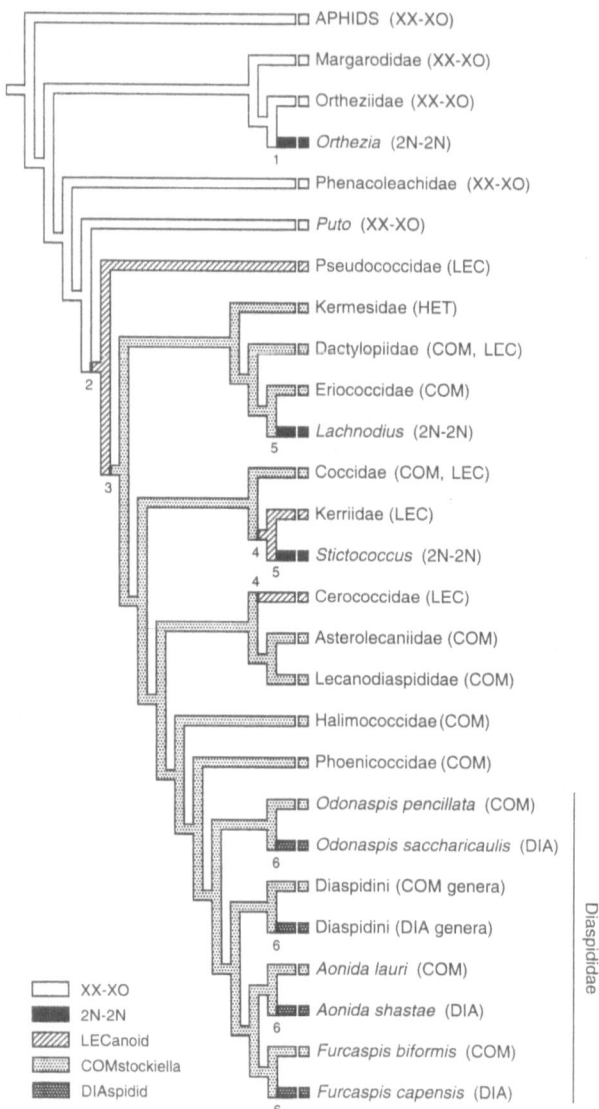

have been expressed only in the male germ line, just prior to meiosis, causing meiotic elimination of the paternal chromosome set. This mutation would have enjoyed an enormous transmission advantage, as explained above, and as first noted by Brown (1963, 1964), who called the resulting sweep to fixation an "automatic frequency response". In effect, PGE is a form of conditional (imprint-cued) meiotic drive. Like other driving genes, a maternally inherited PGE allele gains its transmission advantage at the expense of a victim (the

Fig. 3. A phylogenetic hypothesis for some major events in the evolution of coccoid PGE. The tree represents an informal synthesis of ideas from various published and unpublished sources, and should not be considered authoritative. Chromosome systems are keyed at the *lower left* and indicated after the name of each terminal taxon. Kosztarab and Kozár (1988) and Miller (1990) review coccoid evolution and systematics. Several poorly known families have been omitted for the sake of simplicity. Transitions between chromosome systems were inferred by *MacClade* (Maddison and Maddison 1992). The Diaspididae are represented by four lower-level clades that each include both Comstockiella and diaspidid elimination systems, to illustrate the hypothesis that diaspidid PGE has evolved several times; other tribes and genera (apparently monomorphic for one system or the other) are omitted. Basal taxa and the outgroup (aphids) have conventional XX–XO systems with chiasmate meiosis in males, but a 2N-2N system has evolved in *Orthezia* (transition *1*), possibly representing an instance of the driving-X scenario proposed by Haig (1993a). *Lecanoid system.* Late germline elimination is inferred to have arisen once (transition *2*), presumably from a 2N-2N ancestor. The Pseudococcidae (mealybugs) have been well studied, and except for the genus *Puto*, all nonparthenogenic species have lecanoid PGE. *Puto* is conventionally classified as a pseudococcid; it is shown here as basal, in keeping with the hypothesis that its XX-XO system (Brown and Cleveland 1968) is a relic of the ancestral condition for coccoids, not a rederivation (see Nur 1980). *Comstockiella system.* Partial premeiotic elimination is inferred to have arisen near the base of the clade including the remaining ("higher") coccoids (transition *3*); note that Comstockiella systems occur throughout this clade. Lecanoid systems occur sporadically in several of these families and are assumed to represent multiple losses of premeiotic elimination, as discussed in the text. Families are coded as Comstockiella if any of the included species is known to exhibit this system. Very few species of Kerriidae and Cerococcidae have been studied cytologically; these families are coded as lecanoid (transitions *4*), but further study might show that they, too, include Comstockiella species. Kermesidae are coded HET because male somatic cells contain a heterochromatic chromosome set, but spermatogenesis has not been studied (see Nur 1980). *"Breakout" species.* The 2N-2N systems of *Lachnodius* and *Stictococcus* (transitions *5*) appear to represent independent losses of PGE. Male meiosis is achiasmate in both genera (Brown 1977; Brown and Chandra 1977). Unfortunately, no detailed accounts of spermatogenesis yet exist for either of these 2N-2N lineages (see Nur 1980), so there remains a possibility that the cytology of one or both has been misinterpreted. *Multiple origins of the diaspidid system.* The Comstockiella system is primitive for Diaspididae. Three genera in the tribes Odonaspidini and Aspidotini each contain at least one species with the Comstockiella system and at least one with the diaspidid system, and there are Comstockiella-system and diaspidid-system genera in the tribe Diaspidini. This implies that there have been at least four independent origins of the diaspidid system (transitions *6*), and possibly more (Nur 1990a)

◄──

paternally inherited homologue), which therefore comes under strong selection to resist being eliminated.

How might such resistance be achieved? There are several possible routes, each of which may have been taken in different lineages and at different times. First, paternally inherited genes could completely block PGE, restoring a fair (but possibly achiasmate) meiosis, followed by functional spermiogenesis for all four meiotic products. One problem with this route is that orthodox male meiosis and spermiogenesis will not have occurred for many millions of years in a typical lineage with PGE, and so during this time some necessary genes may have been lost (see Nur 1970). Second, a single paternal chromosome could escape into the sperm-destined (otherwise maternal) haploid chromosome set, either by exchanging places with its maternal homologue, or by restoring a random Mendelian segregation of that particular homologue pair. Simply to become euchromatic and join the maternal set in the sperm (like a B chromosome) would be futile, because doing so would cause aneuploidy. Third, paternal genes could somehow cause the "eliminated" paternal sets to undergo spermiogenesis, rather than disintegrating, so that the male's pool of mature sperm would contain at least some all-paternal sperms along with the usual complement of all-maternal sperms.

Where and when might such resistance be effected? Because meiotic elimination takes place late in germline development, some paternal resistance genes also might be expressed in the germ line (or at least they might have been, during the early evolution of PGE, before the modern pattern of early lecanoid inactivation had been established). Paternal genes in somatic tissues (of the testis, as well as other parts of the body) might also join the resistance. For example, they could produce diffusible factors (such as hormones) that infiltrate germ cells, affecting chromatin states, meiotic processes, or spermiogenesis. Paternal chromosomes are heterochromatic in both the soma and the germ line of males with modern lecanoid and Comstockiella systems (presumably reflecting maternal suppression of paternal "breakouts", as discussed further below), and heterochromatization is clearly controlled, at least in large part, by genes expressed from the maternally inherited chromosomes (Nur 1990a). However, heterochromatic states can be unstable. For example, in some lecanoid genetic backgrounds the paternal chromosomes tend not to remain fully condensed in meiotic prophase (Nur and Brett 1988), and in many taxa heterochromatic states are reversed in some somatic tissues (Nur 1967, 1980, 1990b; Brown and Chandra 1977), possibly to restore benefits of heterozygosity that are unavailable to cells in which the paternal genome is silenced. Such decondensations pose a danger to maternal interests, since they invite paternal counter moves.

An anti-PGE "counter mutation" that somehow causes a breakout need not be fully effective (or even very effective) to increase rapidly in frequency and go to fixation, because an allele that is occasionally transmitted (when paternally inherited) is much fitter than one that is never transmitted (when paternal). Likewise, even a modest transmission of paternal chromosomes will select strongly for maternal counter-counter mutations that squelch the partial breakout or "leak", since the leak inflicts a fitness loss on maternally inherited genes in those males in which it occurs. An alternating succession of maternally expressed PGE mutations and paternally expressed anti-PGE counter mutations might therefore arise and fix. We suggest that such an evolutionary "struggle" between maternal and paternal genes drove the evolution of coccoid PGE systems. In this model (illustrated schematically in Fig. 4), the partial premeiotic eliminations that occur in Comstockiella systems, and the complete embryonic eliminations that occur in diaspidid systems, are both interpreted as advanced maternal answering moves to paternal counter moves that resulted in full or partial breakouts.

The scenario begins with a maternally expressed PGE mutation (M1) that causes meiotic elimination of the paternal chromosome set, perhaps by making it heterochromatic at the onset of meiosis (see Section 4.3, above, and Fig. 2). This maternal move elicits paternally expressed anti-PGE counter moves that restore a fair meiosis (P1). Such counter mutations might arise frequently, because all paternal genes suffer the same twofold loss of fitness. A successful anti-PGE mutation is likely to act earlier in germline development than the eliminator gene whose effect it blocks, because once the paternal set is heterochromatized (to cause its elimination), it is also silenced. The spread of such an anti-PGE counter mutation selects, in turn, for maternally expressed counter-counter mutations that act even earlier, and so on (M2).

At first, maternally inherited genes would have caused PGE without inactivating paternal genes in the soma, but as more paternally inherited genes evolved anti-PGE effects, inactivation of paternal genes in various somatic tissues (and at earlier times in development) would have been required to sustain reliable PGE. A practical limit to this progression toward earlier and more extensive inactivation is reached when the paternal set is made heterochromatic at just the time when it would first begin expression. In lecanoid systems it becomes heterochromatic early in embryonic development, before the germ line is set aside, at the time when transcription is first detected from the zygotic (maternal) genome (Sabour 1972). This is also near the time when, in diaspidid systems, the paternal set is eliminated. According to our model, these temporal associations are not accidental. Paternal genes cannot resist before they are turned on, so there is no selection to impair or eliminate them at an earlier stage of development, at least if they remain turned off later; but if they somehow manage to resist, later in development (P3), then stronger maternal measures (Comstockiella premeiotic elimination, M4, and diaspidid embryonic elimination, M5) may evolve in response.

Male vigor and fitness will necessarily be compromised by the inactivation of an entire chromosome set in most tissues, because this causes males to suffer the effects of deleterious recessive mutations (leaving aside potential issues of dosage compensation). The fact that heterochromatic states are reversed in certain tissues of at least some species (Nur 1967, 1980, 1990b) implies that euchromatic states could be restored in even more tissues. Why are they not? This pattern of widespread inactivation (with reactivation only in particular tissues) could be explained if paternal genes that are now inactivated formerly engaged in successful resistance. Their success could have favored maternally expressed mutations that caused the initial establishment (or restoration) of paternal heterochromatization in those tissues contributing to the resistance, despite a cost to the male's overall fitness, if maternal chromosomes thereby realized a sufficiently increased rate of differential transmission. Interestingly, in at least some lecanoid species, paternal chromosomes are reactivated in testis sheath and cyst cells, and paternal gene expression (presumably

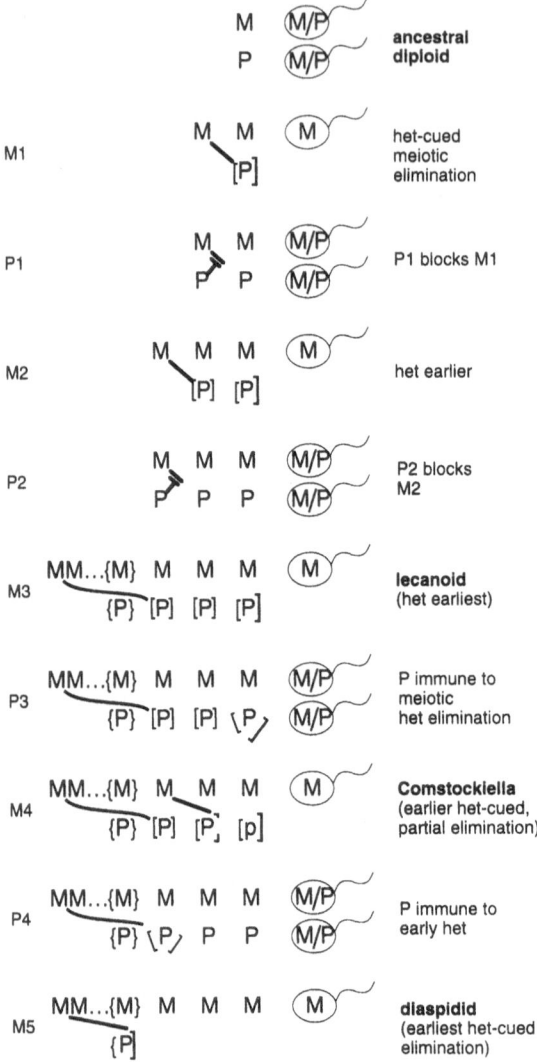

from those tissues) appears to be required for sperm development (Nur 1967, 1990b; Brown and Chandra 1977).

Among species with the Comstockiella variant of the lecanoid system, different numbers of paternal chromosomes (one, several, all but one, or all) are eliminated in the germ line, prior to meiosis. In some species the particular chromosome(s) eliminated may vary from cyst to cyst within an individual testis, and in some the number of chromosomes eliminated also varies from cyst to cyst (Nur 1965, 1980, 1990a; Kitchin 1975). This early elimination could represent a maternal counter-counter move (M4) that preempts a paternal counter move (P3) that, when it first arose, occasionally

Fig. 4. A model for the evolution of coccoid PGE systems. Evolutionary time is represented as proceeding from *top to bottom*. Developmental time proceeds *left to right*, toward spermiogenesis. Sperm contain only maternal genes (M) or Mendelian assortments from both parents (M/P). The M- and P-sets are shown at developmental stages beginning at the time of action of the most recent PGE or counter-PGE mutation. Ancestral coccoids had XX-XO sex determination and recombination in males, who were chromosomally and transmissionally diploid (Brown 1964; Brown and Chandra 1977; Nur 1980). As in *Sciara* (Smith-Stocking 1936), males expressed both paternal and maternal genes in many tissues, so deleterious recessive alleles existed at many loci. Sex determination came under facultative maternal control, and male recombination was lost, as discussed in the text. Successive PGE and counter-PGE mutations (M1–M5, P1–P4) then gave rise to the modern lecanoid, Comstockiella and diaspidid elimination systems. *Lecanoid system.* An initial PGE mutation (M1) eliminates some or all paternal chromosomes in meiosis II, perhaps by heterochromatizing the P-set in meiotic prophase. Heterochromatic P-sets ([P]) are shown with enlarged closing brackets (]) in the stage where they are eliminated. A counter-PGE mutation (P1), expressed from the P-set earlier than M1 (and possibly in tissues other than the germ line), blocks the effect of M1 and restores full or partial transmission of the P-set. M2 then arises, causing earlier heterochromatization of the P-set (possibly in tissues other than the germ line), suppressing P1 and restoring efficient PGE. Analogues of M2 that act even earlier (and affect even more of the soma) may also arise, but they expose their male carriers to the effects of more deleterious recessive mutations expressed from the M-set in tissues where the P-set is inactivated. By acting only slightly earlier than P1 (and only in relevant tissues), M2 exposes relatively few such mutations and is therefore more easily fixed. Recessive mutations are gradually removed from the gene pool as successively earlier-acting (and more widely-acting) M2-like (and P1-like) mutations fix. In lecanoid males, the P-set becomes heterochromatic at the same time that it would first be expressed (and the M-set is first expressed), as the cleavage nuclei migrate to the periphery of the embryo (Sabour 1972). Both sets are shown as inactive prior to this time ({M}{P}). The earliest-acting lecanoid mutations (M3-like) are expressed not from the M-set, but from the diploid maternal genome during oogenesis. *Comstockiella and diaspidid systems.* A counter-mutation (P3) allows the meiotic P-set to become euchromatic and escape elimination. M4 then causes premeiotic (Comstockiella) elimination of some paternal chromosomes, facilitating a reestablishment of heterochromatization of the remainder (p) and their lecanoid elimination in meiosis II. Another counter-mutation (P4) appears and fixes, allowing the P-set to escape later elimination in meiotic prophase or meiosis II (possibly by becoming euchromatic prior to meiotic prophase). Because deleterious recessive mutations are now quite rare (owing to the functionally hemizygous state that prevails in most tissues), the diaspidid mutation (M5) is a viable answer to P4. Expressed from the diploid maternal genome in oogenesis, it causes destruction of the P-set at the end of the rapid cleavages, before the germ line is set aside

spared at least some paternal chromosomes from elimination, by causing them to decondense and thereby compete with their maternal homologs for membership in the sperm-destined chromosome set. A partial premeiotic (Comstockiella) elimination might help to block a paternal escape, if a reduced number of germline paternal chromosomes experienced an effectively higher concentration of some maternally expressed substance needed to keep them securely heterochromatic, silent, and clumped at critical stages of meiosis. If the efficiency of PGE were increased as a consequence, then the genes causing early partial elimination could increase in frequency under selection. Comstockiella elimination also could have evolved to thwart the escape of entire paternal chromosome sets from lecanoid elimination (see below).

Embryonic elimination (M5) represents a decisive maternal move, because no answering paternal counter move is possible in a male with no paternal chromosomes. Mechanistic (developmental) obstacles might often slow or prevent the evolution of embryonic elimination, but it should evolve relatively easily in lineages with histories of embryonic paternal gene inactivation, where male development and physiology will have had time to adapt, gradually, to an effectively haploid somatic genome (as discussed in the legend to Fig. 4).

In 1963, Brown proposed an amazingly modern move-countermove scenario to explain evolutionary transitions between lecanoid and Comstockiella systems. In this model, some mutations go rapidly to fixation by "automatic frequency response"; others that counteract the effects of these previously fixed mutations then arise and fix; and so on. Unfortunately, Brown's model makes two debilitating assumptions. First, it assumes (incorrectly) that when paternal chromosomes seem to disappear in Comstockiella meiotic prophase, they merely become *invisible* and will, in fact, be *transmitted* in sperm. Thus the model seeks to explain a nonevent. Second, the model assumes that paternal chromosomes are heterochromatic in *all* male tissues, and that "fertility factors" therefore must be expressed from such heterochromatic paternal chromosomes (Nelson-Rees 1962). A few years later, Nur (1967) showed that paternal chromosomes become euchromatic and genetically active in some tissues of the testis, thereby explaining more plausibly why the sons of heavily X-irradiated male mealybugs are usually infertile. Although badly misled by these two assumptions, Brown clearly saw that PGE involves conflicts of interest. Had his life not ended tragically at the height of his career (Dempster et al. 1978), he surely would have constructed a successful model.

4.6
Evidence of "Breakouts"

A paternal counter mutation that fully restores diplodiploidy to a population with a prior history of PGE thereby creates a fully diploid species that is embedded phylogenetically within a clade for which PGE is clearly the ancestral condition. Two such apparently revertant (2N-2N) taxa are known in Coccoidea (Nur 1980; transitions 5 in Fig. 3), and our model clearly predicts that others should eventually be found. Of course, such revertants might often be short-lived, since new maternal mutations that restore PGE might well arise.

Do functional sperm ever develop from heterochromatic paternal sets? Uzi Nur (pers. comm.) notes that electron micrographic sections through mature sperm cysts of lecanoid- and Comstockiella-system males occasionally show more than the expected number of sperms (e.g., 17 or 18 instead of 16, in species where 16 is the standard number; see Robison 1990, Fig. 8, for an example). Molecular techniques could be used to screen thousands of sperm or hundreds of offspring at a time for rare instances of paternal allelic transmission (whether by this route or any other).

A partial premeiotic (Comstockiella) elimination could thwart this kind of breakout by making the paternal sets aneuploid. However, this maternal move is unlikely to succeed, over the long run, unless the development of aneuploid spermatids is usually blocked at some later stage. Such a block might well evolve, because every aneuploid sperm that fertilizes an egg wastes a potentially viable zygote. Both maternally and paternally inherited genes stand to benefit from a mechanism that reduces such wastage. Given a mechanism that detects and destroys aneuploid spermatids, the Comstockiella strategy should be relatively simple to implement, and highly effective.

One possible paternal response to this maternal strategy would be to convert the Comstockiella system back to a (leaky) lecanoid system, in which euploid (but heterochromatic) paternal sets are generated in meiosis. In Comstockiella systems where variable numbers of paternal chromosomes are eliminated at meiotic prophase, as few as zero are eliminated in some cysts (Nur 1965; Kitchin 1975), in which "spermatogenesis resembles that of the lecanoid system" (Nur 1990a). There are also some apparently fully lecanoid species (zero in all cysts) with Comstockiella ancestors (Fig. 3). This variation can be interpreted as one reflection of an ongoing evolutionary struggle between paternally expressed breakout mutations and maternally expressed suppressors of such mutations.

The distribution of Comstockiella and diaspidid systems within the Diaspididae strongly implies that diaspidid systems arose on at least four (and probably six or more) occasions from Comstockiella systems, which appear to be primitive for Diaspididae (Nur 1980, 1990a; transitions 6, Fig. 3). The alternative possibility is that Comstockiella systems have been re-derived on several occasions from an ancestral diaspidid embryonic-elimination system, but this seems less likely on mechanistic grounds, since it demands the reactivation of a complex developmental program for germline behaviors that would have been unused for significant evolutionary time (see Bull and Charnov 1985). If diaspidid elimination evolves to squelch paternal leaks, then we might expect to find that diaspidid-system lineages tend to be closely related to relatively leaky Comstockiella-system lineages.

5
PGE and Arrhenotoky in Mites and Bark Beetles

The commonness of arrhenotoky relative to embryonic PGE may be overestimated for animals as a whole, because in many taxa the evidence for arrhenotoky consists largely of haploid chromosome numbers in adult male somas (see Bell 1982; Norton et al. 1993). In relatively few cases are there unequivocal demonstrations that unmated females produce normal males, and only rarely are there the careful cytogenetic studies needed to detect embryonic PGE (which often occurs inside the mother). In some taxa where males are known to be haploid and where unmated females are known not to produce sons, the genetic system has been described as "arrhenotokous gynogenesis", with the implication that fertilization is required to stimulate male development, although paternal chromosomes play no continuing role. Without a cytological study of early embryos, this hypothesis cannot be distinguished from that of embryonic PGE. Thus some taxa presently characterized as arrhenotokous are undoubtedly pseudoarrhenotokous. For example, arrhenotoky is widespread in mites, but the phytoseiid mites appear to be largely pseudoarrhenotokous, with an embryonic PGE system similar in many ways to the coccoid diaspidid system (see Treat 1965; Nelson-Rees et al. 1980; Schulten 1985; Norton et al. 1993; Sabelis and Nagelkerke 1993).

The beetle family Scolytidae has long been known to include many arrhenotokous species (reviewed by Kirkendall 1993). One member of this family, the coffee berry borer *Hypothenemus hampei*, recently has been shown to be pseudoarrhenotokous, with germline elimination that bears striking similarities to the lecanoid system (Brun et al. 1995; Borsa and Coustau 1996; Borsa and Kjellberg 1996). In somatic cells the paternal chromosome set appears "compacted into a darkly staining ball of chromatin", and during the single meiotic division in spermatogenesis "the paternally derived set begins to degenerate while the maternally derived set condenses and divides" (Brun et al. 1995). This remarkable discovery will undoubtedly stimulate a search for other pseudoarrhenotokous scolytids that have been misclassified as arrhenotokous. The family includes species with diploid, haplodiploid, and (now) parahaplodiploid genetic systems, and these systems seem to be correlated with aspects of ecology and population structure (Kirkendall 1993), so the scolytids may present unusually good opportunities to reconstruct the evolutionary histories of PGE systems. They may also present opportunities to compare independently derived PGE mechanisms, both within the family and between the family and other groups such as coccoids.

6
PSR and *Wolbachia* as Direct-acting Eliminators

In the wasp *Nasonia vitripennis*, a zygote's paternal chromosomes may be eliminated by either of two mechanisms that act at the first (gonomeric) mitosis. The first mechanism depends on a B chromosome called *PSR*, for paternal sex ratio (Nur et al. 1988). When a sperm from a male carrying *PSR* fertilizes an egg, the paternal genome (except for *PSR*) fails to complete the first mitosis and is lost by anaphase lag; but *Nasonia* is haplodiploid, like most Hymenoptera, so the resulting haploid embryo develops as a normal male. The

PSR chromosome escapes the destruction inflicted on the paternal set, so an embryo that would otherwise have been diploid and female turns into a haploid male carrying *PSR*. This "ultra-selfish" strategy is expected to maintain *PSR* at high frequencies only in populations where females fertilize more than 50% of their eggs. *Nasonia* females do this (as predicted) in situations where their daughters and sons are likely to mate with each other (see Werren 1991).

The second PGE mechanism of *Nasonia* is effected by intracellular parasitic bacteria of the genus *Wolbachia*. These bacteria cause cytoplasmic incompatibility by interfering with paternal chromosome segregation in the gonomeric mitosis in embryos whose parents are infected with *Wolbachia* of incompatible genotypes, or whose mother is *not* (but father *is*) infected. In many other insects, including *Drosophila*, *Wolbachia* also destroy chromosomes (presumably paternal ones), and the affected embryos die (reviewed by O'Neill et al. 1992, 1997; Rousset et al. 1992; Moran and Baumann 1994; Werren et al. 1995; Lassy and Karr 1996; Werren 1997). In *Nasonia* the paternal genome fails in the same way and at the same stage of development; but instead of killing the female-destined embryo, this elimination converts it to a viable haploid male. *Wolbachia* are transmitted through the egg, not the sperm, so a seemingly "spiteful" conversion of females to males can increase the local frequency of *Wolbachia*-infected females by reducing the frequency of uninfected females (Hurst 1991; Hurst et al. 1997; see Werren 1997). When an infected male mates with a female infected by the same strain of *Wolbachia*, the zygote's paternal chromosomes survive (as in *Drosophila*), indicating that the *Wolbachia*-infected egg cytoplasm provides a substance that rescues the paternal chromosomes from destruction (see Lassy and Karr 1996; Werren 1997).

These two mechanisms of PGE differ in that one involves a supernumerary chromosome (*PSR*), while the other involves infection by a microorganism (*Wolbachia*). In addition, they can exist and function separately or together (Dobson and Tanouye 1996). In incompatible crosses, *Wolbachia* causes the destruction of *PSR* along with the rest of the paternal chromosome set, so *PSR*'s immunity to its own effect does not protect it from the effects of *Wolbachia* (Dobson and Tanouye 1996). However, despite these differences, the two mechanisms of PGE share striking genetic and cytological similarities, suggesting some deeper connection including (perhaps) related origins (Reed and Werren 1995). For example, *PSR* may have arisen from one or several paternal chromosome fragments generated during a cytoplasmic-incompatibility reaction caused by *Wolbachia* (Ryan et al. 1985; Nur et al. 1988; Reed 1993; Reed et al. 1994).

Do *PSR* and *Wolbachia* "imprint" the chromosomes they manipulate? The literature on these systems often invokes imprints, laid down in the male by *PSR* or *Wolbachia*; but these eliminations do not obviously involve chromosomal marks that are *replicated* along with the chromosomes. Some sort of "lesion" must be created during spermatogensis, and it must *persist* through one round of replication, but it need *not* be *replicated*. Presumably, a compatible *Wolbachia*-infected egg cytoplasm provides a substance that "repairs" the "lesion" that otherwise would lead to destruction of the paternal chromosomes. In a male that carries *PSR*, similar lesions could be created during spermatogenesis, with similar consequences in the zygote. Interpretations of *PSR*'s effects are complicated, however, by its physical presence in the affected mitosis.

In short, the paternal genome eliminations induced by *PSR* and *Wolbachia* can be explained by direct effects of these selfish chromosome-manipulating parasites on the chromosomes to be eliminated. There is no need (yet) to postulate a heritable epigenetic mark that would have no effect unless it were read by a mechanism that responded to the information contained in the mark. The purpose of this cautious interpretation is to focus attention on the essential features of Metz's original concept and definition of imprinting. We do not claim that an involvement of imprints in these systems can be ruled out, but merely that none has yet been established. Indeed, as mentioned above, we are intrigued by the possibility that systems of direct PGE, such as these, may open windows into the early evolution of at least some forms of imprint-mediated PGE.

7
Perspective on Imprints and Conflicts of Interest

We have emphasized the relationships between imprints and the mechanisms that use them as cues, but why do imprints exist? Some could be byproducts of gametogenesis. Others could have evolved to provide parent-of-origin information used to direct development in ways that benefit the entire organism and all its genes (as is assumed, uncritically, by many biologists). In either case, imprints would inevitably be, in addition, a source of information that could be exploited by mechanisms that implement selfish strategies driven by intragenomic conflicts of interest. Of course, some imprints may have evolved solely to inform genes that act out such conflicts. However, in that case they might well be opposed by genes elsewhere in the genome that never stand to benefit from the conflict.

Maternal and paternal contributions to the zygote are highly unequal, and these underscore the abundant opportunities for conflict that exist in eukaryotic development. Except for the pronuclei, most components of the zygote are derived primarily from one gamete or the other. The egg provides maternal mRNAs and diverse nutrients, and although paternal mitochondria are introduced from the sperm, maternal mitochondrial DNAs dominate numerically and paternal ones are usually destroyed (see Cosmides and Tooby 1981; Avise 1994; Meusel and Mortiz 1993; Godelle and Rebold 1995; Kaneda et al. 1995; Hurst et al. 1997; Pitnick and Karr 1998); likewise, intracellular bacteria such as *Wolbachia* are transmitted in the egg, not the sperm.

However, paternal factors can be introduced with or within the sperm (Agulnik et al. 1993; Yasuda et al. 1995; Browning and Strome 1996; Herrera et al. 1996; reviewed by Karr 1996). For example, in most animals (including humans) the sperm provides centriolar functions (minus certain components provided by the egg) as proposed by Boveri nearly a century ago (reviewed by Schatten 1994; Gall 1996; Sathananthan 1997). Mice (and possibly rodents generally) are exceptional, in that the egg provides the centriole. Schatten notes that the requirement for both paternal and maternal components to reconstitute zygotic centriolar function should prevent parthenogenesis, and he suggests that this may be the purpose of this system. He proposes that the exception to Boveri's rule in mice reflects an evolutionary replacement of this system with one of complementary (maternal and paternal) genomic imprints. Others have proposed that imprinting evolved to prevent parthenogenesis and ovarian tumors, on the grounds that imprinting lacks any other obvious adaptive function (reviewed and elaborated by Varmuza and Mann 1994a). This proposal has been highly controversial (Haig 1994; Moore 1994; Solter 1994; Varmuza and Mann 1994b; Haig and Trivers 1995), but it serves to remind us that the function(s) of imprinting remain frustratingly obscure. Is "imprinting" a misleadingly simple name for diverse phenomena with diverse causes? Might the various uniparentally contributed zygotic components represent just a few of many distinct sources of conflict that drive the evolution of imprint-cued phenomena such as paternal gene exclusion and differential gene inactivation?

Acknowledgments. We thank J. S. Burner, A. Burt, E. L. Charnov, B. de Saint Phalle, W. J. Dickinson, K. Golic, D. Haig, P. G. Johnston, T. L. Karr, S. Mango, D. L. Miller, N. A. Moran, U. Nur, S. L. O'Neill, M. H. Richards, H. Robertson, J. R. Roth, W. Sullivan, R. Trivers, and J. H. Werren for stimulating discussions, comments on the manuscript, and other forms of help. We dedicate this chapter to Uzi Nur, with gratitude for his teaching, and with deep admiration for his many extremely important contributions to genetics.

References

Abramczuk J, Sawicki W (1975) Pronuclear synthesis of DNA in fertilized and parthenogenetically activated mouse eggs. Exp Cell Res 92:361–372

Agulnik SI, Sabantsev ID, Ruvinsky AO (1993) Effect of sperm genotype on chromatid segregation in female mice heterozygous for aberrant chromosome 1. Genet Res 61:97–100

Avise JC (1994) Molecular markers, natural history and evolution. Chapman and Hall, New York

Bell G (1982) The masterpiece of nature: the evolution and genetics of sexuality. University of California Press, Berkeley

Berry RO (1941) Chromosome behavior in the germ cells and development of the gonads in *Sciara ocellaris.* J Morphol 68:547–583

Blackman, RL (1987) Reproduction, cytogenetics and development. In: Minks AK, Harrewijn P (eds) World crop pests, 2a. Aphids: their biology, natural enemies and control, vol A. Elsevier, Amsterdam, pp 163–195

Borsa P, Coustau C (1996) Single-stranded DNA conformation polymorphism at the *Rdl* locus in *Hypothenemus hampei* (Coleoptera: Scolytidae). Heredity 76:124–129

Borsa P, Kjellberg F (1996) Experimental evidence for pseudo-arrhenotoky in *Hypothenemus hampei* (Coleoptera: Scolytidae). Heredity 76:130–135

Brandriff BF, Segraves GR, Pinkel D (1991) The male-derived genome after sperm-egg fusion: spatial distribution of chromosomal DNA and paternal-maternal genomic association. Chromosoma 100:262–266

Breeuwer JAJ, Werren JH (1990) Microorganisms associated with chromosome destruction and reproductive isolation between two insect species. Nature 346:558–560

Brown SW (1963) The Comstockiella system of chromosome behavior in the armored scale insects (Coccoidea: Diaspididae). Chromosoma 14:360–406

Brown SW (1964) Automatic frequency response in the evolution of male haploidy and other coccid chromosome systems. Genetics 49:797–817

Brown SW (1965) Chromosomal survey of the armored and palm scale insects (Coccoidea: Diaspididae and Phoenicococcidae). Hilgardia 36:189–294

Brown SW (1977) Adaptive status and genetic regulation in major evolutionary changes of coccid chromosome systems. Nucleus 20:145–57

Brown SW, Chandra HS (1977) Chromosome imprinting and the differential regulation of homologous chromosomes. In: Goldstein L, Prescott DM (eds) Cell biology, a comprehensive treatise, vol I. Academic Press, New York, pp 135–189

Brown SW, Cleveland C (1968) Meiosis in the male of *Puto albicans* (Coccoidea-Homoptera). Chromosoma 24:210–232

Brown SW, Nelson-Rees WA (1961) Radiation analysis of a lecanoid genetic system. Genetics 46:983–1007

Brown SW, Wiegmann LI (1969) Cytogenetics of the mealybug *Planococcus citri* (Risso) (Homoptera: Coccoidea): genetic markers, lethals and chromosome rearrangements. Chromosoma 28:255–279

Browning H, Strome S (1996) A sperm-supplied factor for embryogenesis in *C. elegans*. Development 122:391–404

Brun LO, Borsa P, Gaudichon V, Stuart JJ, Aronstein K, Coustau C, and ffrench-Constant RH (1995) "Functional" haplodiploidy. Nature 374:506

Bull JJ (1979) An advantage for the evolution of male haploidy and systems with similar genetic transmission. Heredity 43:361–381

Bull JJ (1983) Evolution of sex determining mechanisms. Benjamin Cummings, Menlo Park

Bull JJ, Charnov EL (1985) On irreversible evolution. Evolution 39:1149–1155

Carmona JA, Sanjur OI, Doadrio I, Machordom A, Vrijenhoek RC (1997) Hybridogenetic reproduction and maternal ancestry of polyploid Iberian fish: the *Tropidophoxinellus alburnoides* complex. Genetics 146:983–993

Chandra HS, Brown SW (1975) Chromosome imprinting and the mammalian X chromosome. Nature 253:165–168

Cimino MC (1972) Egg-production, polyploidization and evolution in a diploid all-female fish of the genus *Poeciliopsis*. Evolution 26:294–306

Cole FR (1969) The flies of North America. University of California Press, Berkeley

Cosmides LM, Tooby J (1981) Cytoplasmic inheritance and intragenomic conflict. J Theor Biol 89:82–129

Crouse HV (1960) The controlling element in sex chromosome behavior in *Sciara*. Genetics 45:1429–1443

Crouse HV (1966) An inducible change in state on the chromosomes of *Sciara*: its effects on the genetic components of the X. Chromosoma 18:230–253

Crouse HV (1979) X heterochromatin subdivision and cytogenetic analysis in *Sciara coprophila* (Diptera, Sciaridae). II. The controlling element. Chromosoma 74:219–239

Davies DR (1974) Chromosome elimination in interspecific hybrids. Heredity 32:267–270

Dawley RM, Bogart JP (eds) (1989) Evolution and ecology of unisexual vertebrates. Bulletin 466, New York State Museum, Albany

Dempster E, Green MM, Nelson-Rees W, St Lawrence P (1978) Spencer Wharton Brown, 1918–1977. Genetics 88 (Suppl.): 137–138

de Saint Phalle B, Sullivan W (1996) Incomplete sister chromatid separation is the mechanism of programmed chromosome elimination during early *Sciara coprophila* development. Development 122:3775–3784

Dobson S, Tanouye M (1996) The paternal sex ratio chromosome induces chromosome loss independently of *Wolbachia* in the wasp *Nasonia vitripennis*. Dev Genes Evol 206:207–214

Donahue RP (1972a) Cytogenetic analysis of the first cleavage division in mouse embryos. Proc Natl Acad Sci USA 69:74–77

Donahue RP (1972b) Fertilization of the mouse oocyte: sequence and timing of nuclear progression to the two-cell stage. J Exp Zool 180:305–316

Du Bois AM (1933) Chromosome behavior during cleavage in the eggs of *Sciara coprophila* (Diptera) in the relation to the problem of sex determination. Z Wiss Biol Abt B – Z Zellforsch Mikrosk Anat 19:595–614

Formusoh ES, Hatchett JH, Black WC IV, Stuart JJ (1996) Sex-linked inheritance of virulence against wheat resistance gene H9 in the Hessian fly (Diptera: Cecidomyiidae). Ann Entomol Soc Am 69:428–434

Fuge H (1994) Unorthodox male meiosis in *Trichosia pubescens* (Sciaridae): chromosome elimination involves polar organelle degeneration and monocentric spindles in first and second division. J Cell Sci 107:299–312

Fuge H (1997) Nonrandom chromosome segregation in male meiosis of a sciarid fly: elimination of paternal chromosomes in first division is mediated by non-kinetochore microtubules. Cell Motil Cytoskel 36:84–94

Gall JG (1996) A pictorial history: views of a cell. American Society for Cell Biology, Bethesda

Gerbi SA (1986) Unusual chromosome movements in sciarid flies. In: Hennig W (ed) Results and problems in cell differentiation. Springer, Berlin Heidelberg New York, pp 71–104

Godelle B, Rebold X (1995) Why are organelles uniparentally inherited? Proc R Soc Lond B 259:27–33

Haig D (1993a) The evolution of unusual chromosome systems in coccoids: extraordinary sex ratios revisited. J Evol Biol 6:69–77

Haig D (1993b) The evolution of unusual chromosome systems in sciarid flies: intragenomic conflict and the sex ratio. J Evol Biol 6:249–261

Haig D (1994) Refusing the ovarian time bomb: three viewpoints and a reply. Trends Genet 10:346–349

Haig D, Grafen A (1991) Genetic scrambling as a defense against meiotic drive. J Theor Biol 153:531–558

Haig D, Trivers R (1995) The evolution of parental imprinting: a review of hypotheses. In: Ohlsson R, Hall K, Ritzen M (eds) Genomic imprinting, causes and consequences. Cambridge University Press, Cambridge, pp 17–28

Hamilton WD (1967) Extraordinary sex ratios. Science 156:477–488

Hamilton WD (1979) Wingless and fighting males in fig wasps and other insects. In: Blum MS, Blum NA (eds) Sexual selection and reproductive competition in insects. Academic Press, New York, pp 167–220

Hartl DL, Brown SW (1970) The origin of male haploid genetic systems and their expected sex ratio. Theor Popul Biol 1:165–190

Hayman DL, Rofe RH (1977) Marsupial sex chromosomes. In: Calaby JH, Tyndale-Biscoe CH (eds) Reproduction and evolution. Australian Academy of Sciences, Canberra, pp 69–79

Heard E, Philippe C, Avner P (1997) X-chromosome inactivation in mammals. Annu Rev Genet 31:571–610

Herrera JA, López-León MD, Cabrero J, Shaw MW, Camacho JPM (1996) Evidence for B chromosome drive suppression in the grasshopper *Eyprepocnemis plorans*. Heredity 76:633–639

Heslop-Harrison JS (1990) Gene expression and parental dominance in hybrid plants. Development Suppl. 1990:21–28

Howlett SK, Bolton VN (1985) Sequence and regulation of morphological molecular events during the first cell cycle of mouse embryogenesis. J Embryol Exp Morphol 87:175–206

Huettner A (1924) Maturation and fertilization in *Drosophila melanogaster*. J Morphol 39:249–265

Hughes-Schrader S (1948) Cytology of coccids. Adv Genet 2:127–203

Hughes-Schrader S, Ris H (1941) The diffuse spindle attachment of coccids, verified by the mitotic behavior of induced chromosome fragments. J Exp Zool 87:429–456

Hurst GDD, Hurst LD, Majerus MEN (1997) Cytoplasmic sex-ratio distorters. In: O'Neill SL, Hoffman AA, Werren JH (eds) Influential passengers: inherited microorganisms and arthropod reproduction. Oxford University Press, Oxford, pp 125–154

Hurst LD (1991) The evolution of intra-populational cytoplasmic incompatibility, or when spite can be successful. J Theor Biol 148:269–277

Kaneda H, Hayashi J, Takahama S, Taya C, Lindahl KF, Yonekawa H (1995) Elimination of paternal mitochondrial DNA in intraspecific crosses during early mouse embryogenesis. Proc Natl Acad Sci USA 92:4542–4546

Karr TL (1996) Paternal investment and intracellular sperm-egg interactions during and following fertilzation in *Drosophila*. Curr Top Dev Biol 34:89–115

Kirkendall LR (1993) Ecology and evolution of biased sex ratios in bark and ambrosia beetles. In: Wrensch DL, Ebbert MA (eds) Evolution and diversity of sex ratio in insects and mites. Chapman and Hall, New York, pp 235–345

Kitchin RM (1970) A radiation analysis of a Comstockiella chromosome system: destruction of heterochromatic chromosomes during spermatogenesis in *Parlatoria oleae* (Coccoidea: Diaspididae). Chromosoma 31:165–197

Kitchin RM (1975) Intranuclear destruction of heterochromatin in two species of armored scale insects. Genetica 45:227–235

Kosztarab M, Kozár F (1988) Scale insects of Central Europe. Junk, Dordrecht

Lassy CW, Karr TL (1996) Cytological analysis of fertilization and early embryonic development in incompatible crosses of *Drosophila simulans*. Mech Dev 57:47–58

Latham KE, McGrath J, Solter D (1995) Mechanistic and developmental aspects of genetic imprinting in mammals. Int Rev Cytol 160:53–98

Lee JT, Jaenisch R (1997) The (epi)genetic control of mammalian X-chromosome inactivation. Curr Opin Genet Dev 7:274–280

Leitch AR, Schwaracher T, Mosgoller W, Bennett MD, Heslop-Harrison JS (1991) Parental genomes are separated throughout the cell cycle in a plant hybrid. Chromosoma 101:206–213

Lyon MF (1993) Epigenetic inheritance in mammals. Trends Genet 9:123–128

Lyon MF, Rastan S (1984) Parental source of chromosome imprinting and its relevance for X chromosome inactivation. Differentiation 26:63–67

Maddison WP, Maddison DR (1992) MacClade: analysis of phylogeny and character evolution, Version 3. Sinauer, Sunderland

Mantovani B, Scali V (1992) Hybridogenesis and androgenesis in the stick-insect *Bacillus rossius-grandii benazzii* (Insecta, Phasmatodea). Evolution 46:783–796

Metz CW (1925) Chromosomes and sex in *Sciara*. Science 61:212–215

Metz CW (1938) Chromosome behavior, inheritance and sex determination in *Sciara*. Am Nat 72:485–520

Meusel MS, Mortiz RFA (1993) Transfer of paternal mitochondrial DNA during fertilization of honeybee (*Apis mellifera* L.) eggs. Curr Genet 24:539–543

Miller DR (1990) Phylogeny. In: Rosen D (ed) World crop pests, 4a. Armored scale insects: their biology, natural enemies and control, vol A. Elsevier, Amsterdam, pp 169–178

Moore T (1994) Refusing the ovarian time bomb: three viewpoints and a reply. Trends Genet 10:346–349

Moran NA, Baumann P (1994) Phylogenetics of cytoplasmically inherited microorganisms of arthropods. Trends Ecol Evol 9:15–20

Nelson-Rees WA (1962) The effects of radiation damaged heterochromatic chromosomes on male fertility in the mealybug, *Planococcus citri* (Risso). Genetics 47:661–683

Nelson-Rees WA, Hoy MA, Roush RT (1980) Heterochromatinization, chromatin elimination and haploidization in the parahaploid mite *Metaseiulus occidentalis* (Nesbitt) (Acarina: Phytoseiidae). Chromosoma 77:263–276

Norton RA, Kethley JB, Johnston DE, O'Connor BM (1993) Phylogenetic perspectives on genetic systems and reproductive modes of mites. In: Wrensch DL, Ebbert MA (eds) Evolution and diversity of sex ratio in insects and mites. Chapman and Hall, New York, pp 8–99

Nur U (1962) A supernumerary chromosome with an accumulation mechanism in the lecanoid genetic system. Chromosoma 13:249–271

Nur U (1963) Meiotic parthenogenesis and heterochromatization in a soft scale, *Pulvinaria hydrangeae* (Coccoidea: Homoptera). Chromosoma 14:123–139

Nur U (1965) A modified Comstockiella chromosome system in the olive scale insect, *Parlatoria oleae* (Coccoidea: Diaspididae). Chromosoma 17:104–120

Nur U (1967) Reversal of heterochromatization and the activity of the paternal chromosome set in the male mealy bug. Genetics 56:375–389

Nur U (1970) Translocations between eu- and heterochromatic chromosomes, and spermatocytes lacking a heterochromatic set in male mealy bugs. Chromosoma 29:42–61

Nur U (1980) Evolution of unusual chromosome systems in scale insects (Coccoidea: Homoptera). In: Blackman RL, Hewitt GM, Ashburner M (eds) Insect cytogenetics. Blackwell, Oxford, pp 97–117

Nur U (1990a) Chromosomes, sex-ratios, and sex determination. In: Rosen D (ed) World crop pests, 4a. Armored scale insects: their biology, natural enemies and control, vol A. Elsevier, Amsterdam, pp 179–190

Nur U (1990b) Heterochromatization and euchromatization of whole genomes in scale insects (Coccoidea: Homoptera). Development Suppl 1990:29–34

Nur U (1990c) Parthenogenesis. In: Rosen D (ed) World crop pests, 4a. Armored scale insects: their biology, natural enemies and control, vol A. Elsevier, Amsterdam, pp 191–197

Nur U, Brett BLH (1988) Genotypes affecting the condensation and transmission of heterochromatic B chromosomes in the mealybug *Pseudococcus affinis*. Chromosoma 96:205–212

Nur U, Werren JH, Eickbush DG, Burke WD, Eickbush TH (1988) A "selfish" B chromosome that enhances its transmission by eliminating the paternal genome. Science 240:512–514

Odartchenko N, Keneklis T (1973) Localization of paternal DNA in interphase nuclei of mouse eggs during early cleavage. Nature 241:528–529

O'Neill SL, Giordano R, Colbert AME, Karr TL, Robertson HM (1992) 16S RNA phylogenetic analysis of the bacterial endosymbionts associated with cytoplasmic incompatibility in insects. Proc Natl Acad Sci USA 89:2699–2702

O'Neill SL, Hoffmann AA, Werren JH (eds) (1997) Influential passengers: inherited microorganisms and arthropod reproduction. Oxford University Press, Oxford

Pitnick S, Karr TL (1998) Paternal products and by-products in *Drosophila* development. Proc R Soc Lond B Biol Sci 265:821–826

Quattro JM, Avise JC, Vrijenhoek RC (1991) Molecular evidence for multiple origins of hybridogenetic fish clones (Poeciliidae: *Poeciliopsis*). Genetics 127:391–398

Reed KM (1993) Cytological analysis of the paternal sex ratio chromosome of *Nasonia vitripennis*. Genome 36:157–161

Reed KM, Beukeboom LW, Eickbush DG, Werren JH (1994) Junctions between repetitive DNAs on the *PSR* chromosome of *Nasonia vitripennis* – association of palindromes with recombination. J Mol Evol 38:352–362

Reed KM, Werren JH (1995) Induction of paternal genome loss by the paternal sex ratio chromosome and cytoplasmic incompatibility bacteria (*Wolbachia*) – a comparative study of early embryonic events. Mol Reprod Dev 40:408–418

Rieffel SAM, Crouse HV (1966) The elimination and differentiation of chromosomes in the germ line of *Sciara*. Chromosoma 19:231–276

Robison WG Jr (1990) Sperm ultrastructure, behavior, and evolution. In: Rosen D (ed) World crop pests, 4a. Armored scale insects: their biology, natural enemies and control, vol A. Elsevier, Amsterdam, pp 205–220

Rousset F, Bouchon D, Pintureau B, Juchault P, Solignac M (1992) *Wolbachia* endosymbionts responsible for various alterations of sexuality in arthropods. Proc R Soc Lond B 250:91–98

Ryan SL, Saul GB 2nd, Conner GW (1985) Aberrant segregation of R-locus genes in male progeny from incompatible crosses in *Mormoniella*. J Hered 76:21–26

Sabelis MW, Nagelkerke K (1993) Sex allocation and pseudoarrhenotoky in phytoseiid mites. In: Wrensch DL, Ebbert MA (eds) Evolution and diversity of sex ratio in insects and mites. Chapman and Hall, New York, pp 512–541

Sabour M (1972) RNA synthesis and heterochromatization in early development of a mealybug. Genetics 70:291–298

Sathananthan AH (1997) Mitosis in the human embryo: the vital role of the sperm centrosome (centriole). Histol Histopathol 12:827–856

Schatten G (1994) The centrosome and its mode of inheritance: the reduction of the centrosome during gametogenesis and its restoration during fertilization. Dev Biol 165:299–335

Schilthuizen M, Stouthamer R (1997) Horizontal transmission of parthenogenesis-inducing microbes in *Trichogramma* wasps. Proc R Soc Lond B 264:361–366

Schrader F (1923) A study of the chromosomes in three species of *Pseudococcus*. Arch Zellforsch 17:45–62

Schulten GGM (1985) Pseudo-arrhenotoky. In: Helle W, Sabelis MW (eds) World crop pests, 1B. Spider mites: their biology, natural enemies and control, vol 1B. Elsevier, Amsterdam, pp 67–71

Schultz RJ (1969) Hybridization, unisexuality and polyploidy in the teleost *Poeciliopsis* (Poeciliidae) and other vertebrates. Am Nat 103:605–619

Schultz RJ (1977) Evolution and ecology of unisexual fishes. In: Heckt MK, Steere WC, Wallace B (eds) Evolutionary biology. Plenum Press, New York, pp 277–331

Schultz RJ (1989) Origins and relationships of unisexual poeciliids. In: Meffe MK, Snelson FF (eds) Ecology and evolution of livebearing fishes (Poeciliidae). Prentice Hall, Englewood Cliffs, New Jersey, pp 69–87

Smith-Stocking H (1936) Genetic studies on selective segregation of chromosomes in *Sciara coprophila* Lintner. Genetics 21:421–443

Solter D (1994) Refusing the ovarian time bomb: three viewpoints and a reply. Trends Genet 10:346–349

Stuart JJ, Hatchett JH (1988) Cytogenetics of the Hessian fly: II. Inheritance and behavior of somatic and germline-limited chromosomes. J Hered 79:190–199

Suomalainen E, Saura A, Lokki J (1987) Cytology and evolution in parthenogenesis. CRC Press, Boca Raton

Surani MAH (1986) Evidences and consequences of differences between maternal and paternal genomes during embryogenesis in the mouse. In: Rossant J, Pedersen RA (eds) Experimental approaches to mammalian embryonic development. Cambridge University Press, Cambridge, pp 401–435

Tinti F, Scali V (1992) Genome exclusion and gametic DAPI-DNA content in the hybridogenetic *Bacillus rossius-grandii benazzii* complex (Insecta Phasmatodea). Mol Reprod Dev 33:235–242

Tinti F, Scali V (1993) Chromosomal evidence of hemiclonal and all-paternal offspring production in *Bacillus rossius-grandii benazzii* (Insecta Phasmatodea). Chromosoma 102:403–414

Tinti F, Scali V (1995) Allozymic and cytological evidence for hemiclonal, all-paternal, and mosaic offspring of the hybridogenetic stick insect *Bacillus rossius-grandii grandii*. J Exp Zool 273:149–159

Treat AE (1965) Sex-distinctive chromatin and the frequency of males in the moth ear mite. J NY Entomol Soc 73:12–18

Varmuza S, Mann M (1994a) Genomic imprinting – defusing the ovarian time bomb. Trends Genet 10:118–123

Varmuza S, Mann M (1994b) Refusing the ovarian time bomb: three viewpoints and a reply. Trends Genet 10:346–349

Werren JH (1991) The paternal-sex-ratio chromosome of *Nasonia*. Am Nat 137:392–402

Werren JH (1997) Biology of *Wolbachia*. Annu Rev Entomol 42:587–609

Werren JH, Zhang W, Guo LR (1995) Evolution and phylogeny of *Wolbachia*: reproductive parasites of arthropods. Proc R Soc Lond B 261:55–71

White MDJ (1973) Animal cytology and evolution, 3rd edn. Cambridge University Press, Cambridge

White MJD (1978) Modes of speciation. Freeman, San Francisco

Whiting PW (1945) The evolution of male haploidy. Q Rev Biol 20:231–260

Wilson EB (1928) The cell in development and heredity, 3rd edn. Macmillan, New York

Wrensch DL, Ebbert MA (eds) (1993) Evolution and diversity of sex ratio in insects and mites. Chapman and Hall, New York

Wright SJ, Longo FJ (1988) Sperm nuclear enlargement in fertilized hamster eggs is related to meiotic maturation of the maternal chromatin. J Exp Zool 247:155–165

Yasuda GK, Schubiger G, Wakimoto BT (1995) Genetic characterization of ms(3)K81, a paternal effect gene of *Drosophila melanogaster*. Genetics 140:219–229

Zissler D (1992) From egg to pole cell: ultrastructural aspects of early cleavage and germ cell determination in insects. Microsc Res Tech 22:49–74

Imprinting and X-Chromosome Inactivation

Mary F. Lyon

1
Introduction

In normal female mammals one of the two X-chromosomes in every somatic cell is inactive i.e. it fails to transcribe RNA (reviews Gartler et al. 1992; Migeon 1994; Lyon 1996). The result of this is that chromosomally XX females and XY males both effectively have a single dosage of the products of X-linked genes. Thus X-chromosome inactivation fulfils the function of dosage compensation of X-linked genes. In eutherian mammals, typically either one of the two X-chromosomes in any cell, the maternally inherited Xm or the paternally inherited Xp, can be inactivated at random. Once the choice is made in each cell, it remains stable in all further cell generations in that individual and hence in the adult there are large clumps of cells with the same X-chromosome active. If the two X-chromosomes bear different alleles of a gene affecting some visible character, such as coat color, the clumps can be seen as a variegated effect. The best-known example of this is the tortoiseshell cat, in which the pattern results from the animal having a gene for ginger coat on one X-chromosome and black or tabby on the other. However, by contrast, in marsupials the same X-chromosome, the paternally derived Xp, becomes inactive in all cells (Cooper et al. 1993; Graves 1996). This preferential inactivation of Xp is seen also in some cells of the extraembryonic lineages of the embryo, which give rise to the placenta and other supporting tissues, in mice and rats (Takagi and Sasaki 1975), and probably, but less clearly, in humans also (Harrison 1989; Goto et al. 1997). Thus, in marsupials and in extraembryonic tissues of eutherians, the X-chromosome shows imprinting (Fig. 1). This imprinting has some similarities and some differences from autosomal imprinting. In order to understand the significance of imprinting in X-chromosome inactivation one must consider the mechanism by which the inactivation is brought about.

In the early female mouse embryo both X-chromosomes are active. X-chromosome inactivation (XCI) is seen first, at the blastocyst stage, in those tissues which show the imprinted type of inactivation, namely the trophectoderm and primitive endoderm (Fig. 1). Random XCI in the embryo

Medical Research Council, Mammalian Genetics Unit, Harwell, Didcot, Oxon OX11 0RD, UK

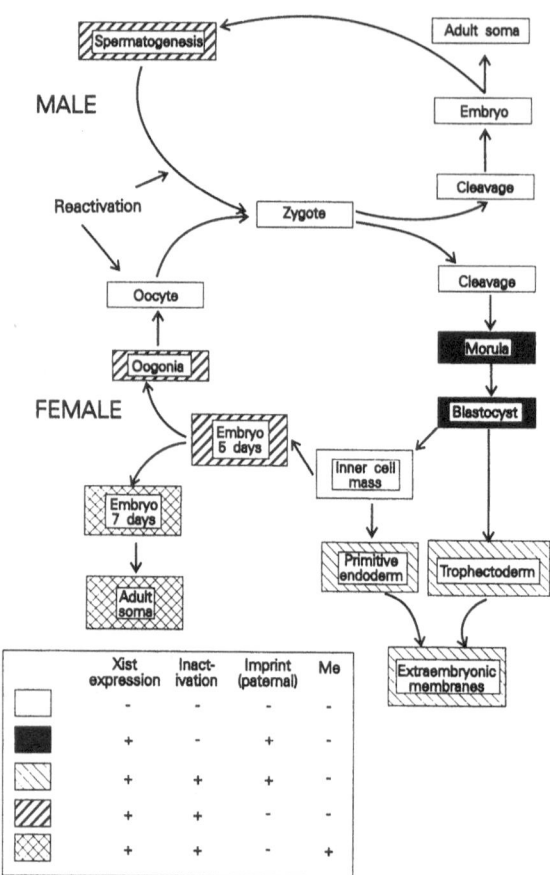

Fig. 1. Cycle of X-chromosome activity in the mouse, showing inactivation in the female embryo, followed by reactivation in the germ cells, and inactivation in the male germ cell, followed by reactivation in the new zygote Also shown is the relation of expression, imprinting and methylation (*Me*) of the *Xist* gene to the cycle of inactivation and reactivation. (Lyon 1993)

itself occurs somewhat later, at 5.5 to 8.5 days, the egg cylinder stage (Gartler et al. 1992; Lyon 1996). The first appearance of XCI is known as its initiation. The inactive X-chromosome, Xi, takes on the characteristic properties that it retains thereafter. It replicates its DNA late in the S phase of the cell cycle, it remains condensed during interphase, it shows hypoacetylation of histones, particularly histone H4 (Jeppesen and Turner 1993), and it lacks transcription. In the somatic cells of eutherians it also exhibits differential methylation of the CpG islands of housekeeping genes, but this methylation is not seen in marsupials, nor in the germ cells nor the extraembryonic tissues of eutherians (review Lyon 1996). The features of the Xi thus resemble those of heterochromatin, which also shows late replication, condensation, and hypoacetylation. Once the Xi has assumed its inactive state it retains this state and the only changes in activity occur in the germ cells. In the female germ cell the Xi is reactivated, so that both X-chromosomes are active, whereas in the male germ cell the single X-chromosome becomes inactive at about the time of meiosis (Fig. 1).

In both human and mouse, rare individuals are found with abnormal numbers of X-chromosomes. These may be XO females, or females or males with additional X-chromosomes, such as XXX or XXXX females or XXY males. In these individuals a single X-chromosome retains the active state and all others take on the properties of Xi. In triploids, however, with three complete sets of chromosomes, either one or two X-chromosomes, and in tetraploids, with four chromosome sets, two X-chromosomes remain active. There is thought to be a counting mechanism which maintains one X-chromosome active per two autosome sets.

2
The X-Inactivation Center

The initiation of XCI originates from the X-inactivation center (XIC) on the X-chromosome. From the XIC a signal spreads along the chromosome bringing about its heterochromatic state. Initiation thus involves counting of X-chromosomes, choice of one to remain active, and induction of spread of the distinctive properties.

The evidence for the existence of the XIC has come from X-chromosome translocations and deletions. When an X-chromosome is involved in a translocation with an autosome inactivation spreads into the attached autosomal segment. However, only one of the two X-chromosomal segments produces this effect and only this segment itself undergoes inactivation, the other segment remaining in the active state. The interpretation of this is that only the segment carrying the XIC can undergo XCI (Rastan and Brown 1990; Rastan 1994). From studies of different translocations and deletions, the XIC has been precisely located in both human and mouse. From this region a gene was cloned which had a unique pattern of expression, being active on the Xi and inactive on the active X-chromosome, Xa. This gene was symbolized *Xist* (*X* inactive specific transcript) (Borsani et al. 1991; Brockdorff et al. 1991; Brown et al. 1991). Experiments in mice have shown that *Xist* is essential for initiation of XCI. A convenient system for studying XCI is provided by embryonic stem (ES) cells. These are cultured cells derived from early mouse embryos. When XX ES cells are maintained in an undifferentiated state, both their X-chromosomes are active. If the cells are allowed to differentiate, however, initiation of XCI occurs (Rastan and Robertson 1985). Penny et al. (1996) made a knockout of the *Xist* gene in XX ES cells by deleting 7 kb of the first exon of the gene. When these cells were allowed to differentiate XCI occurred, indicating that the counting mechanism still recognized the X-chromosome with the deleted *Xist* gene. However, no cells were found with the deleted X-chromosome as the Xi. Cells either had the normal X inactive or both X-chromosomes active. Thus, the deleted chromosome could be chosen for XCI but could not then undergo inactivation. Work with transgenes carrying *Xist* has shown that the sequences for counting and choice of X-chromosomes, though not in the 7-kb deleted region, lie in or close to the *Xist* gene. Lee et al. (1996) made a 450-kb YAC (yeast artificial chromosome) transgene and Herzing et al. (1997) made a 40-kb cosmid transgene and introduced these into XY ES cells where they were integrated into autosomes. In both cases XCI occurred, showing that the counting system recognised the transgene, and either the autosome carrying the transgene or the normal X-chromosome could be inactivated. Thus, the combined knockout and transgene results show that the sequences for counting and choice are within 40 kb of *Xist* and an intact *Xist* RNA is needed to confer inactivity.

Another knockout, this time in XY ES cells, has provided further information. Marahrens et al. (1997) made a deletion of 15 kb of the *Xist* gene in XY ES cells. As expected, these cells survived normally in culture, as XCI is not needed in male cells. The XY ES cells were then introduced into embryos, and adult chimeric mice were obtained which carried the cells with the knockout. The chimeric males were fertile and transmitted the knockout to their female embryos. In male germ cells the single X-chromosome becomes inactive but this result shows that cells with the knockout could develop into mature sperm and that a functional *Xist* gene is not required for normal spermatogenesis. Thus, either *Xist* does not mediate XCI in spermatogenesis or inactivity of the X-chromosome is not a requirement for sperm. However, all the surviving ES cell-derived offspring of these chimeric males were male and thus had received the normal Y chromosome and not the deleted X-chromosome. The authors provided evidence that the female embryos carrying the deleted *Xist* gene died through failure of XCI in the extraembryonic cells. In these animals the deleted X-chromosome was the Xp and hence should have undergone imprinted XCI. It failed to do so, and the normal Xm also failed to undergo XCI. These females thus had double dosage of X-linked genes in the extraembryonic lineages and died. The authors rescued these dying embryos by making chimeras with normal embryos. In these female chimeras the deleted X-chromosome was the Xm and they produced female offspring which survived normally, with the deleted X-chromosome active in all cells. Thus, the *Xist* gene is required for imprinted as well as for random XCI. Furthermore, an Xm cannot undergo imprinted XCI, even when the counting mechanism would require this. This confirms the work of Tada et al. (1993), who found that in female embryos with a supernumerary Xm, i.e., XmXmXp or XmXmY, or in parthenogenetic embryos, which are XmXm (Endo and Takagi 1981), the counting mechanism failed in the extraembryonic membranes. Both Xm chromosomes remained active and the embryos died. In the converse situation, however, in chromosomally XpO embryos, the single Xp remains active in the extraembryonic membranes (Papaioannou and West 1981). The counting mechanism can thus override the imprint on Xp but not on Xm.

3
Function of the *Xist* Gene

The evidence shows clearly that the initiation of XCI requires an intact *Xist* gene. Thus, to understand the imprinting seen in XCI one must know something of the function of the *Xist* gene. Sequencing has shown that the gene has no open reading frame, and does not code for a protein (Brockdorff et al. 1992; Brown et al. 1992). The RNA transcript is retained in the nucleus. In situ hybridization of RNA to the chromosome has shown that, in both human and mouse, the *Xist* RNA forms a clump in the nucleus very close to the Xi (Clemson et al. 1996; Panning and Jaenisch 1996). It is said to coat or paint the Xi, and appears to cover the whole length of this chromosome.

In order to find whether coating with *Xist* RNA brings about the peculiar properties of the Xi Lee and Jaenisch (1997) studied the effect of the *Xist* YAC transgene (mentioned previously) when inserted on mouse Chr 12. The *Xist* RNA coated Chr 12, as on a normal Xi. Histones on Chr 12 were hypoacetylated, as on Xi, the chromosome replicated its DNA late, and three Chr 12 genes at different distances from the inserted transgene were transcriptionally silenced. The presence of the *Xist* gene and its RNA on Chr 12 were thus sufficient to bring about in it three of the properties of a typical Xi. The other properties, differential methylation and condensation, were not studied. Hence, it appears that coating with *Xist* RNA is responsible for the spreading of XCI along the chromosome, and no X-specific sequences along the chromosome are required.

Whether *Xist* is also involved in the maintenance of XCI after its initiation in the embryo is less clear. In female germ cells reactivation of the Xi as the cells approach meiosis is accompanied by loss of expression of *Xist* (McCarrey and Dilworth 1992). Similarly, XCI in cultured human placental cells (chorionic villus cells) can be reversed by artificially hybridizing them with cultured mouse cells and, in this case also, expression of *Xist* is lost (Luo et al. 1995). Reversal of XCI can also be brought about by fusing adult mouse somatic cells with certain lines of embryonal carcinoma (EC) cells, derived originally from early embryos. This reversal is accompanied by change of expression of *Xist* (Mise et al. 1996). However, in two studies, deletion of *Xist* from human adult cells did not result in reactivation of the affected X-chromosome (Brown and Willard 1994; Rack et al. 1994) and treatment of cultured cells with the demethylating agent 5-azacytidine led to reactivation of some X-linked genes but expression of the *Xist* gene continued (Hansen et al. 1996). Thus, in human adult cells, *Xist* expression is not necessary or sufficient to maintain the inactive state of the X-chromosome. It may be that maintenance of XCI in eutherian adult cells is complex. Differential methylation of CpG islands is thought to play a part in maintenance of XCI, and late replication may be involved also (reviewed Riggs 1990; Riggs and Pfeifer 1992; Lyon 1996). In germ cells and in extraembryonic cells differential methylation does not occur. Loss of expression of *Xist* may therefore be sufficient to bring about reactivation in these cells, but not in adult cells where methylation is present.

3.1
Imprinting of the *Xist* Gene

In view of the evidence for the essential role of *Xist* in initiation of XCI, it is reasonable to consider whether the imprinting of the X-chromosome is mediated by the imprinting of *Xist*. To seek evidence of imprinting, the expression of *Xist* in very early embryos has been studied. Kay et al. (1993) showed that in mouse embryos *Xist* is expressed very early, being detectable by PCR at the four-cell stage. In embryos with recognizably different *Xist* alleles on Xm and Xp, however, only Xp was expressed at early stages, and these authors did not see expression of the maternal allele of *Xist* until the late blastocyst stage. To investigate differential expression of *Xist* further, this group later studied experimentally produced embryos with abnormal parental origin of their chromosomes (Kay et al. 1994). Androgenetic embryos have two complete sets of paternally derived chromosomes whereas gynogenotes and parthenogenotes both have two complete chromosome sets of maternal origin. In androgenotes both Kay et al. (1994) and Latham and Rambhatla (1995) found that *Xist* was already expressed at the four- to eight-cell stage. In gynogenotes Kay et al. (1994) saw no expression of *Xist* until the morula to blastocyst stage, whereas Latham and Rhambhatla could detect expression as early as eight-cells, but at a much lower level than in androgenotes. Thus, in both normal and experimental mouse embryos, the expression of *Xist* shows an imprinting effect in early development, with earlier and stronger expression of the paternal allele.

An unexpected finding in the androgenotes was that *Xist* appeared to be expressed in almost all embryos in cleavage stages. Androgenotes can be chromosomally of three types: XX, XY, or YY. The YY embryos would be expected to die very early. The surviving embryos would thus be a mixture of XX and XY. As all or nearly all of these expressed *Xist,* this must mean that XY as well as XX embryos were expressing, and hence that the counting mechanism was not functional at early stages. It must develop at some later stage. Conversely, the imprinting mechanism, although present in early stages, must disappear before the onset of random XCI in the embryo proper. In very early human embryos, the picture is somewhat different, as expression of *Xist* has been found in both male and female embryos and hence on Xm as well as Xp (Daniels et al. 1997; Ray et al. 1997). Thus, as in the mouse it appears that the counting mechanism is not present in early stages, but in contrast the imprinted XCI seen in placental trophoblast later is apparently not preceded by imprinted expression of *Xist* in cleavage stages.

3.2
Differential Methylation of *Xist*

In view of the evidence for differential methylation of cytosine residues in CpG dinucleotides in DNA of autosomal imprinted genes, and also in view of the differential methylation of CpG islands of housekeeping genes on the inactive X-chromosome, studies have been made of the methylation of the *Xist* gene itself.

Norris et al. (1994) studied the methylation of sites in the promoter and 5' region of the gene. In males, all the sites studied were fully methylated. In females both methylated and unmethylated alleles were present. In females carrying the T(X;16)16H translocation the same X-chromosome, that involved in the translocation, is active in all cells. Norris et al. used these females to show that the methylated allele of *Xist* was that on the Xa, the translocated X-chromosome, whereas the allele on the inactive normal X-chromosome was unmethylated in all cells. In male germ cells, in which the X-chromosome becomes inactive, the methylation of *Xist* was lost at around the time of entry into meiosis, which is near the time at which the X becomes inactive, and this demethylation persisted in the mature sperm. Thus methylation of *Xist* was correlated with lack of activity of the gene and demethylation with activity.

In order to study whether differential methylation provided the imprint on *Xist* Ariel et al. (1995) and Zuccotti and Monk (1995) compared methylation at certain 5' sites in the gene in sperm and oocytes. Some sites were methylated in oocytes and remained so in the early embryo, but were unmethylated in sperm. Thus, differential methylation is a strong candidate to form the imprinting mark on *Xist*, and through *Xist* on the whole X-chromosome.

3.3
Developmental Changes in *Xist* Expression

Although *Xist* is expressed in early cleavage stages of mouse embryos, its level of expression is very low. Latham and Rambhatla (1995) found only a few hundred copies per embryo in early stages, whereas after initiation of XCI Buzin et al. (1994) found about 2000 copies per cell. Moreover, the low level of expression in early embryos did not bring about XCI; Latham and Rambhatla showed that several X-linked genes were expressed in androgenotes at eight-cell to blastocyst stages despite the expression of *Xist*. Further insight into levels of *Xist* RNA has come from in situ hybridization studies. In ES cells before differentiation, *Xist* is expressed from both X-chromosomes of female cells and the single X-chromosome of male cells. By in situ hybridization this expression is seen as a small dot overlying each X-chromosome (Lee et al. 1996; Panning and Jaenisch 1996). When differentiation, followed by XCI, occurs, expression is increased on one X-chromosome of female cells and the RNA then expands to coat the chromosome. On the other X-chromosome of female and the single

X-chromosome of male cells expression is extinguished (Fig. 2). The timing of expansion of the *Xist* RNA correlates well with the time of initiation of XCI. In ES cells the characteristic large clump of RNA appeared several days after the cells were allowed to differentiate (Panning et al. 1997; Sheardown et al. 1997), and in embryos it was seen in postimplantation stages at 5.5 to 8.5 days, when initiation of XCI was occurring in the embryo proper. In both ES cells and embryos at the relevant stage three types of cells were visible. First, cells with two dots of *Xist* RNA, overlying the two X-chromosomes, second, cells with one dot and one large clump of RNA, and third, cells with a large clump of RNA only (Fig. 2). With the passage of time the first two types of cells disappeared, and all cells showed a single large clump of RNA, the characteristic pattern in adult somatic cells (Panning et al. 1997; Sheardown et al. 1997). In male ES cells and embryos there were two types of cells, those with a single dot of *Xist* RNA, and those with no signal. In un-differentiated ES cells, before XCI, female cells all showed the double dot pattern and male cells the single dot. The interpretation of these data is that before XCI all X-chromosomes show a low level of *Xist* expression, and at initiation of XCI the level of *Xist* RNA increases markedly over the future Xi. In a transient stage the dot expression over the Xa persists, giving cells with one dot and a clump of RNA. Finally this is followed by extinction of *Xist* expression on the Xa in both female and male cells, giving the characteristic adult pattern (Fig. 2). Sheardown et al found that at the intermediate stage the X-linked gene *Pgk1* was still being expressed from both alleles in XX cells. Hence, at this stage, XCI was not complete. Thus, there are at least two steps in the change in *Xist* expression, first the increase in RNA over Xi, and second the ending of expression from Xa (Panning et al. 1997; Sheardown et al.

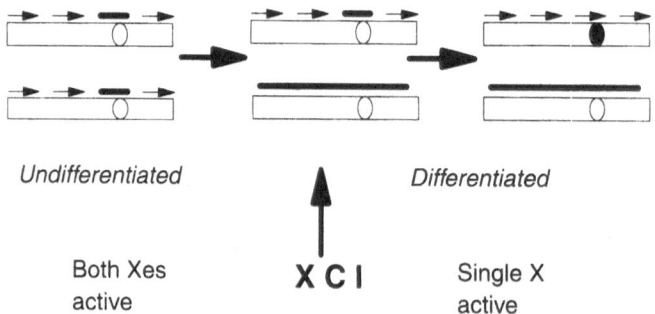

Undifferentiated   Differentiated

X C I

Both Xes   Single X
active   active

Fig. 2. Expression of the *Xist* gene, coating of the X-chromosome with *Xist* mRNA, and transcription of X-linked genes before and after XCI. The position of the *Xist* locus is denoted by an *ellipse*, *Xist* RNA by a *heavy line*, and transcription from the X-chromosome by *arrows*. The sequence of events is shown as in embryonic stem cells, according to Panning et al. (1997). In undifferentiated ES cells there is pinpoint expression of *Xist* mRNA from both X-chromosomes. At the onset of XCI *Xist* RNA on the future Xi is stabilized and expands to coat the X-chromosome, while pinpoint expression continues on the Xa. Soon afterwards the pinpoint expression on the Xa disappears, and the *Xist* gene on the Xa becomes inactive (shown as *solid ellipse*)

1997). Detailed studies of *Xist* RNA production showed that the increase in *Xist* RNA over Xi is not due to increased transcription. The rate of transcription remains constant but the RNA becomes stabilized, with a much longer half-life. There may be factors that stabilize *Xist* RNA on the Xi (Panning et al. 1997) or that destabilize it in the early embryo before XCI (Sheardown et al. 1997). The possibility of alternative splicing giving rise to the stable and unstable RNA types cannot be ruled out. Whatever the mechanism of stabilization of *Xist* RNA, another unanswered question is how the stabilizing or destabilizing factors are restricted in their distribution in the nucleus so that the Xi and Xa accumulate RNA differently. Yet another question is how the stabilized RNA is constrained to lie close to the Xi, rather than spreading diffusely in the nucleus. Panning et al. (1997) suggest that the RNA may be associated with additional factors.

An unexpected finding in Sheardown et al.'s work concerned pre-implantation embryos. In XX 8–16-cell embryos most or all cells showed a large clump of *Xist* RNA, and in some cells a small dot also, rather than the two dots that might have been expected in cells in which both X-chromosomes are active. In somewhat older embryos at the blastocyst stage, some cells still showed a clump of RNA, whereas others showed dots only. In 4–5-day hatched blastocysts, the presumptive inner cell mass (ICM) cells, which later form the embryo proper, showed two dots of *Xist* RNA. Some presumptive trophectoderm cells, in which imprinted XCI would be expected to occur, showed the clump and dot pattern of *Xist* RNA. The clump of RNA seen in these early embryos presumably arose from the Xp, since XY embryos, having the Xm only, never showed clumps. That the Xp should show stronger expression than Xm in early embryos agrees with this group's earlier work on imprinting of *Xist*. However, it is surprising that *Xist* RNA should build up at a stage when both X-chromosomes are active. It is also surprising that so many cells of 8–16-cell embryos showed the clump of RNA. Some of these cells must have been destined to become ICM cells, in which random XCI would later occur, and hence at some stage the clump of RNA over Xp must have reverted to the dot type. Sheardown et al. suggest that the buildup of RNA on the Xp may form the basis of the preferential XCI of Xp in extraembryonic cell lineages and that the reversion of RNA on the Xp to the dot type in ICM cells is connected with the erasure of imprinting, leading to random XCI in the embryonic lineage.

3.4
Effect of Lack of Methylation on Expression of *Xist*

In order to investigate the effect of changes in methylation on the developmental changes in *Xist* expression and on XCI Beard et al. (1995) and Panning and Jaenisch (1996) studied embryos and ES cells homozygous for knockouts of the gene for DNA methyltransferase (*Dnmt*). In undifferentiated XY ES cells mutant for *Dnmt* low-level dot expression of *Xist* occurred, as in normal ES cells. After differentiation, however, some cells showed a high level of expression of *Xist*, which formed a clump over the X-chromosome. These cells failed to express two X-linked genes *Pgk1* and *Mecp2*. Similarly, in embryos at 8.5 or 9.5 days gestation, a stage by which normally XCI is complete, some cells of XY embryos showed high level, clump, expression of *Xist*. In female XX embryos some cells showed two clumps of *Xist* RNA, accompanied by lack of expression of *Pgk1* and *Mecp2*. Thus, it appeared that lack of *Dnmt* and hence lack of methylation of *Xist* could lead to its inappropriate expression and this, in turn, to X-chromosome inactivation. In undifferentiated cells, however, before the normal time of XCI, lack of *Dnmt* did not alter expression of *Xist*. It is not clear why only some *Dnmt* mutant cells showed inappropriate expression of *Xist*. Beard et al. (1995) showed that the *Xist* gene is relatively resistant to lack of methylation, when compared with the autosomal genes *H19* and *Igf2r*, and it is possible that sufficient methylation persisted to keep the gene repressed in some cells. However, several other explanations are also possible (Panning and Jaenisch 1996).

These results show that XCI can occur without normal DNA methylation but the Xa may not retain its active state. In the *Xist* gene itself methylation appears to have two roles, one in forming the parental mark providing the imprint on the gene, and the other in maintaining its repressed state on the Xa.

4
Implications of Results with *Xist* for the Mechanism of XCI

Of the three types of XCI, imprinted, random and in male germ cells, *Xist* is clearly of key importance in the first two. The situation in male germ cells is unclear. The *Xist* RNA coats the X- and Y-chromosomes as it does in somatic cells (Ayoub et al. 1997). However, as mentioned earlier, spermatogenesis occurs normally in males with a deleted *Xist* gene (Marahrens et al. 1997). It is not clear whether *Xist* is not involved in the loss of X-chromosome activity in male germ cells, or whether this loss of activity is not needed for normal spermatogenesis. It is also possible that *Xist* is involved in XCI in male germ cells but is not the only factor.

4.1
Choice of X-Chromosome for XCI

Initiation of XCI in somatic cells involves choice and counting of X-chromosomes. In the mouse the choice of X for inactivation is affected by the *Xce* (X-controlling element) locus (Johnston and Cattanach 1981). This locus maps to the same point as the XIC. Four alleles of it are known which affect the chance of the X-chromosome on which they lie becoming the active one. They are graded in strength from Xce^a (weakest) to Xce^d (strongest) such that in heterozygotes the stronger allele is the one more likely to be on the X-chromosome that becomes the Xa. Thus, in Xce^c/Xce^a heterozygotes about 70% of cells have the X-chromosome bearing Xce^c as the Xa. The Xce effect is known from its phenotype and there is no underlying cloned gene at present. It maps very close to $Xist$ and the question arises whether the Xce effect is produced by variation at the $Xist$ locus. Brockdorff et al. (1991) found that the level of expression of $Xist$ was inversely related to the strength of the Xce allele on that chromosome. Hendrich et al. (1993) found a sequence difference in the $Xist$ gene in strains carrying different alleles of Xce, but could not say if this difference had a functional effect. On the other hand, Simmler et al. (1993) found a crossover between Xce and $Xist$, indicating that these are two separate gene loci with Xce lying 3' to $Xist$. In this region Courtier et al. (1995) found different levels of methylation on the Xa in adult male mice carrying different Xce alleles. The weaker alleles, Xce^a and Xce^b, were more highly methylated than the strong alleles Xce^c and Xce^d. Such methylation differences could be involved in the initiation of XCI, but their exact significance is not clear.

Apparent effects on choice of X-chromosome for inactivation have also been seen in humans. These effects are recognized by the occurrence of families in which females heterozygous for X-linked genes show skewing of inactivation away from the expected 50:50 pattern. In many cases this is not due to primary skewing of choice of X-chromosome for XCI, but rather due to random XCI followed by cell selection against one or other cell line. However, Plenge et al. (1997) found two families with females having highly skewed inactivation patterns in which these females carried a mutation in the $XIST$ promoter. The mutation resulted in reduced transcription, and all the females carrying it had preferential inactivation of the X-chromosome carrying the mutation. It is not clear how reduced transcription of $XIST$ should produce such an effect, but it obviously supports the other evidence on the importance of $XIST/Xist$ in counting and choice of X-chromosomes.

4.2
Counting of X-Chromosomes

The mechanism of counting is at present unknown but recent work has somewhat narrowed the possibilities. Counting is clearly involved in random XCI but there is an element of doubt whether it also takes part in imprinted XCI. The evidence that both Xm remain active in the extraembryonic membranes in sex chromosome aneuploid mice such as XmXmXp and XmXmY, as well as in XmXm parthenogenotes (Tada et al. 1993), suggests that counting is not involved, but that the gametic imprint alone determines activity. However, that Xp remains active in XpO females suggests that counting is present. If so, then the fact that two Xm can remain active restricts the possible options for the mechanism. Various suggestions have been made to explain how one chromosome can be singled out, including a single site in the nucleus to which the Xa can attach, and a limited quantity of some factor (Gartler and Riggs 1983). Such explanations would be unlikely if two Xm can remain active. Other suggestions include the cooperative binding of some factor (Gartler and Riggs 1983; McBurney 1988). A major event at the initiation of XCI is the stabilization of *Xist* RNA on the Xi. This could be due to activation of a stabilizing factor (Panning et al. 1997) or inhibition of a destabilizing factor (Sheardown et al. 1997), or alternative splicing could be involved. The counting process protects the future Xa from this change. The genes concerned, and whether they are autosomal or X-linked, are not known. In normal embryos an essential requirement for XCI is the presence of an Xa and it is formally possible that no specific autosomal genes take part, but that interactions among the X-chromosomes produce the effects. One possibility is that at a certain stage in development changes occur which favor the production of a stabilizing factor which binds at a site in the XIC. On the future Xa the binding site is blocked, either by an imprinting mark or a developmental mark. The blocking of this site then accelerates the production in unlimited quantities of the stabilizing factor, which then binds to all other X-chromosomes present. For such a model to work, production of the stabilizing factor would need to be much more rapid than the insertion of the block in the binding site. It could explain the activity of two Xm if the binding sites were blocked by an imprinting mark, and the activity of Xp in XpO, as it would acquire a developmental mark. The mark might involve methylation. In *Dnmt* mutants there might be enough residual methylation to insert a mark, and since the mark is needed to accelerate production of the stabilizing factor, and hence the occurrence of an Xi, X-inactivation could still occur. However, if the mark was weaker in these mutants, reversal of XCI might occur once there was a high level of stabilizing factor in the cell. As suggested by Panning et al. (1997), the state in which the Xa is still producing unstable *Xist* RNA but the stabilizing factor cannot bind may be rather unstable, and in a separate step *Xist* transcription on the Xa ceases. This model is a purely speculative one. Many other explanations are possible and further work will be needed to reveal the true basis of counting. At the least, the mechanism is likely to involve a site for a blocking factor, in or near the *Xist* gene but not in the regions deleted in the knockouts, and also a feedback or a cooperative binding system. The blocking factor may well involve methylation.

4.3
Spreading of XCI Along the X-Chromosome

Once the *Xist* RNA has become stabilized, it brings about the characteristic properties of the chromatin of the Xi, but as with counting and choice of X-chromosome, the mechanism of this is unknown. Firstly, it is not clear how the RNA is constrained to lie over the Xi rather than dispersing diffusely in the nucleus. It has been suggested that some association of the RNA with a protein is involved (Panning et al. 1997). Association of RNA with protein is involved in the chromatin changes that occur in dosage compensation in *Drosophila* (Kuroda and Meller 1997; Willard and Salz 1997). When a *Xist* transgene was inserted in an autosome the RNA coated the autosome. Hence the coating effect apparently does not require DNA sequences peculiar to the X-chromosome (Lee and Jaenisch 1997). However, the travel of the silencing effect along the autosome was not as complete as in the X-chromosome. This is in line with earlier work on mouse X-autosome translocations in which travel of inactivation into the attached autosomal material appeared not to be complete (Rastan 1983; Russell 1983). This suggests that X-chromosomal DNA may tend to enhance the spreading of XCI. The X-chromosome is known to be particularly rich in repetitive sequences of the LINE type (Boyle et al. 1990). Whether this is significant for travel of XCI is not clear.

A further question concerns how *Xist* RNA brings about the altered chromatin of the Xi. Hypoacetylation of histones on the Xi is found in marsupial XCI as well as eutherian (Wakefield et al. 1997) together with late replication of DNA. A change in acetylation, in the opposite direction, is also seen in dosage compensation in Drosophila (Baker et al. 1994). Thus, it is possible that *Xist* RNA mediates hypoacetylation which, in turn, leads to late replication, but the reverse with late replication leading to hypoacetylation is also possible.

As discussed above, the differential methylation of CpG islands seen in the Xi in somatic tissues of eutherians is thought to be involved in maintenance of XCI, and late replication also probably takes part by forming a feedback loop in which early replication is required for transcription, and transcription for early replication (Riggs 1990; Riggs and Pfeifer 1992). Also, as discussed above, the *Xist* RNA probably also takes part in maintaining the inactive state, as can be seen by loss of *Xist* expression accompanying reactivation of the X-chromosome in female germ cells and extraembryonic cells (McCarrey and Dilworth 1992; Luo et al. 1995).

5
Evolution of X-Chromosome Inactivation

It is interesting to consider the differences between eutherian, marsupial, and monotreme XCI and speculate on the evolution of the phenomenon for the possible insight this may give into the interrelation of XCI and imprinting. Very little is known of XCI in monotremes. Part of the X-chromosome replicates asynchronously in some tissues, which suggests that a limited form of XCI may occur (reviewed Graves 1996). In marsupials, as previously mentioned, XCI is wholly paternal. In addition, it is less complete and stable than in eutherians. Reversal of inactivation of some genes occurs in some tissues or in culture (Cooper et al. 1993). This lower stability is thought to be associated with the lack of differential methylation of CpG islands in marsupials.

From the fact that paternal XCI is seen in both marsupials and some cells of eutherians, it is suggested that paternal XCI was the primitive type (Graves 1996) and random XCI a later development in eutherians. It is also suggested that XCI in male germ cells may have occurred first (Lifschytz and Lindsley 1972). This may have provided a gametic imprint. Carryover of this imprint into the embryo after fertilization may have provided a basis for somatic XCI.

It is also possible that the counting of X-chromosomes, like random XCI, may have been a later development in eutherians. It is not clear whether counting occurs in marsupials. A gametic imprint, leading to activity of Xm and inactivity of Xp would give a single active X-chromosome in normal XmY males and XmXp females. In eutherians it is known from sex chromosome aneuploids (XXY, XXX, etc.) and polyploids that counting occurs, but in marsupials fewer of these are available for study. Little is known about initiation of XCI in general in marsupials. In the earliest embryos that have been studied, XCI was already present (reviewed Lyon 1996). It is possible that the complex changes in X-chromosome activity seen in embryos of eutherians do not occur, and that the Xp enters the new zygote inactive and remains so throughout life, except for reactivation in the germ cells. In view of the possible differences between marsupials and eutherians, and the importance of *Xist* in initiation, it is interesting that *Xist* has not yet been found in marsupials (Wakefield et al. 1997). It seems unlikely that it has been lost, as it is such a key factor in eutherian XCI. However, in the course of evolution its function and hence its sequence may have been so modified that it is not easily recognizable.

6
Comparison with Autosomal Imprinting

Cattanach and Beechey (1990) drew attention to the similarities between X-chromosome and autosomal imprinting. On the X-chromosome there is a gene susceptible to imprinting, *Xist*, the activity of which results, by a *cis*-acting effect, in inactivity not only of neighboring genes but also of those far distant on the chromosome. Similar patterns of an imprinted gene whose activity results in inactivity of neighboring genes are seen also in autosomes. The activity of *H19* results in inactivity of the nearby gene *Igf2* (Leighton et al. 1995). On mouse chromosome 17, a sequence producing an antisense RNA with its promoter in an intron of *Igf2r* is needed for imprinting of *Igf2r* (Wutz et al. 1997). A further similarity among these cases is that an untranslated RNA is involved in producing the inactive state. Another major similarity is in the role of differential methylation. In both XCI and autosomal imprinting, methylation seems to form the gametic mark on the key gene and may also serve to stabilize the inactive state at a later stage. The examples appear to differ in the way in which the expression competition between the active and inactive genes is brought about. A major difference also is the long range action of *Xist*, affecting a whole chromosome, and the very localized action of the autosomal genes. X-chromosome inactivation, and the long-range action of *Xist*, may be a special case of a fairly general imprinting mechanism (Cattanach and Beechey 1990). Possibly in the evolution of mammals, XCI began as a localized form of imprinting, with a small region of the Xp inactive and later evolved to become a chromosomal phenomenon. Later still, as discussed above, random XCI evolved in eutherians. Such suggestions would fit with ideas on the evolution of the mammalian X and Y chromosomes, and the role of XCI as a means of dosage compensation for X-linked genes. The suggestion (Graves 1996) is that originally the X- and Y-chromosomes were equal in size and carried homologous genes. As evolution progressed genes were gradually lost from the Y-chromosome. On a few occasions there were also additions of autosomal genes to the X- and Y-chromosomes, and these additions were again followed by loss of genes from the Y-chromosome. Thus, originally only a few genes on the X-chromosome had no homologues on the Y-chromosome and thus needed dosage compensation, and now nearly all X-linked genes require this. In present-day monotremes, it appears that only a small part of the X-chromosome may undergo XCI, and this, too, is in line with the idea that the earliest XCI may have been a localized inactivation (Graves 1996).

Acknowledgments. I am very grateful to Ms. Theresa Kent for secretarial help and to Dr. Bruce Cattanach for valuable comments on the manuscript.

References

Ariel M, Robinson E, McCarrey JR, Cedar H (1995) Gamete-specific methylation correlates with imprinting of the murine *Xist* gene. Nat Genet 9:312–315

Ayoub N, Richter C, Wahrman J (1997) *Xist* RNA is associated with the transcriptionally inactive XY body in mammalian male meiosis. Chromosoma 106:1–10

Baker BS, Gorman M, Marin I (1994) Dosage compensation in *Drosophila*. Annu Rev Genet 28:491–521

Beard C, Li E, Jaenisch R (1995) Loss of methylation activates *Xist* in somatic but not in embryonic cells. Genes Dev 9:2325–2334

Borsani G, Tonlorenzi R, Simmler MC, Dandolo L, Arnaud D, Capra V, Grompe M, Pizzuti A, Muzny D, Lawrence C, Willard HF, Avner P, Ballabio A (1991) Characterization of a murine gene expressed from the inactive X chromosome. Nature 351:325–329

Boyle AL, Ballard SG, Ward DC (1990) Differential distribution of long and short interspersed element sequences in the mouse genome: chromosome karyotyping by fluorescence in situ hybridization. Proc Natl Acad Sci 87:7757–7761

Brockdorff N, Ashworth A, Kay GF, Cooper P, Smith S, McCabe VM, Norris DP, Penny GD, Patel D, Rastan S (1991) Conservation of position and exclusive expression of mouse *Xist* from the inactive X chromosome. Nature 351:329–351

Brockdorff N, Ashworth A, Kay GF, McCabe VM, Norris DP, Cooper PJ, Swift S, Rastan S (1992)

The product of the mouse *Xist* gene is a 15 kb inactive X-specific transcript containing no conserved ORF and located in the nucleus. Cell 71:515–526

Brown CJ, Willard HF (1994) The human X-inactivation centre is not required for maintenance of X-chromosome inactivation. Nature 368:154–156

Brown CJ, Ballabio A, Rupert JL, Lafreniere RG, Grompe M, Tonlorenzi R, Willard HF (1991) A gene from the region of the human X inactivation centre is expressed exclusively from the inactive X chromosome. Nature 349:33–44

Brown CJ, Hendrich BD, Rupert JL, Lafreniere RG, Xing Y, Lawrence J, Willard HF (1992) The human *XIST* gene: analysis of a 17-kb inactive X-specific RNA that contains conserved repeats and is highly localised within the nucleus. Cell 71:527–542

Buzin CH, Mann JR, Singer-Sam J (1994) Quantitative RT-PCR assays show *Xist* RNA levels are low in mouse female adult tissue, embryos and embryoid bodies. Development 120:3529–3536

Cattanach BM, Beechey CV (1990) Autosomal and X-chromosome imprinting. Development Suppl: 63–72

Clemson CM, McNeil JA, Willard H, Lawrence JB (1996) *XIST* RNA paints the inactive X chromosome at interphase: evidence for a novel RNA involved in nuclear/chromosome structure. J Cell Biol 132:259–275

Cooper DW, Johnston PG, Watson JM, Graves JAM (1993) X-inactivation in marsupials and monotremes. Sem in Dev Biol 4:117–128

Courtier B, Heard E, Avner P (1995) *Xce* haplotypes show modified methylation in a region of the active X chromosome lying 3' to *Xist*. Proc Natl Acad Sci 92:3531–3535

Daniels R, Zuccotti M, Kinis T, Serhal P, Monk M (1997) *XIST* expression in human oocytes and preimplantation embryos. Am J Hum Genet 61:33–39

Endo S, Takagi N (1981) A preliminary cytogenetic study of X chromosome inactivation in diploid parthenogenetic embryos from LT/Sv mice. Jpn J Genet 56:349–356

Gartler SM, Riggs AD (1983) Mammalian X-chromosome inactivation. Annu Rev Genet 17:155–190

Gartler SM, Dyer KA, Goldman MA (1992) Mammalian X-chromosome inactivation. In: Friedmann T (ed) Molecular genetic medicine. Academic Press, New York vol 2. pp 121–160

Goto T, Wright E, Monk M (1997) Paternal X-chromosome inactivation in human trophoblastic cells. Mol Hum Reprod 3:77–80

Graves JAM (1996) Mammals that break the rules: genetics of marsupials and monotremes. Annu Rev Genet 30:233–260

Hansen RS, Canfield TK, Fjeld AD, Gartler SM (1996) Role of late replication timing in the silencing of X-linked genes. Hum Mol Genet 5:1345–1353

Harrison KB (1989) X-chromosome inactivation in the human cytotrophoblast. Cytogenet Cell Genet 52:37–41

Hendrich BD, Brown CJ, Willard HF (1993) Evolutionary conservation of possible functional domains of the human and murine *XIST* genes. Hum Mol Genet 2:663–672

Herzing LBK, Romer JT, Horn JM, Ashworth A (1997) *Xist* has properties of the X-chromosome inactivation centre. Nature 386:272–275

Jeppesen P, Turner BM (1993) The inactive X chromosome in female mammals is distinguished by a lack of histone H4 acetylation, a cytogenetic marker for gene expression. Cell 74:281–289

Johnston PG, Cattanach BM (1981) Controlling elements in the mouse IV. Evidence of non-random X-inactivation. Genet Res 37:151–160

Kay GF, Penny GD, Patel D, Ashworth A, Brockdorff N, Rastan S (1993) Expression of *Xist* during mouse development suggests a role in the initiation of X chromosome inactivation. Cell 72:171–182

Kay GF, Barton SC, Surani MA, Rastan S (1994) Imprinting and X chromosome counting mechanisms determine *Xist* expression in early mouse development. Cell 77:639–650

Kuroda MI, Meller VH (1997) Transient *Xist*-ence. Cell 91:9–11

Latham KE, Rambhatla L (1995) Expression of X-linked genes in androgenetic, gynogenetic, and normal mouse preimplantation embryos. Dev Genet 17:212-222

Lee JT, Jaenisch R (1997) Long-range *cis* effects of ectopic X-inactivation centres on a mouse autosome. Nature 386:275-279

Lee JT, Strauss WM, Dausman JA, Jaenisch R (1996) A 450 kb transgene displays properties of the mammalian X-inactivation center. Cell 86:83-94

Leighton PA, Ingram RS, Eggenschwiler J, Efstriadis A, Tilghman SM (1995) Disruption of imprinting caused by deletion of the *H19* gene region in mice. Nature 375:34-39

Lifschytz E, Lindsley DL (1972) The role of X-chromosome inactivation during spermatogenesis. Proc Natl Acad Sci 69:182-186

Luo S, Torchia BS, Migeon BR (1995) XIST expression is repressed when X inactivation is reversed in human placental cells: a model for study of *XIST* regulation. Somat Cell Mol Genet 21:51-60

Lyon MF (1993) Epigenetic inheritance in mammals. Trends Genet 9:123-128

Lyon MF (1996) Molecular genetics of X-chromosome inactivation. Adv Genome Biol 4:119-151

Marahrens Y, Panning B, Dausman J, Strauss W, Jaenisch R (1997) *Xist*-deficient mice are defective in dosage compensation but not spermatogenesis. Genes Dev 11:156-166

McBurney M (1988) X chromosome inactivation: a hypothesis. Bio Essays 9:85-88

McCarrey JR, Dilworth DD (1992) Expression of *Xist* in mouse germ cells correlates with X-chromosome inactivation. Nat Genet 2:200-203

Migeon BR (1994) X-chromosome inactivation: molecular mechanisms and genetic consequences. Trends Genet 10:230-235

Mise N, Sado T, Tada M, Takada S, Takagi N (1996) Activation of the inactive X chromosome induced by cell fusion between a murine EC and female somatic cell accompanies reproducible changes in methylation pattern of the *Xist* gene. Exp Cell Res 223:193-202

Norris DP, Patel D, Kay GF, Penny GD, Brockdorff N, Sheardown SA, Rastan S (1994) Evidence that random and imprinted *Xist* expression is controlled by preemptive methylation. Cell 77:41-51

Panning B, Jaenisch R (1996) DNA hypomethylation can activate *Xist* expression and silence X-linked genes. Genes Dev 10:1991-2002

Panning B, Dausman J, Jaenisch R (1997) X chromosome inactivation is mediated by *Xist* RNA stabilization. Cell 90:907-916

Papaioannou VE, West JD (1981) Relationships between the parental origin of the X chromosomes, embryonic cell lineage and X chromosome expression in mice. Genet Res Camb 37:183-197

Penny GD, Kay GF, Sheardown SA, Rastan S, Brockdorff N (1996) Requirement for *Xist* in X chromosome inactivation. Nature 379:131-137

Plenge RM, Hendrich BD, Schwartz C, Arena JF, Naumova A, Sapienza C, Winter RM, Willard HF (1997) A promoter mutation in the *XIST* gene in two unrelated families with skewed X-chromosome inactivation. Nat Genet 17:353-356

Rack KA, Chelly J, Gibbons RJ, Rider S, Benjamin D, Lafrenière RG, Oscier D, Hendriks RW, Craig IW, Willard HF, Monaco AP, Buckle VJ (1994) Absence of the XIST gene from late-replicating isodicentric X chromosomes in leukaemia. Hum Mol Genet 3:1053-1059

Rastan S (1983) Non-random X-chromosome inactivation in mouse X-autosome translocation embryos — location of the inactivation centre. J Embryol Exp Morphol 78:1-22

Rastan S (1994) X chromosome inactivation and the *Xist* gene. Curr Opin Genet Dev 4:292-297

Rastan S, Brown SDM (1990) The search for the mouse X-chromosome inactivation centre. Genet Res 56:99-106

Rastan S, Robertson EJ (1985) X-chromosome deletions in embryo-derived (EK) cell lines associated with lack of X-chromosome inactivation. J Embryol Exp Morphol 90:379-388

Ray PF, Winston RML, Handyside AH (1997) *XIST* expression from the maternal X chromosome in human male preimplantation embryos at the blastocyst stage. Hum Mol Genet 6:1323-1327

Riggs AD (1990) DNA methylation and late replication probably aid cell memory, and type I DNA reeling could aid chromosome folding and enhancer function. Philos Trans R Soc Lond B 326:285–297

Riggs AD, Pfeifer GP (1992) X-Chromosome inactivation and cell memory. Trends Genet 8:169–174

Russell LB (1983) X-autosome translocations in the mouse: their characterization and use as tools to investigate gene inactivation and gene action. In: Sandberg AA (ed) Cytogenetics of the mammalian X chromosome, Part A. Basic mechanisms of X chromosome behaviour. Alan R Liss, New York, pp 205–250

Sheardown SA, Duthie SM, Johnston CM, Newall AET, Formstone EJ, Arkell RM, Nesterova TB, Alghisi G-C, Rastan S, Brockdorff N (1997) Stabilization of *Xist* RNA mediates initiation of X chromosome inactivation. Cell 91:99–107

Simmler MC, Cattanach BM, Rasberry C, Rougeulle C, Avner P (1993) Mapping the murine *Xce* locus with $(CA)_n$ repeats. Mammal Genome 4:523–530

Tada T, Takagi N, Adler I-D (1993) Parental imprinting on the mouse X chromosome: effects on the early development of XO, XXY and XXX embryos. Genet Res Camb 62:139–148

Takagi N, Sasaki M (1975) Preferential inactivation of the paternally derived X chromosome in the extraembryonic membranes of the mouse. Nature 256:640–642

Wakefield MJ, Keohane AM, Turner BM, Graves JAM (1997) Histone underacetylation is an ancient component of mammalian X chromosome inactivation. Proc Natl Acad Sci 94:9665–9668

Willard HF, Salz HK (1997) Remodelling chromatin with RNA. Nature 386:228–229

Wutz A, Smrzka OW, Schweifer N, Schellander K, Wagner EF, Barlow DP (1997) Imprinted expression of the *Igf2r* gene depends on an intronic GpG island. Nature 389:745–749

Zuccotti M, Monk M (1995) Methylation of the mouse *Xist* gene in sperm and eggs correlates with imprinted *Xist* expression and paternal X-inactivation. Nat Genet 9:316–320

The Mechanisms of Genomic Imprinting

Bernhard Horsthemke[1], Azim Surani[2], Tharapell James[3], and Rolf Ohlsson[4]

1
Introduction

Genomic imprinting can be regarded as one of many variants of epigenetic modes of gene regulation in eukaryotic cells. There is no a priori reason, therefore, to invoke fundamentally novel mechanisms to explain the imprinting phenomenon in mammals. For example, the different factors involved in imprinting, such as the stable propagation of different chromatin states that repress or permit gene transcription, are well-known entities in a wide range of eukaryotic cells (see Pirotta Chap. 10, and Geramisova and Corces Chap. 11, this vol.). From this point of view, we should perhaps consider the mechanism(s) of imprinting as being the unusual result of combinatory events that occur regularly on an evolutionary scale.

When attempting to explain the imprinting mechanism(s), we face a number of pertinent issues: what is the character of an imprint and how is it propagated and interpreted? Does imprinting mark an allele for silencing or activation (or both)? Is there a common mechanism, or does a range of imprinting mechanisms exist which are particular for a subset of imprinted genes? A rational approach to these topics includes a dissection of the imprinting mechanism into discrete processes; such as (1) the resetting of the imprint during gametogenesis, (2) its propagation (somatic inheritance), and (3) interpretation in the soma. To be able to understand these subprocesses, it might be useful to first consider some common characteristics.

2
Many Imprinted Loci Are Organized in Subchromosomal Domains

The accumulated list of imprinted genes and their physical locations reveals that many imprinted genes are clustered to form replicon-sized

[1] Institut für Humangenetik, Universitätsklinikum Essen, Essen, Germany
[2] Wellcome/CRC Institute, Cambridge, UK
[3] Department of Biochemistry, University of Dublin, Dublin, Ireland
[4] Department of Animal Development and Genetics, Uppsala University, Uppsala, Sweden

subchromosomal domains (see Beechey, Appendix, this Vol.). Such regions may be functionally related to the imprinting phenomenon as suggested by the demonstration that the clusters of imprinted genes behave as single replicative units during the S phase. Both the major clusters at human chromosomes 11p15.5 and 15q11–q13 replicate asynchronously, such that the paternal allele replicates early and the maternal allele late (Kitsberg et al. 1993; Knoll et al. 1994). Conversely, neighboring domains can replicate asynchronously in opposite manners, i.e., the maternal allele replicates before the paternal allele (Bickmore and Carothers 1995; LaSalle and Lalande 1995). For the the cluster on distal chromosome 7 in the mouse, which is homologous with the human cluster at 11.p15.5, the *H19* locus appears to mediate a transition between imprinted and non-imprinted DNA replication patterns (Greally et al. 1998).

Taken together, the results show that epigenetic imprints mark subchromosomal regions to be placed in different nuclear environments (representative for the early and late S phase, respectively) depending on their parental origin. Although the obvious implication of these observations is that the transcriptional activity of imprinted genes gains or loses impetus depending on the replication timing, any cause or consequence relationship has not been documented. For example, it may be that any gene may replicate early because it is transcriptionally active. On the other hand, there is no strict distinction between gene activity and replication timing, since the transcriptionally active paternal *Igf2* allele replicates early and the transcriptionally active maternal *H19* allele replicates late.

The possibility that specific *cis* elements control the replication timing during the S phase and thereby dictate transcriptional competence is supported by the demonstrations that the control of asynchronous replication and the imprinting behavior of the imprinted genes in the 15q11–q13 cluster is governed by local *cis* element(s) in the *SNRPN* (Gunaratne et al. 1995) and *H19* loci (Greally et al. 1998). As a caveat to these links between a genetically defined imprinting center and the generation of replication asynchronies, it has been reported that the 15q11–q13 cluster of PWS patients with an imprinting mutation replicates asynchronously (White et al. 1996).

Whereas the 15q11–q13 cluster seems to be guided by one single imprinting center, the genes within the mouse distal chromosome 7/human chromosome 11p15.5 cluster obey different imprinting signals, as exemplified by the fact that the deletion of the *H19* locus including an upstream region, affects imprinting of *Igf2* and *Ins2* in *cis*, whilst leaving the imprinting pattern of the neighboring cluster member *Mash2* untouched (Leighton et al. 1995b). In addition, both *Igf2* and *H19* behave as a single imprinting unit in transgenic animals when included in a 130-kb YAC (Ainscough et al. 1997). The absence of an imprinting center of the type that has been described for the 15q11–q13 region in humans, is puzzling. Either one does not exist, or it has not yet been uncovered. It is perhaps relevant here that breakpoints associated with the

Beckwith-Wiedemann syndrome (BWS) are overrepresented in one region within and one region outside the cluster, which presumably perturbs the imprinting status of the *IGF2* gene (Reik and Maher 1997). It is not known, however, if the imprinting patterns of the entire cluster are affected in patients exhibiting this genetic defect. We are thus left with the impression that, even within clusters of imprinted genes, subdomains may also exist.

Not only imprinted but also nonimprinted regions display asynchronous replication timing (Bickmore and Carothers 1995). While it cannot currently be ruled that such nonimprinted domains harbor as yet unidentified imprinted genes, or nonimprinted genes which are expressed monoallelically (Holländer et al. 1998), it is clear that asynchronous replication timing is not sufficient per se for manifesting the imprinting phenomenon. The fact that nonimprinted genes, such as the *PAR3* gene, coexist with imprinted loci within the 15q11–q13 cluster, for example, demonstrates that local *cis* elements must be involved to somehow allow the recognition of long-range-acting epigenetic marks.

3
Parent of Origin-Specific Epigenetic Marks and Imprintable Elements

Any parent of origin-specific mark must be epigenetic in its nature. From a practical point of view, this means that the parental imprint must not modify the primary sequence of the genome, although it will somehow mark the imprinted allele to be active or inactive in the offspring. In particular, the imprint should be erasable, in order to allow a resetting of the parental identity of the imprintable allele for the next generation.

3.1
Epigenetic Marks

There are two main known epigenetic systems which are reversible and that are associated with regulation of gene expression: CpG methylation and histone acetylation. Whereas genome-wide CpG methylation is a relatively late evolutionary acquisition in animal cells, first occurring in primitive vertebrates (Tweedie et al. 1997), the acetylation of histones to regulate genome-wide gene activity is universal in eukaryotic cells (Grunstein 1997). Specifically, the complex pattern of acetylation of the histones at individual lysine residues is intimately associated with chromatin structure by regulating the interaction between nucleosomes and hence the accessability for *trans*-acting factors to *cis* elements which are pivotal for gene transcription or silencing (Grunstein 1997). Examples relevant to the topic of this chapter include the specific association of hypoacetylated histones to the inactive X chromosome (Jeppesen and Turner 1993) and isoforms of acetylated histones to heterochromatin of yeast (Grunstein 1997). These and other data have led to the suggestion that the regional pattern of histone acetylation serves as an epigenetic memory during M->G0/G1 transition to preserve proper sets of active/inactive genes from one cell cycle to the next (Jeppesen 1997).

There are some links between genomic imprinting and histone acetylation: trichostatin A, an inhibitor of histone deacetylase, prevents the silencing of the paternal *H19* allele during early mouse development (Svensson et al. 1998), although it remains to be established whether this is a direct or an indirect effect. In addition, inhibitors of histone deacetylation, but not inhibitors of CpG methylation, regulate replication timing (Bickmore and Carothers 1995). These data imply, but do not prove, that the replication asynchronies displayed by the imprinted loci are reflected in local differences in the distribution of acetylated histones. Although not directly comparable to such a condition, the ability of the paternal genome to compete out the maternal genome for a maternally derived reservoir of acetylated histones during the initial round of replication at the one cell stage of the mouse zygote (Adenot et al. 1997) is intriguing.

It has not been documented, however, whether or not any differential distribution of acetylated histones is faithfully replicated on the daughter strands during the S phase (irrespective of the transcriptional activity of any involved locus), which is a fundamental property of the gametic imprint in somatic cells. Although it could be argued that the differential distribution of acetylated histone isoforms on the active and inactive X chromosomes replicates an epigenetic status, this is likely to be an indirect consequence of the *Xist* function (see Lyon, Chap. 4, this vol.). Although it is formally possible that the positioning of acetylated histones around crucial *cis* elements might transiently consitute a gametic imprint that predates the methylation imprint (see Sect. 4), the regional or localized histone acetylation status is not likely to constitute a parental mark in somatic cells.

CpG methylation, which is a well-known parameter of the regulation of gene transcription (see Razin and Shemer, Chap. 9, this vol.), qualifies for all formulated requirements mentioned in the previous paragraphs. Thus, all imprinted genes studied this far display allele-specific differences in their methylation pattern (Razin and Shemer, Chap. 9, this Vol; Reik and Walter 1998). Moreover, conceptuses which were homozygous for a perturbed DNA methyltransferase gene displayed abnormal allelic expression patterns of imprinted loci and died prenatally (Li et al. 1993). For example, both parental *H19* alleles were active, whereas both parental *Igf2* alleles were inactive. This study was followed up by a demonstration that the restoration of the normal DNA methyltransferase gene in embryonic stem cells was not sufficient to establish a normal imprinting pattern (Tucker et al. 1996). This suggests that de novo CpG methylation patterns laid down at the time of blastocyst implantation and onwards are not sufficient for the proper manifestation of normal imprinting patterns.

3.2
Imprintable *cis* Elements

The accumulated genetic evidence has narrowed down three regions within the *SNRPN* (Lyko et al. 1998), *Igf2r* (Wutz et al. 1997), and *H19* (Elson and Bartolomei 1997; Ripoche et al. 1997) loci, which are implicated in the control of imprinting in *cis*. With the possible exception of the *H19* imprinting control region, the sequences of the *SNRPN* and *Igf2r* imprinting control regions fit the hypothesis that the imprintable elements consist of CpG-rich tandem repeats (Neumann et al. 1995). The absence of any significant sequence similarity between the imprintable *cis* elements suggests that particular features in the structure of the DNA and not the sequence per se are responsible for the effects. This is analogous to the epigenetically controlled centromere functions, which are poorly preserved at the sequence level although sequence repeats are a common motif (Karpen and Allshire 1997).

Although the inheritance pattern of a deleted region in genetically manipulated mouse conceptuses may uncover important *cis* elements that regulate parent of origin-specific expression patterns, it does not rule out that other regions within the imprinting domain also play a role. As noted above, deletion experiments have revealed that the upstream region of the *H19* gene behaves as an imprinting control region for the physically linked *Igf2* and *H19* genes, for example. Transgenes under the control of the imprinting control region upstream of the *H19* promoter, however, only partially manifest parent of origin-specific expression patterns (Elson and Bartolomei 1997). Such results are compatible with the existence of additional *cis* elements which may be strategically positioned to facilitate the propagation of *cis*-specific effects from imprinting control regions.

Bestor (Bestor 1990; Yoder et al. 1997) and Barlow (Barlow 1993) have independently promoted the hypothesis that genomic imprinting is a by-product of a host defense system. A point in case is the observation that the differential methylation pattern, which is one hallmark of parental imprints, appears to be a general character of repetitive elements (Yoder et al. 1997). Another analogy is that repetitive elements (Garrick et al. 1998) and imprinting control regions, such as those found in the *H19* upstream region (Lyko et al. 1997) and the 15q13–q11 imprinting center (Lyko et al. 1998), silence nearby genes in *cis*. The transposition or rearrangement of repetitive sequences to generate differential methylation patterns and gene silencing activities in parent of origin-specific manners, is an intriguing although speculative possibility.

4
The Resetting of the Parental Identity

A fundamental aspect of the imprinting phenomenon in mammals is the resetting of the parental identity in the germ line. In practical terms, this means that the maternal imprint is erased and replaced by a paternal imprint (or vice versa) in primordial germ cells (PGCs). Minimally, this must represent a two-step process. For PGCs of both sexes, it is crucial that the chromosomes lose their parental marks, such that new parent of origin-specific imprints can be subsequently laid down (Tada et al. 1998). It is expected that this process will be intrinsic to germ cells and that it functionally abolishes the parental imprint. Indeed, using embryonic germ cell-lymphocyte hybrids, erasing of allele-specific methylation imprints and reactivation of the silent alleles could be observed (Tada et al. 1997, 1998). The underlying mechanism(s) that generates epigenetically neutral homologous chromosomes from the parental genomes is, however, unclear (see also Sect. 5.2.2).

The acquistion of a new parental identity is a fundamental step in the genomic imprinting phenomenon. There are good reasons to suspect that it involves progressive modification events. For example, maternal modifications during oocyte growth are required for expression and repression of maternal alleles during embryogenesis (Obata et al. 1998). Since the gametes display differentially methylated regions, which are propagated in the soma (see below), the de novo methylation step which establishes differences in methylation pattern in male and female gametes may represent an end-point step in the resetting of the parental identity (Xu and Bestor 1997). Such a germline-specific enzymic activity has yet to be identified, since the DNA methyltransferase gene in mammals, the *Dnmt1* gene, is responsible for the maintenance of the methylation status during DNA replication in somatic cells (Yoder and Bestor 1996). It is now known, however, that the *Dnmt1* gene also encodes a de novo methylation activity and that there are alternative forms of the enzyme in the germline, due to sex-specific alternative splicing (Mertineit et al. 1998). The sex-specific exons in the 5' region regulate the location and activity of the DNA methyltransferase, which can be found in the nucleus of growing oocytes and in discrete foci of spermatogonium and the leptotene/ zygotene stages. The *Dnmt1* gene is, therefore, a candidate for producing the de novo methyltransferase that is involved in the resetting of the parental imprint in the germ line. The role, if any, of a second DNA methyltransferase gene, *Dnmt2* (Yoder and Bestor 1998), in genomic imprinting is unclear.

Given the absence of any extensive sequence specificity, beyond the CpG motif, for the action of the DNA methyltransferase activity (Yoder and Bestor 1996), it is likely that the de novo methylation of imprinting control regions is a consequence of the chromatin structure that inhibits or facilitates the interaction between the de novo methyltransferase and particular *cis* elements (Xu and Bestor 1997). The same sequence may adopt different chromatin structures in male and female PGCs due to the presence of sex-specific chromatin configurations. This conclusion finds some support in the demonstration that meiotic crossingover frequencies for subchromosomal regions, including imprinted genes, are sex-specific (Paldi et al. 1995). In addition, it has been suggested that the transcriptional activation in the male or female germline may set the stage for a differential recruitment of DNA methyltransferases (M. A. H. Surani, submitted for publication).

In one sense, this situation parallels the establishment of the stable propagation of active or inactive chromatin states during early *Drosophila* development. At this stage, any inactivated gene will remain inactive due to Polycomb-mediated assembly of repressive chromatin (see Pirotta, Chap. 10, this vol.). Since the potential for a gene to become active during later development is at odds with such a situation, there is a need to maintain an open chromatin configuration (and escape Polycomb-mediated maintenance of repressed state) without inducing premature transcriptional activation. In these instances, there must exist factors that do not transcriptionally activate such genes but prevent the assembly of repressive chromatin (see (Vashee et al. 1998) for an interesting solution to this dilemma). By analogy, the imprinting control regions may specifically attract sex-specific germline factors which prevent and/or facilitate its interaction with the de novo methyltransferase, in order to establish differentially methylated imprinting control regions without involving any transcriptional activation.

This line of thinking may have some bearing on the genetic data in association with the Prader Will and Angelman syndromes (see below and Hall, Chap. 6, this vol.). Based almost entirely on genetic evidence, Horsthemke and colleagues have proposed that the "imprintor" transcript of chromosome 15q11–13 functions in the germline by acting on a switch initiator site at the exon 1 region of the *SNRPN* gene, to presumably induce local changes in chromatin structure (see Sect. 6.3).

5
Manifestation of the Imprinted Status

5.1
Propagation of the Imprint

In humans, the allele-specific repression of the *H19* gene can be observed in the liver of 90-year-old individuals (Ekström et al. 1995), demonstrating the extraordinary stability of the imprinted state in somatic cells. This may be achieved by a series of progressive events that are likely to involve differential CpG methylation within imprinting control regions (see chapter by Razin and Shemer, this vol.; Reik and Walter 1998). Given the genome-wide modifications of the methylation patterns during early development, the propagation of gamete-specific differentially methylated regions (DMRs) poses particular problems. Firstly, the methylated (imprinted) allele must be protected from demethylation activities during preimplantation development. Secondly, the nonimprinted allele must resist de novo methylation during postimplantation development to maintain the epigenetic difference in the imprinting control region.

Using the bisulite protocol of genomic sequencing, two reports independently documented a differentially methylated region which resisted the extensive demethylation activities during preimplantation development and which mapped within the imprinting control region upstream of the *H19* promoter (Olek and Walter 1997; Tremblay et al. 1997). These authors suggest that the parental imprint may be the result of a combination of differential DNA methylation patterns, laid down during gametogenesis, together with allele-specific recognition of methylated *cis* elements by *trans*-acting factors, which preserve the gamete-specific differences during preimplantation development. Similar arguments can be put forward to explain the specific protection of the nonimprinted allele against de novo methylation during postimplantation development. The demonstration that Sp1-binding sites protect CpG islands from becoming methylated during development (Brandeis et al. 1994) is in keeping with these possibilities. From this, it follows that the temporal expression of factors which protect against demethylation and de novo methylation, respectively, should be very tightly regulated during early development. In addition, such factors should either be absent or inactive during early gametogenesis in both sexes, to allow resetting of parental identity.

The *Igf2r* gene, which provides an interesting parallel to the *H19/Igf2* paradigm, harbors two DMRs. Whereas region 2, which is a CpG island and an imprinting control region, displays persistent methylation on the maternal allele during development, region 1 acquires a methylation imprint on the paternal allele later in development (Stöger et al. 1993). Similarly, a DMR within a CpG island 200 bp from the *U2afbp-rs* gene acquires a maternal-specific methylation. Although both the female and male gametes display a hypermethylated pattern in this region, this methylation pattern is preserved only on the maternal allele during mouse preimplantation development (Shibata et al. 1997). The establishment of this DMR may therefore reflect the existence of a protective factor(s) which is specific for the female gamete. The *Snrpn* gene displays a more complex methylation pattern; although two different DMRs maintain a persistent methylation pattern during development, the DMR2 is methylated in sperm and DMR1 is methylated in the egg (Shemer et al. 1997).

Although numerous DMRs have been determined within imprinted loci, many of these seem to have been acquired postzygotically, presumably as a consequence to the primary imprint. This is exemplified by the region 2 DMR of the *Igf2* gene which is remethylated on the expressed, paternally derived allele during postimplantation development (Feil et al. 1994; Moore et al. 1997;

Reik and Walter 1998). The reason for this acquisition is not known; although it has been claimed that it is important for the transcriptional activity of the paternal *Igf2* allele (Hu et al. 1997). It has also been argued that the reactivation of the maternal *Igf2* allele in a double *H19* enhancer mouse model is not associated with a maternal-specific gain of methylation (data not shown in Webber et al. 1998).

Whilst these examples are straightforward by and large, the temporal disappearance of an/*Hpa* II-sensitive site within region 2 of the *Igf2r* gene between the 2 and 8 cell stage preimplantation conceptuses (Shemer et al. 1996) may be the result of dynamic modifications and complex equilibriums. Given the plasticity of development, with a range of developmental rates particularly during early development, we should perhaps expect many more examples of variations in methylation patterns outside imprinting control regions but within imprinted loci. Another and related parameter to consider deals with genotype-specific modifiers. This notion is highlighted by the observation that the methylation and allele-specific expression of the transgene TKZ751 depends on the genetic background (Allen et al. 1990). It is an open question, therefore, whether or not DMR regions within or outside imprinting control regions are exempt from modifiers that regulate expressivity and penetrance of epigenetic modifications (see also Martin and Sapienza, Chap. 12, this vol.).

5.2
Reading the Imprint

The net result of the primary imprint is the generation of one active and one more or less inactive allele, either throughout the conceptus or in particular cell types, in parent of origin-specific manners. Does the parental mark play a direct or an indirect role in this process? The answer to this question may differ in each individual case, since some imprinted genes, such as *Snrpn*, are already expressed monoallelically during preimplantation development of the mouse (Szabo and Mann 1995), whilst the functional imprinting status is manifested only subsequently for other imprinted loci, such as *Igf2r* (Lerchner and Barlow 1997) and *H19* (Svensson et al. 1998). Moreover, the functional imprinting status for several genes can be cell type-specific within the same normal conceptus (see Appendix, this vol.). Whether or not this means that the imprinting status in such instances is never manifested or that the repressed allele is reactivated (as may be the case for biallelic expression of *Igf2* in the choroid plexus of rat; Overall et al. 1997) is an open question.

When scrutinizing these considerations, we face a fundamental question: what aspect of gene expression is affected by the parental imprint? Formally, any given aspect of gene expression, such as nuclear architecture (see Marshall and Sedat, Chap. 14 and Paldi and Youvenot, Chap. 13, this vol.), chromatin structure (see Razin and Shemer, Chap. 9, Geramisova and Corces Chap. 11, and Pirotta, Chap. 10, this vol.), gene transcription (see Franklin, Chap. 8, this vol.), RNA processing, stability, and mRNA transport out of the nucleus and/or any process outside the nucleus that somehow allows the parental origin of the gene product to be examined by a regulatory machinery, may manifest the imprinting process. There is currently no precedence, however, for the transduction of the epigenetic imprint into primary or secondary gene products. Given the regional control of the imprinting phenomenon, encompassing clusters of imprinted genes, for example, the least provocative scenario for the function of the parental imprint involves the interpretation of its information into the formation of active and repressive chromatin structures, which manifest the imprinting status and the transcriptional activities (see also Paldi and Youvenot Chap. 13, Razin and Shemer Chap. 9, Lyon, Chap. 4, Pirotta Chap. 10, and Franklin Chap. 8, this vol. for complementary approaches to this topic).

5.2.1
The Chromatin Connection

The manifestation of epigenetically repressed states of chromatin may, in general, take the form of the cytologically identifiable and highly condensed heterochromatin (Holmquist 1975) at one extreme and to the specifically silenced single-copy homeotic genes during postembryonic development (Franke et al. 1994) at the other end of a spectrum. The inactive X-chromosomes of the mammalian female (see Lyon Chap. 4, this vol.) and the condensed sex chromosomes of certain coccid insects (see Herrick and Seger, Chap. 3, this vol.) share many cytological features with constitutive heterochromatin, although there are distinct biochemical differences at the level of higher order organization of repressed chromatin.

The nonhistone heterochromatin-associated protein, heterochromatin protein 1 (HP1), with the ability to modulate epigenetic silencing in Drosophila (James and Elgin 1986), can be found at all major heterochromatic sites identified by both genetic and cytologic methods (James et al. 1989). Since the discovery of HP1, additional proteins that associate with heterochromatin have been identified. Apart from a subset of chromatin proteins that are known to share a chromobox domain, which is essential for the correct type of protein-protein association and/or polymerization of HP1 (Paro and Hogness 1991), other types of heterochromatic proteins have been identified. Of particular interest, in the context of this chapter, is MeCP2, which specifically binds clusters of methylated DNA in a sequence nonspecific manner and appears to be specifically localized in heterochromatin (Nan et al. 1997).

It is likely that protein-protein interactions leads to supercondensation of heterochromatic regions (see also Pirotta, Chap. 10, this vol.). Molecular and genetic analyses have suggested that the establishment of silenced chromatin may require some S phase-specific function and that the assembly and maintenance of silenced chromatin can also be influenced at other phases of the cell cycle. In this regard, it is interesting that that the ORCs are associated with Sir1p (Triolo and Sternglanz 1996) and HP1 (Pak et al. 1997) that manifest repressed chromatin in yeast and *Drosophila*. This finding also reinterprets the heterochromatin specificity of the localization of heterochromatic proteins. ORC units are widely distributed over the chromosomes and not just in heterochromatin. Standard immunocytochemistry may not show the location of HP1 unless there is sufficient amount of heterochromatin. From this, it follows that heterochromatic proteins may participate in the formation of local, repressive chromatin that presumably emanates from imprinting control regions. It is conceivable, therefore, that imprinting control regions, either by default or by requiring a parental mark, nucleate the formation of repressive chromatin the spreads into adjacent loci. The differentially methylated regions may dictate the formation of such localized "heterochromatin"-like structures. For example, the methylated *H19* allele may recruit MeCP2 to effectuate the progressive silencing of the paternal allele during early mouse development. Such a possibility is at odds, however, with the observation that the *H19* imprinting control region appears to include a silencer function which operates in the methylation-free context of *Drosophila* (Lyko et al. 1997).

Except for accessibility to various chromatin probes (see for example Feil et al. 1997), very little is known about the differences in the higher-order organization of chromatin between the repressed and the active allele. It has been suggested that heterochromatinization of the inactive paternal genome of the mealy bug (*Planococcus lilacinus*) involves nuclease-resistant chromatin which can already be observed in sperm, even if the sperm chromatin is not heterochromatic (Khosla et al. 1996). This potential imprint includes repetitive sequences which show some similarity with nuclear matrix attachment regions. Indeed, such matrix-associated regions have been postulated to play a role in the imprinting behavior of the *Igf2* and *H19* genes by organising subchromosomal domains (Banerjee and Smallwood 1995). Moreover, by using a biochemical approach, it has been claimed that several regions within the *Igf2/H19/L23* subdomain (see Appendix, this vol.) genes could specifically associate with the nuclear matrix (Greally et al. 1997). As can be seen in Fig. 1, however, both parental alleles of the *Igf2* and *H19* genes can be visualized as extranuclear loops in salt-extracted mouse cells, subjected to DNA FISH analysis. Since resistance to salt extration is generally considered as an experimental verification for matrix-specific association, we conclude that there is not any good evidence for allele-specific structures of this kind, in this instance at least.

Another avenue to resolve the assertions that higher-order chromatin structure plays a role in manifesting the imprinting status, may have been provided by the observations that noncoding transcripts have been documented to associate with chromatin to contribute to the formation of both active and repressive chromatin at its site of production. For example, *Xist* RNA has been documented to mediate the formation of repressive chromatin in mammalian

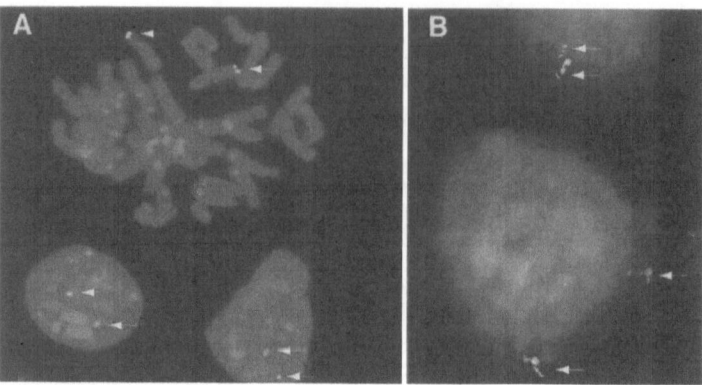

Fig. 1A, B. DNA FISH analysis of the *Igf2* and *H19* genes in mouse cells, subjected to high salt extraction. Such a treatment removes DNA from the nuclear cage except at specific attachment points specified by so-called MAR elements. The panels show superimposed images resulting from FISH analysis using *Igf2* and *H19* probes and DAPI counterstain. A DNA FISH analysis of metaphase chromosome and interphase cells. B DNA FISH analysis of salt-extracted cells with halos extending into the cytoplasm. The centromere region, which has a lighter tint in DAPI stains, appears to be associated with the nuclear cage in contrast to the extranuclear position of both parental alleles for a 100-kb region encompassing *Igf2-H19*. Magnifications are 600-fold (A) and 400-fold (B). These data were generated by Ms Tongwei Tang (Uppsala University), who tragically passed away on 23 January 1998

females (see Lyon, Chap. 4, this vol.). Conversely, it has been reported that dosage compensation via the hyperactivity of the single X-chromosome in male *Drosophila* might be achieved by another noncoding regulatory RNA molecule (Willard and Salz 1997). Although the exact mechanisms by which these transcripts regulate chromatin formation remains to be established, the accumulating number of noncoding, antisense transcripts within imprinted loci, is noteworthy (Moore et al. 1997; Wutz et al. 1997). These transcripts are themselves expressed in a parent of origin-dependent manner and has in case of the *Igf2r* antisense transcript been shown to remain associated to its site of production, even in metaphase chromosomes (D. Barlow, pers. inf.). Moreover, the "imprintor" transcript of chromosome 15q11–13 has, based on genetic evidence, been postulated to act on a switch initiator site at the exon 1 region of the *SNRPN* gene by inducing local changes in chromatin structure (see Sect. 6.2). It deserves to be pointed out, however, that these two instances are not directly comparable, since the former very likely takes place in somatic cells whereas the latter is postulated to be germline-specific.

The manifestation of silenced chromatin found in mating type loci, the telomeres and the ribosomal DNA of yeast, in PEV and the heritable inactivation of homeotic genes in flies (see Pirotta, Chap 10, this vol.), and X-chromosome inactivation and genomic imprinting in mammals all share several features (Singh and James 1995). While it is possible that silencer regions commonly operate by recruiting histone deacetylases to promoter regions, there can be a thermodynamic explanation for the presence of deacetylated histones in repressed chromatin structures (De Rubertis et al. 1996). Moreover, the role of various other proteins in silencing is not fully known (see also Pirotta, Chap. 10, and Geramisova and Corces Chap. 11, this vol.). It is not yet possible, therefore, to formulate a unified view on the status of the repressed chromatin in imprinted genes.

Since active genes can be found on both parental alleles both in individual imprinted loci and clusters of imprinted genes, a regional heterochromatinization of the classical type (such as X-inactivation) is unlikely to manifest genomic imprinting. Based on ectopic expression of homeotic genes and homeotic transformations of floral organs in methylation-defective plants (Yoder and Bestor 1996), it can be suggested that heritable methylation patterns reinforce and/or supplant chromosomal protein-mediated heritable gene control, such as those derived from the polycomb and trithorax groups of genes (see Pirotta, Chap. 10, this vol.). In addition, it cannot be excluded that heterochromatin proteins, such as HP1 or MeCP2, may participate in the formation of local, repressive chromatin, in some instances. With the identification of *cis*-acting silencer elements within two different imprinting control regions in transgenic flies, the potential link between established mechanisms of formation of repressive chromatin in *Drosophila* and genomic imprinting in mammals is now amenable for genetic analysis.

5.2.2
The Silencers of Imprinting Control Regions

Using the transgenic fly system, Lyko et al. (1998) observed strong silencing of the *lacZ* and mini-*white* reporter genes in constructs containing the *SNRPN* exon 1 region. The putative silencing element was narrowed down to a 215-bp region overlapping the *SNRPN* promoter. Silencing was bidirectional and effective over more than 1 kb. These findings suggest that a silencer element participates in the repression of the maternal *SNRPN* allele, or the long-range regulation of the imprinted 15q11–q13 domain. If the latter were true, the silencing observed in *Drosophila* would reflect one direction of the imprint switch, i.e., the generation of a repressed chromosomal environment on 15q11–q13 (maternal imprint). Erasure of this imprint in primordial germ cells must then require an activating factor which interacts with the silencer to displace the silencing proteins. A maternally derived deletion of this target sequence, as observed in fathers of PWS imprinting mutation patients would, therefore, result in a failure to erase the maternal imprint in the male germline and, consequently, in the transmission of a paternal chromosome with a (grand)maternal imprint. The failure to observe a block of the paternal to maternal imprint switch by a paternally derived deletion of the silencer in mothers of AS patients can be explained by the fact that *SNRPN* is likely to play a role in PWS and therefore cannot be deleted in normal individuals.

Similarly, the imprinting control region upstream of the *H19* promoter was demonstrated to possess silencing activity in the *Drosophila* transgenic approach (Lyko et al. 1997). The absence of DNA methylation in *Drosophila* implies that the activity of the silencer must function by default, i.e., be active on the unmethylated, maternally inherited allele. In mouse, however, the paternal silencer region is methylated and yet represses the paternal *H19* allele. If the silencing activity of the *H19* imprinting control region in the *Drosophila* transgenic assay applies to the mammalian context, the activity of the *H19* silencer element appears not to be regulated by the parental imprint per se, although its consequences may be. From this, it follows that the silencer function directed to the *H19* locus is neutralized when maternally inherited. This could be achieved by factors competing with chromatin constitutents which organize the repressive chromatin. If so, such factors should be associated to the imprinting control/silencer region by default and be sensitive to DNA methylation, either directly or indirectly. This is not appliccable to the *SNRPN* silencer, however, since the repression of the 15q11–13 cluster takes place on the maternally derived, nonimprinted allele (see Sect. 6.2).

Whether or not the localization of silencers and imprinting control region is the exception or the rule is not known. It does illustrate, however, an interesting aspect of the imprinting phenomenon and clearly influences our current perception of the imprinting mechanism. At the same time, we should be aware that imprinting may mark an allele to be silent **or** active. Turning the argument around, imprinting may prevent silencing (*Igf2r?*) or promote activation (*p57kip2?*). Clearly, there is a need for much more information on this topic.

6
Models of Imprinting

It is currently difficult to envisage a common mechanistic model that applies to all known imprinted loci and yet is compatible with all published results. The expression competition model is, however, attractive, since it may apply to at least two different instances: the *Igf2/H19* and *Igf2r* loci. A second model, which is based almost entirely on genetic data in association with the Prader-Willi and Angelman syndromes in humans, is also presented.

6.1
The Expression Competition Model

The *IGF2* and *H19* loci are expressed in essentially identical patterns at the cellular level in several tissues during prenatal development (Poirier et al.

1991; Ohlsson et al. 1994). The reason for this was partly resolved when examining conceptuses carrying a deletion of the *H19* endoderm-specific enhancers: both maternal-specific *H19* expression and paternal-specific *Igf2* expression were lost in endodermally derived tissues (Leighton et al. 1995a). These results strongly suggested that the endodermal enhancers acted not only on the *H19* promoter, but also on the *Igf2* promoters, which are separated by more than 90 kb from the enhancer elements. This deduction is in agreement with the expression competition model, which states that the epigenetic status of the *Igf2* and *H19* promoters determines the activity of the endodermal enhancers (Bartolomei et al. 1993). More specifically, the model states that the paternally derived *Igf2* promoters outcompete the *H19* promoter on the same chromosome, (and vice versa for the maternally derived chromosome), for the same enhancer function (Fig. 1A). The model received experimental support when it was demonstrated that the deletion of the entire *H19* locus upregulated the normally repressed maternal *Igf2* allele (Leighton et al. 1995b). It was clear from these studies that the *H19* locus and/or its environment dictates the expression of the *Igf2* gene in *cis*, in a parent of origin-specific manner. Additional experiments subsequently verified that a region upstream of the *H19* promoter performs as an imprinting control region (Elson and Bartolomei 1997; Ripoche et al. 1997).

The *Igf2r* locus provides an interesting parallel to the *Igf2/H19* example. Whereas the maternal allele exclusively transcribes the *Igf2r* gene during postimplantation development (Lerchner and Barlow 1997), a novel antisense transcript, which has no open reading frame and is derived from the immediate vicinity of the *Igf2r* imprinting control region, is transcribed exclusively from the paternal allele (Wutz et al. 1997). It has been proposed that these genes compete for a common enhancer, in a scenario analogous to the enhancer competition model (Wutz et al. 1997). An interesting issue, which compounds this interpretation, is that the antisense transcript remains associated with the *Igf2r* locus throughout the cell cycle, even in metaphase chromosomes (D. Barlow, pers. comm.). It cannot be excluded, therefore, that the *Igf2r* antisense transcript is recognized by the local chromatin to influence its function (see above). From this, it follows that the repression of the paternal *Igf2r* allele could result from either an enhancer competition situation, or an X-inactivation-like process, or both.

Given the strong genetic data, there is little reason to question the basic features of the enhancer competition model. It is important to note, however, that the generality of the model has not been documented. We are currently unable to discriminate between the possibilities, therefore, that the expression competion model is of general validity (where the endodermal enhancer can be replaced with other lineage-specific enhancers) or represents a specific adaptation to the endodermal lineage. These alternatives are compounded by the observation that biallelic expression of *H19* can coexist with high levels of monoallelic *Igf2* expression in both human (Adam et al. 1996) and murine

(Svensson et al. 1998) extraembryonic tissue. Since other results also appear to be incompatible with an expression competition model (Svensson et al. 1995; Franklin et al. 1996; Looijenga et al. 1997), we should not rule out other interpretations of data that concerns the mechanism of *Igf2/H19* imprinting (see Geranisova and Corces, Chap. 11, this vol.).

6.2
The SNRPN Imprinting Center and the Imprintor and Switch Initiator Model

6.2.1
Structure and Function of the 15q11–q13 Imprinting Center

Mutations affecting the acquisition, maintenance, or reading of imprints can be expected to interfere with normal development. Aberrant imprinting was first detected in a small group of patients with Angelman syndrome (AS) and Prader-Willi syndrome (PWS). In contrast to the majority of AS and PWS patients, who have a deletion, UPD or single gene defect (Table 1), some patients (including several sibs) were found to have apparently normal chromosomes of biparental origin, but abnormal methylation throughout the imprinted chromosome domain (Glenn et al. 1993; Buiting et al. 1994; Reis et al. 1994). The PWS patients have a maternal methylation pattern on both chromosomes and do not express any of the paternally active genes, wheras the AS patients have a paternal methylation pattern on both chromosomes and express *SNRPN* and *IPW* biallelically (Saitoh et al. 1996). The presence of a maternal epigenotype on a paternal chromosome and vice versa is highly suggestive of an imprinting defect. Aberrant imprinting may result from a mutation that acts either in *cis* or in *trans*. Since all of these patients exhibit a

Table 1. Clinical and genetic findings in PWS and AS

	PWS	AS
Clinical signs	Neonatal hypotonia	Microcephalus
	Hypogonadism	Jerky movements
	Hyperphagia and obesity	No speech
	Short stature	Abnormal EEG
	Small hands and feet	Severe mental retardation
	Craniofacial dysmorphism	Paroxysms of laughter
	Mental retardation	Hypopigmentation
	Hypopigmentation	
Genetic lesions (frequency)	Paternal deletion 15q11–13 (70%)	Maternal deletion 15q11–13 (70%)
	Maternal disomy (29%)	Paternal disomy (1%)
	Imprinting defect (1%)	Imprinting defect (4%)
		Single gene defect (25%)

classical phenotype, the mutation does not disturb the imprinting process in general, but affects 15q11–q13 only. A *cis* effect resulting from a mutation or subtle chromosomal rearrangement within the critical PWS/AS region was suggested in affected sibs, because the sibs were found to share the same paternal and maternal chromosomes. Detailed molecular studies then revealed inherited microdeletions in the *D15S63/SNRPN* region in several cases (Reis et al. 1994; Sutcliffe et al. 1994; Buiting et al. 1995; Dittrich et al. 1996; Saitoh et al. 1996; Ohta et al. 1997; Fig. 2). Based on these findings, it was suggested that the deletions affect a single genetic element termed the imprinting center (IC), which regulates in *cis*, the chromatin structure, DNA methylation, and gene expression throughout a 2-Mb imprinted domain within 15q11–q13 (Buiting et al. 1995; Horsthemke et al. 1995).

Mutations of the IC do not exert a simple position effect on neighboring loci. In the AS families, the deletions were found on the maternal chromosome of the patients and on the paternal chromosome of the phenotypically normal mothers. In the PWS families, the deletions were found on the paternal chromosome of the patients and on the maternal chromosome of the phenotypically normal fathers and paternal grandmothers (Reis et al. 1994; Sutcliffe et al. 1994; Buiting et al. 1995; Dittrich et al. 1996; Saitoh et al. 1996; Ohta et al. 1997). These data suggest that IC mutations block the imprint switch during gametogenesis. They can be transmitted silently through the germline of one sex and manifest themselves only after transmission through the germline of the opposite sex. Failure of switching the imprint results in a paternal chromosome with a maternal epigenotype or a maternal chromosome with a paternal epigenotype. Of course, we do not know whether the imprints are not erased or are erased and reestablished.

A closer look at the microdeletions suggests that the IC has a bipartite structure: a centromeric part, which is deleted in AS patients, and a telomeric part, which is deleted in PWS patients. We have recently identified a transcript spanning the IC region (Dittrich et al. 1996). A DNA sequence close to *D15S63* (PW71) and a related sequence 30 kb upstream are alternative start sites of this novel transcript, which consists of a distinct 5' untranslated region (exons BD1-3) spliced to exons 2 to 10 of *SNRPN*, skipping exon 1 (Fig. 2). The transcripts have different 5' exons (BD1B and BD1A), but share the exons BD2 and BD3. BD1B' and BD1B* are alternative exons of the BD1B transcript. The novel transcripts are made from the paternal chromosome only and are present at a very low copy number. They are found mainly in brain, heart, and germline containing tissues.

In seven of eight AS IC deletion patients studied, exon BD3 is deleted (Fig. 2). In one family (AS-H), a 6-kb deletion immediately distal to exon BD3 may interfere with the expression or splicing of the BD transcript. Another patient (AS-C2) has a point mutation in the splice donor site of exon BD2, although this may be a neutral polymorphism. These data suggest that the BD transcript or a regulatory sequence close to BD3 is involved in the paternal → maternal

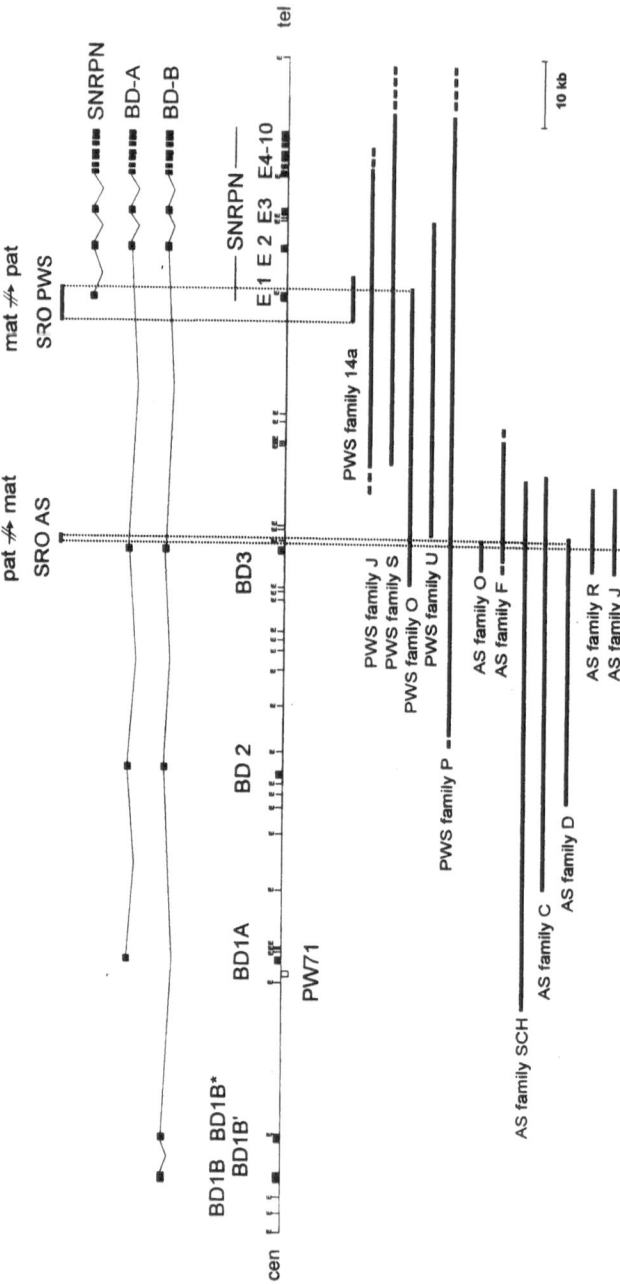

Fig. 2. The 15q11–q13 imprinting center. The extent of inherited microdeletions affecting the imprint switch during gametogenesis is shown by *bars*. *SRO* Smallest region of deletion overlap; *E EcoRI* sites; *open box* PW71; *hatched boxes* BD exons; *black boxes* SNRPN exons

imprint switch in the female germline. In each of five PWS IC deletion patients studied, *SNRPN* exon 1 is deleted. This suggests that the *SNRPN* transcript or a regulatory sequence close to *SNRPN* exon 1 is involved in the maternal → paternal imprint switch in the male germline.

Based on these findings, we have proposed a model for imprint switching (Dittrich et al. 1996). In this model (Fig. 3), the IC consists of an imprintor

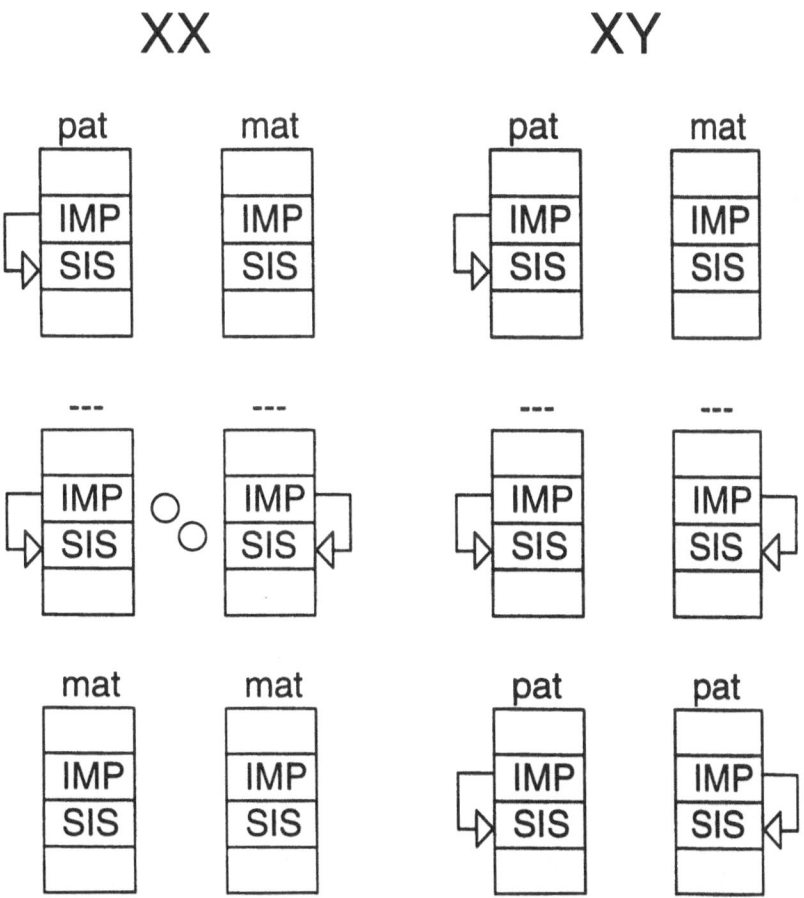

Fig. 3. Model for imprint switching on chromosome 15. The imprinted domain on chromosome 15 is indicated as a *box*. Imprint switching requires imprintor (*IMP*) activity in *cis* (arrow) and a germline-specific factor in *trans* (*circles*). Mutations of the imprintor a prevent the acquisition of a maternal imprint in the female germline, and a paternal imprint develops by default. Children inheriting this chromosome will have two chromosomes 15 with a paternal imprint and develop AS. Mutations of the switch initiation site (*SIS*) block the erasure or switch of the maternal imprint in the paternal germline. Children inheriting this chromosome will have two chromosomes 15 with a maternal imprint and develop PWS. *pat* Paternal imprint; *mat* maternal imprint; – erased imprint

(IMP) and an imprint switch initiation site (SIS). The imprintor encodes the BD transcript. It is transcribed from the paternal chromosome only and acts in *cis* on the switch initiation site, possibly by introducing a local change in the chromatin structure. This change may be brought about by the process of transcription or by the interaction of the BD transcripts with the DNA. In the XX germline, a *trans*-acting factor specific for female germline cells is involved in completing the switch and/or initiating the bidirectional spreading of the maternal imprint on the paternally derived chromosome, which shuts off the imprintor activity. In the absence of the maternal *trans*-acting factor (XY germline), the maternal imprint on the maternally derived chromosome is lost; starting from the switch initiation site (spreading of paternal imprint). This may occur by default or involve other factors.

While there are good indications that the BD transcript is the imprintor transcript involved in establishing the maternal imprint in the female germline, the role of the *SNRPN* exon 1 region deleted in the PWS imprinting mutation families is less clear. As *SNRPN* is inactive on the maternal chromosome, *SNRPN* transcription is unlikely to initiate the maternal→paternal imprint switch during male gametogenesis, although it may be necessary to maintain the paternal imprint. One possibility is that the *SNRPN* exon 1 region is the imprint switch initiation site. Based on DNA methylation and replication data, we have proposed that the paternal copy of the imprinted domain within 15q11–q13 may have a euchromatin-like chromatin structure, from which the PWS genes are transcribed, whereas the maternal copy may have a heterochromatin-like chromatin structure, from which the AS gene(s) are transcribed (Buiting et al. 1995). It is possible that the switch initiation site represents a nucleation center for the assembly of repressive chromatin assembly and disassembly, which is made accessible by the IC transcript, as discussed above.

6.2.2
Sporadic Imprinting Defects: Implications for Imprint Switching and Maintenance

All affected PWS or AS sib pairs studied to date have an IC mutation. In the majority of sporadic patients, however, we could not detect such a mutation, although we performed an extensive search including Southern blot and heteroduplex analysis of the critical IC regions and sequence analysis of the BD exons and *SNRPN* exon 1. Interestingly, several AS and PWS patients were found to share a maternal and paternal haplotype, respectively, with an unaffected sib. Furthermore, in four isolated AS patients the aberrantly imprinted chromosome was of grandmaternal origin, rather than of grandpaternal origin, as in all the IC mutation cases (Bürger et al. 1997; Buiting et al. 1998; Table 2). These patients are unlikely to have a familial *cis*-acting mutation

Table 2. Grandparental origin of aberrantly imprinted chromosome. (Buiting et al. 1998)

Origin	Angelman		Prader-Willi	
	IC mutation	No IC mutation	IC mutation	No IC mutation
Mat. grandfather	5	5		
Mat. grandmother		5		
Pat. grandfather				
Pat. grandmother			6	9

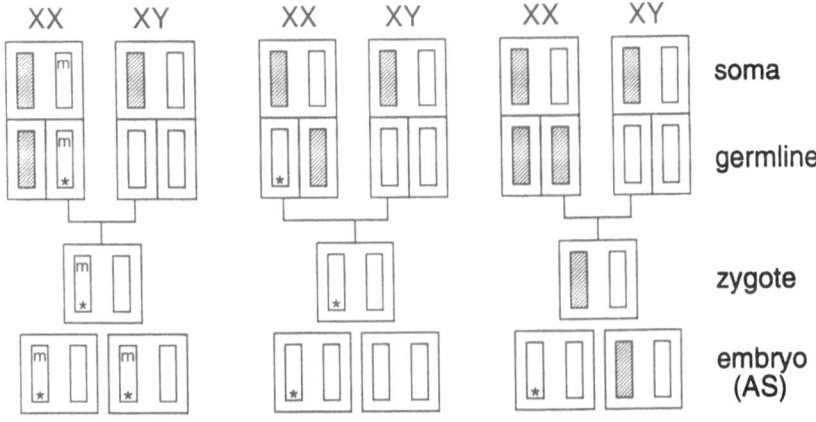

Fig. 4. Pre- and postzygotic imprinting defects. Aberrant imprints on chromosome 15 may result from an inherited IC mutation (*left diagram*), a de novo imprinting defect (*middle*) or a postzygotic imprint maintenance failure (*right diagram*). In this example, the patients have a paternal epigenotype on the maternal chromosome (marked by an *asterisk*) and develop Angelman syndrome. *Hatched box* maternal epigenotype; *open box* paternal epigenotype; *m* IC mutation

affecting the imprint switch as shown in Fig. 4 (left), and other mechanisms have to be considered.

The aberrant imprints in these cases may result from an imprint reset failure in a single germ cell (Fig. 4, middle) or from a postzygotic imprint maintenance failure (Fig. 4, right). The mutation is either (1) an as yet undetected structural mutation in the IC region, (2) an epimutation (Holliday 1987) resulting from an imprinting error, or (3) a paramutation (Brink 1973) resulting from interchromosomal transfer of epigenetic states. The latter process is mechanistically related to homologous recombination and probably involves hemimethylated chromatids (Holliday 1987; Colot and Rossiguol 1996). In this context it is of interest to note that the maternal and the paternal chromosomes 15 pair in the late S phase of the cell cycle (LaSalle and Lalande 1996).

This pairing occurs specifically at the imprinted 15q11–q13 region and may contribute to increased recombination.

If the sporadic imprinting defects arise prezygotically (Fig. 4, middle), we will have to accept the idea that the female germline is capable of making a paternal imprint, because in four AS patients the aberrantly (paternally) imprinted maternal chromosome was inherited from the maternal grandmother (Table 2). So far, we have not observed the reciprocal situation in PWS, i.e., PWS patients in whom the aberrantly (maternally) imprinted paternal chromosome was inherited from the paternal grandfather. Since in the PWS patients, the maternal imprint on the grandmaternal chromosome may not have been erased in the paternal germline, rather than having been established de novo, we do not know whether the paternal germline is capable of making a maternal imprint. Although the number of PWS patients studied is too low to rule out such a case, the nonreciprocal findings in AS and PWS suggest that the paternal imprint is the default imprint, which develops in the paternal and the maternal germline, if an XX-specific factor is missing or fails. Alternatively, the imprinting defect may be the result of a mutation in *trans*. It should be noted, however, that a *trans*-acting mutation affecting the parental germline is unikely in AS and PWS patients having an unaffected sib, because such a mutation should affect all germ cells of a carrier parent. A *trans*-acting mutation impairing imprint maintenance during postzygotic development of the patient remains a possibility, but in this case one might expect to find two affected AS or PWS sibs with different maternal or paternal haplotypes, respectively. So far, such families have not been observed. It is difficult to imagine that a *trans*-acting mutation affects only one imprinted region, but *trans*-acting mutations in Polycomb-group genes affecting the inheritance of epigenetic states through somatic cell divisions in *Drosophila* (Orlando and Paro 1995) and *Arabidopsis* (Goodrich et al. 1997) have been described.

7
Concluding Remarks

Despite considerable efforts to underpin the imprinting process since the first imprinted genes were identified during the early 1990s, we still lack a coherent view on its underlying mechanisms. It is a paradox that the common denominator of the imprinting mechanisms is their epigenetic complexity. We cannot discount, therefore, the possibility that each individual imprinted gene has evolved particular and even unrelated principles that repress or derepress parental alleles. The clusters of imprinted genes, on the other hand, may provide a general paradigm for the imprinting mechanism. This assertion is reinforced by the fact that some of the newest members of the class of imprinted genes have been found simply by looking for them within already known clusters of imprinted genes (although it is currently not possible to predict the generality of this aspect of the imprinting phenomenon).

The clusters of imprinted genes exemplify that the imprinting mechanism is not likely to have evolved in a one-step process, but suggest the sequential generation of individual *cis* elements that are of importance in the imprinting process and its manifestation. For example, the imprinting within the cluster might have been initiated with a particular genetic event that created the preconditions for establishing imprintable *cis* elements. During evolution, the parent of origin-specific imprints and/or their functions might have spread in *cis* to affect allelic expression patterns of physically linked genes (see also Trivers and Burt, Chap. 1, this vol.). The existence of two or more imprinting control regions within the *Igf2/Ipl* cluster (see Sect. 2) is evidence for such a stepwise acquisition of imprintable *cis* elements.

This outline of the generation of clusters of imprinted genes assumes that the clusters preexist in a nonimprinted version in nonmammalian vertebrates. Indeed, it has been demonstrated that the gene order and organization of some genes in the domain harboring the *H19-Ipl* imprinting cluster is strikingly similar in zebrafishes and mammals (Postlethwait et al. 1998). Alternatively, the genes of the clusters may have attained their physical location and imprinting status independently of each other, due to recombinatory hot spots within subchromosomal domains, for example. A third possibility suggests that we are looking at the wrong end of the stick; perhaps the effects of genomic imprinting are simply a vehicle to facilitate the spreading of parasitic DNA elements, that are more commonly known as selfish DNA. This possibility assumes that the juxtapositioning of selfish DNA to crucial target genes exploits inborn tensions of the diploid genome that enable its propagation. In any of these scenarios, the reasons for selecting particular genes to be imprinted may be very complex and entirely unrelated from one gene to another.

All of these considerations serve to illustrate that the complexity of the mechanisms of genomic imprinting may reflect the complexity of their evolution. It cannot be ruled out that the adaptive conditions that have selected the plethora of imprinting mechanisms may differ on an evolutionary scale for each individual locus, even for individual steps of the imprinting mechanism. Our understanding of imprinting mechanisms is likely, therefore, to benefit from interdisciplinary accounts of the imprinting phenomenon. In particular, it will be fundamental to examine the imprinting phenomenon in well-defined contexts that address individual subprocesses of the underlying mechanisms. Such approaches may yield the desired molecular handles, such as the identification of *cis* elements and *trans*-acting factors that are involved in the establishment and manifestation of the imprinting phenomenon.

Acknowledgments. We wish to thank Gary Franklin, J. Walter, Alex Olek, Ben Tycko, Oskar Smrzka, Denise Barlow, Laurence Hurst, Vincenzo Pirotta, David Haig, and Wolf Reik for discussions and sharing data prior to publication. In addition, we want to acknowledge R. D. Nicholls, A. Reis, S. Malcolm, and B. Korn for collaboration, S. Groß and C. Lich for expert technical assistance, all colleagues who provided us with patient samples for their inavaluable contribution, and E. Passarge for continuous support. Work in the authors' laboratory was supported by the Deutsche Forschungsgemeinschaft, the Human Frontier Science Program, The Wellcome/ CRC Trust, the Swedish Cancer Research Foundation and the Swedish Natural Science Foundation.

References

Adam G, Cui H, Miller S, Flam F, Ohlsson R (1996) Allele-specific in situ hybridization (ASISH) analysis: a novel technique which resolves differential allelic usage of *H19* within the same cell lineage during human placental development. Development 122:839–847

Adenot P, Mercier Y, Renard J-P, Thompson E (1997) Differential H4 acetylation of paternal and maternal chromatin precedes DNA replication and differential transcriptional activity in pronuclei of 1-cell mouse embryos. Development 124:4615–4625

Ainscough J, Koide T, Tada M, Barton S, Surani M (1997) Imprinting of *Igf2* and *H19* from a 130-kb YAC transgene. Development 124:3621–3632

Allen N, Norris M, Surani M (1990) Epigenetic control of transgene expression and imprinting by genotype-specific modifiers. Cell 61:853–861

Banerjee S, Smallwood A (1995) A chromatin model of *IGF2/H19* imprinting. Nat Genet 11:237–238

Barlow D (1993) Methylation and imprinting: from host defence to gene regulation. Science 260:309–310

Bartolomei MM, Webber AL, Brunkow ME, Tilghman SM (1993) Epigenetic mechanisms underlying the imprinting of the mouse *H19* gene. Genes Dev 7:1663–1673

Bestor TH (1990) DNA methylation: evolution of a bacterial immune function into a regulator of gene expression and genome structure in higher eukaryotes. Philos Trans R Soc Lond B 326:179–187

Bickmore W, Carothers A (1995) Factors affecting the timing and imprinting of replication on a mammalian chromosome. J Cell Sci 108:2801–2809

Brandeis M FD, Keshet I, Siegfried Z, Mendelsohn M, Nemes A, Temper, Razin A, Cedar H (1994) Sp1 elements protect a CpG island from de novo methylation. Nature 371:435–438

Brink R (1973) Paramutation. Annu Rev Genet 7:129–152

Buiting K, Dittrich B, Robinson WP, Guitart M, Abeliovich D, Lerer I, Horsthemke B (1994) Detection of aberrant DNA methylation in unique Prader-Willi syndrome patients and its diagnostic implications. Hum Mol Genet 3:893–895

Buiting K, Saitoh S, Gross S, Dittrich B, Schwartz S, Nicholls R, Horsthemke B (1995) Inherited microdeletions in the Angelman and Prader-Willi syndromes define an imprinting centre on human chromosome 15. Nat Genet 9:395–400

Buiting K, Dittrich B, Groß S, Lich C, Färber C, Buchholz T, Smith E, Reis A, Bürger J, Nöthen M, Barth-Witte U, Janssen B, Abeliovich D, Lerer I, van den Ouweland A, Halley D, Schrander-Stumpel C, Smeets H, Meinecke P, Malcolm S, Gardner A, Lalande M, Nicholls R, Friend K, Schulze A, Matthijs G, Kokkonen H, Hilbert P, Van Maldergem L, Glover G, Carbonell P, Willems P, Gillessen-Kaesbach G, Horsthemke B (1998) Sporadic imprinting defects in Prader-Willi syndrome and Angelman syndrome: implications for imprint switch models, genetic counseling and prenatal diagnosis. Am J Hum Genet (in press)

Bürger J, Buiting K, Dittrich B, Groß S, Lich C, Sperling K, Horsthemke B, Reis A (1997) Different mechanisms and recurrence risks of imprinting defects in Angelman syndrome. Am J Hum Genet 61:88–93

Colot V, Rossiguol JL (1996) Interchromosomal transfer of epigenetic states in ascobolus: Transfer of DNA methylation is mechanistically related to homologous recombination. Cell 86:855–864

De Rubertis F, Kadosh D, Henchoz S, Pauli D, Reuter G, Struhl K, Spierer P (1996) The histone deacetylase RPD3 counteracts genomic silencing in *Drosophila* and yeast. Nature 384:589–591

Dittrich B, Buiting K, Korn B, Rickard S, Buxton J, Saitoh S, Nicholls RD, Poustka A, Winterpacht A, Zabel B, Horsthemke B (1996) Imprint switch mutation on human chromosome 15 may involve alternative transcripts of the *SNRPN* gene. Nat Genet 14:163–170

Ekström TJ, Cui H, Li X, Ohlsson R (1995) Promoter-specific *IGF2* imprinting status and its plasticity during human liver development. Development 121:309–316

Elson D, Bartolomei M (1997) A 5′ differentially methylated sequence and the 3′-flanking region are necessary for *H19* transgene imprinting. Mol Cell Biol 17:309–317

Feil R, Walter J, Allen N, Reik W (1994) Developmental control of allelic methylation in the imprinted mouse *Igf2* and *H19* genes. Development 120:2933–2943

Feil R, Boyano M, Allen N, Kelsey G (1997) Parental chromosome-specific chromatin conformation in the imprinted *U2af1-rs1* gene in the mouse. J Biol Chem 272:20893–20900

Franke A, Messmer S, Mohrle A, Orlando V, Zink D, Paro R (1994) Mechanisms of heritable gene silencing during *Drosophila* development. Biochem Soc Trans 22:561–565

Franklin GC, Adam GIR, Ohlsson R (1996) Genomic imprinting and mammalian development. Placenta 17:3–14

Garrick D, Fiering S, Martin D, Whitelaw E (1998) Repeat-induced gene silencing in mammals. Nat Genet 18:56–59

Glenn C, Nicholls R, Robinson W, Saitoh S, Niikawa N, Schinzel A, Horsthemke B, Driscoll DJ (1993) Modification of 15q11-q13 DNA methylation imprints in unique Angelman and Prader-Willi patients. Hum Mol Genet 2:1377–1382

Goodrich J, Puangsomlee P, Martin M, Long D, Meyerowitz E, Coupland G (1997) A Polycomb-group gene regulates homeotic gene expression in *Arabidopsis*. Nature 386:44–51

Greally J, Guinness M, McGrath J, Zemel S (1997) Matrix-attachment regions in the mouse chromosome 7F imprinted domain. Mamm Genome 8:805–810

Greally J, Starr D, Hwang S, Song L, Jaarola M, Zemel S (1998) The mouse *H19* locus mediates a transition between imprinted and non-imprinted DNA replication patterns. Hum Mol Genet 7:91–95

Grunstein M (1997) Histone acetylation in chromatin structure and transcription. Nature 389:349–352

Gunaratne P, Nakao M, Ledbetter D, Sutcliffe J, Chinault A (1995) Tissue-specific and allele.specific replication timing control in the imprinted human Prader-Willi syndrome region. Genes Dev 9:808–820

Holländer G, Zuklys S, Morel C, Mizoguchi E, Mobisson K, Simpson S, Terhorst C, Wishart W, Golan D, Bhan A, Burakoff S (1998) Monoallelic expression of the interleukin-2 locus. Science 279:2118–2121

Holliday R (1987) The inheritance of epigenetic defects. Science 238:163–170

Holmquist G (1975) Hoechst 33258 fluorescent staining of *Drosophila* chromosomes. Chromosoma 49:333–356

Horsthemke B, Dittrich B, Buiting K (1995) Parent-of-origin specific DNA methylation and imprinting mutations on human chromosome 15. In Parental imprinting: causes and consequences. Cambridge University Press, Cambridge, pp 295–308

Hu J, Vu T, Hoffman A (1997) Genomic deletion of an imprint maintenance element abolishes imprinting of both insulin-like growth factor II and *H19*. J Biol Chem 272:20715–20720

James TC, Elgin SC (1986) Identification of a nonhistone chromosomal protein associated with heterochromatin in *Drosophila melanogaster* and its gene. Mol Cell Biol 6:3862–3872

James TC, Eissenberg JC, Craig C, Dietrich V, Hobson A, Elgin SC (1989) Distribution patterns of HP1, a heterochromatin-associated nonhistone chromosomal protein of Drosophila. Eur J Cell Biol 50:170–180

Jeppesen P (1997) Histone acetylation: a possible mechanism for the inheritance of cell memory at mitosis. Bioessays 19:67–74

Jeppesen P, Turner B (1993) The inactive X chromosome in female mammals is distinguished by a lack of histone H4 acetylation, a cytogenetic marker for gene expression. Cell 74:281–289

Karpen G, Allshire R (1997) The case for epigenetic effects on centromere identity and function. Trends Genet 13:489–496

Khosla S, Kantheti P, Brahmachari V, Chandra H (1996) A male-specific nuclease-resistant chromatin fraction in the mealybug *Planococcus lilacinus*. Chromosoma 104:386–392

Kitsberg D, Selig S, Brandeis M, Simon I, Keshet I, Driscoll DJ, Nicholls RD, Cedar H (1993) Allele-specific replication timing of imprinted gene regions. Nature 364:459–463

Knoll JHM, Cheng S-D, LaLande M (1994) Allele-specificity of DNA replication timing in the Angelman/Prader Willi syndrome imprinted chromosomal region. Nat Genet 6:41–46

LaSalle J, Lalande M (1995) Domain organisation of allele-specific replication within the GABRB3 gene cluster requires biparental 15q11–13 contribution. Nat Genet 9:386–394

LaSalle J, Lalande M (1996) Homologous association of oppositely imprinted chromosomal domains. Science 272:725–728

Leighton P, Saam J, Ingram R, Stewart C, Tilghman S (1995a) An enhancer deletion affects both *H19* and *Igf2* expression. Genes Dev 9:2079–2089

Leighton PA, Ingram RS, Eggenschwiler J, Efstratiadis A, Tilghman SM (1995b) Disruption of imprinting caused by deletion of the *H19* gene region in mice. Nature 375:34–39

Lerchner W, Barlow D (1997) Paternal repression of the imprinted mouse *Igf2r* locus occurs during implantation and is stable in all tissues of the post-implantation mouse embryo. Mech Dev 61:141–149

Li E, Beard C, Jaenisch R (1993) Role for DNA methylation in genomic imprinting. Nature 366:362–365

Looijenga L, Verkerk A, De Groot N, Hochberg A, Oosterhuis J (1997) *H19* in normal development and neoplasia. Mol Reprod Dev 46:419–439

Lyko F, Brenton JD, Surani MA, Paro R (1997) An imprinting element from the mouse *H19* locus functions as a silencer in *Drosophila*. Nat Genet 16:171–173

Lyko F, Buiting K, Horsthemke B, Paro R (1998) Identification of a silencing element in the human 15q11–q13 imprinting center by using transgenic *Drosophila*. Proc Natl Acad Sci USA 95:1698–1702

Mertineit C, Yoder J, Taketo T, Laird D, Trasler J, Bestor T (1998) Sex-specific exons control DNA methyltransferase in mammalian germ cells. Development 125:889–897

Moore T, Constancia M, Zubair M, Bailleul B, Feil R, Sasaki H, Reik W (1997) Multiple imprinted sense and antisense transcripts, differential methylation and tandem repeats in a putative imprinting control region upstream of mouse *Igf2*. Proc Natl Acad Sci USA 94:12509–12514

Nan X, Campoy F, Bird A (1997) MeCP2 is a transcriptional repressor with abundant binding sites in genomic chromatin. Cell 88:471–481

Neumann B, Kubicka P, Barlow DP (1995) Characteristics of imprinted genes. Nat Genet 9:12–13

Obata Y, Kaneko-Ishino T, Koide T, Takai Y, Ueda T, Domeki I, Shiroishi T, Ishino F (1998) Disruption of primary imprinting during oocyte growth leads to the modified expression of imprinted genes during embryogenesis. Development 125:1553–1560

Ohlsson R, Hedborg F, Holmgren L, Walsh C, Ekström TJ (1994) Overlapping patterns of *IGF2* and *H19* expression during human development: biallelic *IGF2* expression correlates with a lack of *H19* expression. Development 120:361–368

Ohta T, Buiting K, Kokkonen H, Saitoh S, McCandless S, Cassidy S, Driscoll D, Horsthemke B, Nicholls R (1997) Molecular analysis in two large AS imprinting mutation (IM) families and identification of microdeletion junctions in AS and PWS IM families. Am J Hum Genet 61 (Suppl):A1850

Olek A, Walter J (1997) The pre-implantation ontogeny of the *H19* methylation imprint. Nat Genet 17:275–276

Orlando V, Paro R (1995) Chromatin multiprotein complexes involved in the maintenance of transcription patterns. Curr Opin Genet Dev 5:174–9

Overall M, Bakker M, Spencer J, Parker N, Smith P, Dziadek M (1997) Genomic imprinting in the rat: linkage of *Igf2* and *H19* genes and opposite parental allele-specific expression during embryogenesis. Genomics 45:416–420

Pak DT, Pflumm M, Chesnokov I, Huang DW, Kellum R, Marr J, Romanowski P, Botchan MR (1997) Association of the origin recognition complex with heterochromatin and HP1 in higher eukaryotes. Cell 91:311–323

Paldi A, Gyapay G, Jami J (1995) Imprinted chromosomal regions of the human genome display sex-specific meiotic recombination frequencies. Curr Biol 15:1030–1035

Paro R, Hogness DS (1991) The Polycomb protein shares a homologous domain with a heterochromatin-associated protein of *Drosophila*. Proc Natl Acad Sci USA 88:263–267

Poirier F, Chan C-T, Timmons P, Robertson E, Evans M, Rigby P (1991) The murine *H19* gene is activated during embryonic stem cell differentiation in vitro and at the time of implantation in the developing embryo. Development 113:1105–1114

Postlethwait J, Yan Y, Gates M, Horne S, Amores A, Brownlie A, Donovan A, Egan E, Force A, Gong Z, Goutel C, Fritz A, Kelsh R, Knapik E, Liao E, Paw B, Ransom D, Singer A, Thomson M, Abduljabbar T, Yelick P, Beier D, Joly J, Larhammar D, Talbot W et al. (1998) Vertebrate genome evolution and the zebrafish gene map. Nat Genet 18:345–349

Reik W, Maher E (1997) Imprinting in clusters: lessons from Beckwith-Wiedemann syndrome. Trends Genet 13:330–334

Reik W, Walter J (1998) Imprinting mechanisms in mammals. Curr Opin Genet Dev 7:154–164

Reis A, Dittrich B, Greger V, Buiting K, Lalande M, Gillessen-Kaesbach G, Anvret M, Horsthemke B (1994) Imprinting mutations suggested by abnormal DNA methylation patterns in familial Angelman and Prader-Willi syndromes. Am J Hum Genet 54:741–747

Ripoche M, Kress C, Poirier F, Dandolo L (1997) Deletion of the *H19* transcription unit reveals the existence of a putative imprinting control element. Genes Dev 11:1596–1604

Saitoh S, Buiting K, Rogan P, Buxton J, Driscoll D, Arnemann J, König R, Malcolm S, Hors-themke B, Nicholls R (1996) Minimal definition of the imprinting center and fixation of a chromosome 15q11–13 epigenotype by imprinting mutations. Proc Natl Acad Sci USA 93:7811–7815

Shemer R, Birger Y, Dean WL, Reik W, Riggs AD, Razin A (1996) Dynamic methylation adjustment and counting as part of imprinting mechanisms. Proc Natl Acad Sci USA 93:6371–6376

Shemer R, Birger Y, Riggs A, Razin A (1997) Structure of the imprinted mouse Snrpn gene and establishment of its parental-specific methylation pattern. Proc Natl Acad Sci USA 94:10267–10272

Shibata H, Ueda T, Kamiya M, Yoshiki A, Kusakabe M, Plass C, Held W, Sunahara S, Katsuki M, Muramatsu M, Hayashizaki Y (1997) An oocyte-specific methylation imprint center in the mouse *U2afbp-rs/U2af1-rs1* gene marks the establishment of allele-specific methylation during preimplantation development. Genomics 44:171–178

Singh PB, James TC (1995) Chromobox genes and the molecular mechanisms of cellular determination. In: Ohlsson R, Hall K, Ritzen M (eds) Genomic imprinting: causes and consequences. Cambridge University Press, Cambridge, pp 71–108

Stöger R, Kubicka P, Liu C-G, Kafri T, Razin A, Cedar H, Barlow D (1993) Maternal-specific methylation of the imprinted mouse *Igf2r* locus identifies the expressed locus as carrying the imprinting signal. Cell 73:61–71

Sutcliffe JS, Nakao M, Christian S, Örstavik KH, Tommerup N, Ledbetter DH, Beaudet AL (1994) Deletions of a differentially methylated CpG island at the *SNRPN* gene define a putative imprinting control region. Nat Genet 8:52–58

Svensson K, Walsh C, Fundele R, Ohlsson R (1995) *H19* is functionally imprinted in the mouse fetal choroid plexus and leptomeninges. Mech Dev 51:31–37

Svensson K, Mattsson R, James T, Wentzel P, Pilartz M, MacLaughlin J, Miller S, Olsson T, Eriksson U, Ohlsson R (1998) The paternal allele of the *H19* gene is silenced in a stepwise manner during early mouse development: the acetylation status of histones may be involved in the generation of variegated expression patterns. Development 125:61–69

Szabo P, Mann J (1995) Allele-specific expression and total expression levels of imprinted genes during early mouse development: implications for imprinting. Genes Dev 9:3097–3108

Tada M, Tada T, Lefebvre L, Barton S, Surani M (1997) Embryonic germ cells induce epigenetic reprogramming of somatic nucleus in hybrid cells. EMBO J 16:6510–6520

Tada T, Tada M, Hilton K, Barton S, Sado T, Takagi N, Surani M (1998) Epigenotype switching of imprintable loci in embryonic germ cells. Dev Genes Evol 207:551–562

Tremblay K, Duran K, Bartolomei MS (1997) A 5′ 2-kilobase-pair region of the imprinted mouse *H19* gene exhibits exclusive paternal methylation throughout development. Mol Cell Biol 17:4322–4329

Triolo T, Sternglanz R (1996) Role of interactions between the origin recognition complex and SIR1 in transcriptional silencing. Nature 381:251–253

Tucker K, Beard C, Dausmann J, Jackson-Grusby L, Laird P, Lei H, Li E, Jaenisch R (1996) Germ-line passage is required for establishment of methylation and expression patterns of imprinted but not of nonimprinted genes. Genes Dev 10:1008–1020

Tweedie S, Charlton J, Clark V, Bird A (1997) Methylation of genomes and genes at the invertebrate-vertebrate boundary. Mol Cell Biol 17:1469–1475

Vashee S, Melcher K, Ding W, Johnston S, Kodadek T (1998) Evidence for two modes of cooperative DNA binding in vivo that do not involve direct protein-protein interactions. Curr Biol 8:452–458

Webber A, Ingram R, Levorse J, Tilghman S (1998) Location of enhancers is essential for the imprinting of *H19* and *Igf2* genes. Nature 391:711–715

White L, Rogan P, Nicholls R, Wu B, Korf B, Knoll J (1996) Allele-specific replication of 15q11-q13 loci: adiagnostic test for detection of uniparental disomy. Am J Hum Genet 59:423–430

Willard H, Salz H (1997) Remodelling chromatin with RNA. Nature 386:228–229

Wutz A, Smrzka O, Schweifer N, Schellander K, Wagner E, Barlow D (1997) Imprinted expression of the *Igf2r* gene depends on an intronic CpG island. Nature 389:745–749

Xu G, Bestor T (1997) Cytosine methylation targetted to pre-determined sequences. Nat Genet 17:376–378

Yoder JA, Bestor TH (1996) Genetic analysis of genomic methylation patterns in plants and mammals. Biol Chem 377:605–610

Yoder J, Bestor T (1998) A candidate mammalian DNA methyltransferase related to *pmt1p* of fission yeast. Hum Mol Genet 7:279–284

Yoder J, Walsh C, Bestor T (1997) Cytosine methylation and the ecology of intragenomic parasites. Trends Genet 13:335–340

Human Diseases and Genomic Imprinting

Judith G. Hall

1
Introduction

It has become increasingly clear over the past few years that genomic imprinting has a great deal of relevance to human diseases. As discussed by Trivers and Burt, Chap. 1, and Horsthemke et al. (Chap. 5, this vol.), imprinted genes are involved in various aspects of growth and behavior that are laid down during prenatal development. The imprinted status can, however, be manifested differentially during development, such that the parent of origin specific monoallelic expression patterns are stage- and tissue-specific (see Horsthemke et al. Chap. 5, this vol.). Moreover, the comparison of human and mouse imprinted genes suggest that genomic imprinting is also species and even strain-specific (Hall, 1990). Efforts to understand the role of imprinting in human disorders are challenging, therefore, and may entail tissue specific and age-specific analysis (Hall, 1997; see Table 1).

It is often the case that poorly understood normal processes are uncovered as a result of their manifestation of disease phenotype when something goes awry. Genomic imprinting is no exception. Because of parent of origin specific expression patterns of growth – regulating genes (see Tycko, Chap. 7, this vol.), unusual inheritance patterns can be observed and tissue specific abnormalities of growth and expression can be seen, rather than true congenital anomalies with malformations. There are many types of parent of origin effects in human diseases; imprinting effects are observed as disease processes when there is lack of expression (because of deletions, mutation, or deficiency) of a gene usually expressed monoallelically from a specific parent, or when there is over-expression (because of duplication, relaxation, or loss of control) of a normally monoallelically expressed gene (Langlois et al. 1995).

Department of Pediatrics, University of British Columbia and BC's Children's Hospital, Vancouver, Canada

Table 1. Links between human disease processes and genomic imprinting

Disease processes	Abnormality produced

1. Chromosomal deletions	Uncovers imprinting
	   Syndromes – Prader-Willi
	                 – Angelman
	                 – Beckwith Wiedemann
	   Cancer – loss of tumor suppressor
	                 – Wilms tumor

2. Uniparental disomy	Uncovers imprinting by lack of specific parental contribution
	Mental retardation
	  – Prader Willi-mat UPD15
	  – Angelman-pat UPD15
	Abnormal growth
	  – Undergrowth – mat UPD7
	  – Overgrowth – mat UPD15
	Nondisjunction leading to UPD by trisomy rescue
	  – Results from trisomy – therefore maternal age related
	  – Vestigial trisomic – may produce abnormal tissue growth and cancer
	Autosomal recessive disorders
	  – When UPD chomosome carries abnormal allele

3. Abnormal regulation of expression of imprinted genes	– Insulin gene control
	  – Yolk sac – uniparental paternal expression
	  – Newborn pancreas – biparental expression – controlled by paternal chromosome 6
	  – Adult pancreatic expression – controlled somewhere else than chromosome 6
	– Loss of imprinting (relaxation)
	  – Overgrowth (Beckwith Wiedemann)
	  – Hemihypertrophy
	  – Cancer

4. Mutation in an imprintable gene (leads to nonfunction when the nonimprinted parental gene is supposed to express)	– Tumor growth – paraganglions (expression of paternal allele) – Syndrome – Albright osteodystrophy (under growth, brain growth, calcium metabolism)
5. Disturbances in early embryonic growth	– Placental overgrowth – placental mole when only paternal contribution (androgenetic) – Ovarian teratoma-pathenogenesis (gynogenetic) – Triploid phenotype dependent on from which parent the two chromosome sets are

6. Abnormal behavior	– Mouse models
	   Hyperactive
	   Hypoactive
	– Syndromes
	  – Prader Willi hypoactivity
	  – Angelman hyperactivity
	– Turner syndrome – socialization

2
Chromosome Deletion Syndromes

2.1
Prader-Willi Syndrome

The Prader-Willi syndrome (PWS) is a well-known and relatively easily recognized syndrome. In the newborn, hypotonia and failure to thrive are observed. At 6 months to 1 year of age, affected individuals begin to eat voraciously. Without rigorous calorie restriction they usually put on weight rapidly and become quite obese. Affected individuals also have hypothalamic, hypogonadotrophic hypogonadism. Genitalia are small and puberty is delayed. Affected individuals tend to have almond-shaped eyes, narrow bitemporal diameter, and a somewhat long face. They often develop small hands and feet as young children and moderate developmental delay is also observed.

More than a decade ago, chromosome deletions of the 15q11–13 region were recognized in 60% of PWS patients. Shortly thereafter it was discovered that these deletions were always deletions of the paternally inherited chromosome. This suggested that there was a critical region on chromosome 15, which, when deleted, resulted in the PWS. Shortly thereafter, another group of individuals with PWS were observed to be related to the lack of a paternal chromosome 15, since both copies of chromosome 15 had come from the mother, i.e., maternal uniparental disomy. This is obviously a second way to be deficient in the PWS critical region on the paternal chromosome 15. For practical purposes, the phenotypes of these two types of PWS are indistinguishable; although since the region of the deletion on chromosome 15 often involves a pigment gene, patients with PWS due to deletions may have blond hair and light complexions.

A small percentage of Prader-Willi patients may be due to mutations of an imprinting control region. The Prader-Willi critical region has recently been investigated by numerous laboratories and found to involve several genes which are only expressed from the paternally inherited chromosome. Among these are *SNRPN* (a developmentally regulated protein component of spliceosome expressed predominantly in neuronal tissue), *ZNF127*, *1PW*, *NDN* and *PAR1*, and *PAR5* (see Beechey, Appendix and Horsthemke et al. Chap. 5, this vol.). The distance between these imprinted genes is 1–1.5 Mb, suggesting that additional imprinted genes may exist in this critical region.

2.2
Angelman Syndrome

Angelman syndrome (AS) is characterized by severe developmental delay and unusual behavior. Affected individuals are almost never able to walk. They have outbursts of inappropriate laughter and a very happy disposition. AS patients also tend to have ataxic movement and exhibit episodes in which they clap or wave their hands. Affected individuals also have a characteristic abnormality on their EEG and often have seizures. They tend to have a large mouth and midfacial hypoplasia. Individuals with Angelman syndrome were also recognized to have deletions of chromosome 15q11–13 which involved an area overlapping the Prader-Willi deletion. The deletion in Angelman syndrome is always on the maternally inherited chromosome, suggesting there is also an Angelman critical region. Just as in Prader-Willi syndrome, this critical region may be missing because of a deletion of the maternally inherited chromosome, or through paternal uniparental disomy of chromosome 15. Approximately 60–70% of the patients are due to a deletion, while only 5% are due to uniparental paternal disomy, and the rest appear to result from abnormalities of the control of the critical region or mutations in a specific gene.

UBE3A (a gene encoding a protein which functions in ubiquitination) has been identified in the Angelman critical region and exhibits differential allelic expression (Matsuura et al. 1997). Interestingly, it has tissue-specific expression such that there is biparental expression in most tissues, but only uniparental maternal expression in the brain (Donlon 1997; Kishino et al. 1997; Vu and Hoffman 1997). The maternal allele is normally expressed only in the brain, and, if there is a deletion or lack of expression of the maternal allele in the brain, the Angelman syndrome phenotype is the result.

The involved region of chromosome 15q11–13 has been extensively studied. It has parental specific methylation patterns, parental specific replication patterns, and several control elements which would all appear to be involved in the differential expression (Glenn et al. 1997). The exact mechanisms of control are still not yet understood, but the maternal chromosomal abnormalities seem to more often involve control elements, while the paternal chromosome abnormalities seem to more often involve deficiency (see Horsthemke et al. Chap. 5, this vol.). A specific PCR test based on methylation can be used to identify the parental origin of a particular chromosome 15 (Buchholz et al. 1997; Zeschnigk et al. 1997). Determining the mechanism leading to Angelman and Prader-Willi syndrome is important, e.g., deletion (including translocation) *versus* uniparental disomy *versus* control gene alterations. The recurrence risks vary depending on which mechanism is responsible and whether one of the parents carries an abnormality. The highest recurrence risk is associated with control alterations of an imprinting center or gene control element, the lowest for uniparental disomy (Bürger et al. 1997). Once the mechanism has been established, prenatal diagnosis of a second affected pregnancy is possible.

The Prader-Willi/Angelman deletions have drawn attention to the possibility that regions of other chromosomes may involve imprinting effects. The question is now raised with all chromosomal anomalies (deletions, duplications, trisomies, isochromosomes, etc.) as to whether they are more likely to occur on the chromosome inherited from one or the other parent and whether there are different parent-of-origin phenotypic effects. A great deal of work needs to be done in humans to answer this question in a tissue- and/or stage-specific way.

3
Uniparental Disomy

The concept that one could inherit both of a pair of chromosomes from only one parent was first suggested by Engel (reviewed in Engel, 1997). Molecular genetic techniques now allow investigators to trace chromosomes, parts of chromosomes, and genes from parent to child and in a single individual from tissue to tissue. Uniparental disomy describes the situation where an individual has the normal number of chromosomes but two chromosomes of a pair (or parts of those chromosome) have been inherited from only one parent. There is usually no contribution at all of that chromosome (or region) from the other parent. Full chromosomes or parts of chromosome can be involved in this type of process. It may even exist in a mosaic form (i.e., present in some cells but not in others). The chromosomes present in the individual with uniparental disomy may be the two different chromosomes from that parent (heterodisomy), or they may be identical copies of the same chromosome (isodisomy). Uniparental disomy used to be thought to be extremely rare. Recently, however, it has been recognized to be surprisingly common and to be responsible for a variety of abnormalities (Ledbetter and Engel 1995).

Uniparental disomy is thought to occur primarily through trisomy rescue. Ten to 15% of all recognized human conception occur as trisomies, most of which, however, are lethal and miscarry early in pregnancy (Dimmick and Kalousek 1992). The only way that such a conception can survive is for one of the three chromosomes to be lost. If that happens, one third of the time the individual will convert from a trisomy with three chromosomes to a uniparental disomy with two chromosomes from one parent. This is because the trisomy has two chromosomes from one parent and one from the other. Trisomies most often have two maternal chromosomes and one paternal chromosome, but they can also occur with two paternal chromosomes and one maternal chromosome. If there was nondisjunction in the first meiotic division, then there would be two different chromosomes (heterodisomy) transmitted from one parent that become part of the trisomy and then the disomy. If the nondisjunction occurred in the second meiotic division, then there may be isodisomy with two identical chromosomes. Isodisomy may also occur with somatic nondisjunction. Because of crossingover during meiosis, there may be isodisomy for only part of a chromosome.

3.1
Uncovering Imprinted Genes

If a particular gene must be inherited from a particular parent in order for normal development to occur, then uniparental disomy involving the presence of two chromosomes from the other parent will mean that that critical region or gene is not present. In the case of Prader-Willi, there is a critical region that must be inherited from the father. When there are only two maternal copies of chromosome 15, the paternally derived Prader-Willi critical region and its gene products will not be present. It is clear from uniparental disomy work in mice that there are many areas of the mouse genome where the presence of uniparental disomy leads to developmental abnormalities. In humans, six such regions have been identified to date (paternal chromosome 6q, maternal chromosome 7q, paternal chromosome 11q, maternal chromosome 14q, and both maternal and Paternal chromosome 15q). (Hall 1997; see Table 2) More are likely to be defined.

3.2
Producing Autosomal Recessive Disorders

If one parent is a carrier for an autosomal recessive disorder and an isodisomy (exactly the same copy of the chromosome or chromosome region) is present, then, since the specific chromosome carries an abnormal allele and the individual with uniparental disomy has no normal allele, the individual will mani-

Table 2. Uniparental isodisomy and recessive disorders

Chromosome	Transmission	Recessive disorders	Reference
5	Paternal	Spinal muscular atrophy	Brzustowicz et al. (1994)
6	Paternal	Complement deficiency	Welch et al. (1990)
7	Maternal	Cystic fibrosis	Voss et al. (1989)
	Maternal	Osteogenesis imperfecta	Spotila et al. (1992)
	Paternal	Congenital chloride diarrhea	Hoglund et al. (1994)
8	Paternal	Lipoprotein lipase deficiency	Benlain et al. (1996)
9	Maternal	Cartilage hair hypoplasia	Sulisalo et al. (1994)
11	Maternal	Beta thalassemisa	Beldjord et al. (1992)
13	Paternal	Retinoblastoma	Cavenee et al. (1983)
14	Maternal	Rod monochromacy	Pentao et al. (1992)
15	Maternal	Bloom syndrome	Woodage et al. (1994)
16	Paternal	Alpha thalassemia	Ngo et al. (1993)
	Maternal	Familial Mediterranean Fever	Korenstein et al. (1994)
X	Paternal	Hemophilia transmitted from father to son	Vidaud et al. (1989)
	Maternal	Duchenne muscular dystrophy in a female	Quan et al. (1997)

fest the autosomal recessive disorder. Thus, amazingly enough, only one of the parents will be a carrier for the autosomal recessive disorder. This phenomenon was first described in humans in an individual affected with cystic fibrosis. A long list of other genes have subsequently been recognized to be present in a recessive form secondary to uniparental disomy (Table 2).

3.3
Vestigial Aneuploidy

If uniparental disomy is derived from a trisomy, then there may be some residual trisomic cells which survive. When prenatal diagnosis is done, 2-3% of all cases are found to have some trisomic cells, among either amniocytes or chorionic villus cells. This percentage is higher with chorionic villus sampling than with amniocentesis. These cases appear to represent a trisomy conception that has converted to disomy. If there are some trisomic cells present in the individual, they can lead to malformations and growth abnormalities. Care must be taken to determine whether such abnormalities can be identified.

3.4
Prenatal Diagnosis

Obviously, if trisomy is found in some cells when prenatal diagnosis is performed, there is concern for vestigial aneuploidy in the child. Amniocentesis and cord blood studies are frequently done and these are usually normal. Nevertheless, there may be trisomic cells left in some tissues. Depending on the particular trisomy, surviving cells might be expected to be present in different tissues. In such situations, careful analysis of multiple tissues may help to define prognosis. It is also important to determine whether uniparental disomy has developed, since that may put the fetus at risk for imprinting abnormalities, depending on which chromosome is involved. At this time, concern should be raised for chromosomes 6, 7, 11, 14, 15, and 16 to make sure that uniparental disomy has not developed.

3.5
Advanced Maternal Age

Interestingly, advanced maternal age is a risk factor for uniparental disomy when trisomy rescue is involved. The forms of AS and PWS seen with uniparental disomy are also associated with advanced maternal age. Chromosome deletions in these disorders are not associated with advanced maternal age. Translocation carrier parents, because they may predispose to trisomy, are also at risk for uniparental disomy and subsequently for the potential complications and sequelae of perturbed imprinting, producing autosomal recessive diseases, and vestigial aneuploidy.

4
Growth Disorders

The early mouse work on imprinting indicated that imprinted genes have both positive and negative effects on growth. This suggested that imprinted genes were possibly growth factors, growth factor receptors, or factors that partici- pate in the regulation or downstream function of such protein products. Con- sequently, imprinted genes would be expected to play an important role in embryonic and fetal development, and in disorders of overgrowth, under- growth, and uncontrolled growth such as cancer (see Tycko, Chap. 7, this vol.).

4.1
Intrauterine Growth Retardation

In mice it is quite clear that imprinted genes may play an important role in intrauterine growth. In humans uniparental disomy of maternal chromosome 7 is associated with moderately severe intrauterine growth retardation. This suggests that an important gene for normal growth is on the paternal chromo- some 7. The MEST gene on chromosome 7 has been identified as having only paternal expression (Nishita et al. 1996; Kobayashi et al. 1997) although it has been ruled out as a culprit in the Russell-Silver syndrome (RSS) (Kalscheuer, pers. inform.). RSS is associated with growth retardation but with normal head size. Recently, uniparental maternal disomy has been observed in individuals described as having RSS phenotype. As many as 15% of Russell-Silver syn- drome individuals have been found to have maternal uniparental disomy of chromosome 7 (Kotzot et al. 1995). Interestingly, RSS patients often have hemihypotrophy (i.e., one side of the body even smaller than the other). This may reflect mosaicism for uniparental disomy.

4.2
Undergrowth and Congenital Diabetes

Uniparental paternal disomy of chromosome 6 has been associated with tran- sient neonatal diabetes mellitus and intrauterine growth retardation. This is a form of diabetes where insulin is not produced during the newborn period, but then spontaneously starts to be produced after 6–12 months. Several cases have been associated with uniparental paternal disomy of chromosome 6, suggesting that chromosome 6 regulates insulin production in the newborn period (Temple et al. 1996). Most recent studies suggest that duplication of paternal chromosome 6, or uniparental paternal disomy of chromosome 6, may be responsible (Arthur et al. 1997). Interestingly, the gene for insulin is on chromosome 11 and must have different types of control *in utero*, and in the newborn period compared to later life. It is known that in the yolk sac of embryonic mice only the paternal insulin allele is expressed, suggesting that there is tissue-specific, stage-specific, and parent of origin-specific control of insulin expression (Giddings et al. 1994; see also Paldi and Jouvenot, Chap. 13, this vol.).

4.3
Overgrowth and Hemihypertrophy (BWS)

There are a large number of overgrowth syndromes known to occur in children (Jones 1988). They all seem to be associated with permissiveness for tumorigenesis. One of the most striking and common overgrowth syndrome is the Beckwith-Wiedemann syndrome (BWS), in which there is either relaxation of the maternal *IGF2* gene or uniparental paternal disomy (Reik and Maher 1997; see Tycko, Chap. ••, this vol.). Many BWS patients have hemihypertrophy, suggesting that the overgrowth phenomenon may be present in only half the cells of the body implying that local factors such as receptors are involved rather than circulating factors. Interestingly, individuals with Beckwith-Wiedemann syndrome are also at risk for Wilms tumor; a malignancy which is well known to have parent-of-origin effects and loss of heterozygosity (Wiedemann 1983; Ward 1997).

BWS involves fetal overgrowth, particularly of internal organs such that there is often an omphalocele because of the large intraabdominal organs. Additional cardinal signs include a large tongue and ear pits and creases. Because of a large pancreas, excess insulin may be secreted and hypoglycemia can be seen in newborn infants. Usually during childhood, the oversized proportions of affected individuals become more normal, such that by the time they are adults they do not appear unusual. This suggests that control of the involved genes is specific for prenatal and early postnatal developmental periods. In the past, it had been recognized that BWS tended to run in the families, so that there were nonaffected transmitting females. The condition was said to have a nonpenetrant or variably penetrant autosomal dominant pattern of inheritance. There were also cases of BWS with partial paternal chromosome 11 duplication. There are numerous examples of monozygotic twins affected with BWS where only one of the monozygotic twin is affected. The discordance suggests that somatic events may be responsible for producing Beckwith-Wiedemann syndrome. These discordant twins tended to be female monozygotic twins. Recently, several genes have been implicated in producing Beckwith-Wiedemann syndrome including the imprinted *IGF2*, *H19*, *p57^KIP2*, and *KVLQT* genes (Catchpoole et al. 1997; Joyce et al. 1997; O'Keefe et al. 1997; see Tycko, Chap. ••, this vol.). *KVLQT1* shows tissue-specific imprinting with only paternal expression in most tissues, but biparental expression in the heart (Lee et al. 1997). The risk of recurrence for BWS will depend on the genetic rearrangement that produces it. Clearly, it can have as much as 50% recurrence risk.

An interesting X-linked disorder resembles BWS clinically; Simpson-Golabi-Behmel syndrome is characterized by overgrowth, coarse faces, anomalies of skeleton, heart, kidney, gastrointestinal tract, and central nervous system. The gene responsible (*GPC3*) is a member of the glypican family of heparin sulfate proteoglycans (Lindsay et al. 1997). It seems possible that the two conditions share a common pathway that impinges on the concentration of free IGF-II ligand in the vicinity of cell surface receptors.

5
Behavior

Pronuclear transplantation work in mice suggests that neither androgenetic cells with only a paternal chromosome complement nor gynogenetic pronuclear or parthenogenic cells with only a maternal complement can survive unless they are mixed with normal cells. When these abnormal uniparental cells are mixed with normal cells, however, embryos tend to have tissue selection for the gynogenetic and androgenetic cells. Androgenetic cells end up in bone, muscle, and dermis (mesenchymal tissues), while gynogenetic cells end up in central nervous system and the epidermis (ectodermal tissues). (Ferguson-Smith 1996). These findings would suggest that the maternal genetic contribution to the brain may have a disproportionate effect on the behavior of an individual (see Horsthemke et al. Chap. 5, this vol.).

The mouse models of uniparental disomy suggested that hyperactivity and hypoactivity are seen with imprinted genes, as illustrated by the AS and PWS. In the newborn period, individuals with PWS are extremely hypotonic and fail to feed, so that they often have failure to thrive and without special supportive efforts would be likely to die. AS, on the other hand, is associated with a very peculiar type of hyperactivity with puppet-like movements, clapping of hands, inappropriate laughter, and ataxic movements. The exact mechanism of these behavioral abnormalities is not clear. In the case of AS, however, it appears that at least one of the responsible gene(s) is imprinted only in the central nervous system. Several psychiatric disorders seem to demonstrate parent-of-origin affects. Time will tell whether these involve genomic imprinting.

Recent studies in Turner syndrome suggest there may be a genomically imprinted gene(s) on the X chromosome (Skuse et al. 1997). Turner syndrome occurs when there is absence or abnormality of one of the sex chromosomes. Half of Turner syndrome individuals have only one X chromosome. The other half have various combinations of deletions and isochromosomes of the X chromosome or remnants of the Y chromosome. Approximately 80% of Turner syndrome women and girls have lost their paternal X chromosome. It has been known for some time that women and girls with Turner syndrome have unusual behavior. They act immaturely, have a particular spatial perceptual problem, and/or may lack ambition and social skills. A recent study has suggested that this unusual behavior correlates with loss of the paternal X chromosome. Thus, those Turner syndrome individuals who maintain a paternally derived X chromosome seem to have normal skills, while those who retain their maternal X chromosome do not. This would suggest that normal males who always receive their X chromosome from their mother are less socially adept because of imprinting of some gene on the X chromosome (McGuffin and Scowfield 1997).

6
Other Disorders

There are a number of other conditions in humans which show a very strong suggestion of imprinting effects. These observations have come from careful evaluations of pedigrees. One such disorder is pseudo-pseudohypoparathyroidism or Albright's hereditary osteodystrophy (Davies and Hughes 1993). A marked excess of maternal transmission has been observed suggesting that full expression of the disorder occurs with maternal transmission but only partial expression when the gene is inherited from the father. There appear to be at least two linkage groups associated with Albright's hereditary osteodystrophy, one involving the GS protein. Paragangliomas are also imprinted such that expression occurs with inheritance from father. There are numerous other disorders with some suggestion of imprinting effects. Careful studies will lead to clarifying the situation but the clinician should be suspicious when an unusual inheritance pattern is present. Many disorders which have been described as multifactorial may indeed turn out to have at least one of the genes which is genomically imprinted in a complex pathway (Hall 1990; see Table 1).

7
Complexity of the Control of the Genomic Imprinting Process

Since genomic imprinting is tissue-, stage-, species-, and possibly strain-specific, it is very difficult to study, particularly in human beings where embryonic and fetal tissues as well as placental tissue at various stages are not readily available. In particular, in human beings, to study only blood lymphoctyes will certainly misrepresent the situation present in other tissues, so it does not seem appropriate to study only blood when looking for imprinting effects. There are a number of complex or multifactorial disorders that show some suggestion of parent-of-origin effects such as neural tube defects, diabetes, schizophrenia, and manic depressive disorder. These disorders are almost surely heterogeneous and also have complex biochemical pathways. It seems quite possible that one or two genes along a pathway may be imprinted while the others are not. Obviously, sorting out such genomic imprinting effects is a complex process.

Imprinting clearly involves differential methylation of DNA (see Razin and Shemer, Chap. 9, and Horsthemke et al. Chap. 5, this vol.). Although the imprinted allele has increased methylation when nonexpressing, there are however, islands of methylation in expressing imprintable genes as well. The process of methylation requires folic acid. Folic acid deficiency has recently been recognized to be extremely important in the genesis of a variety of birth defects as well as adult diseases. Many humans are folic acid-deficient and therefore may not methylate as efficiently as if they were not deficient. In particular, it appears that the process of meiosis and gametogenesis are important steps in establishing normal parent-of-origin imprinting effects (Tycko et al. 1997); it is possible that these periods constitute stages that are sensitive to folic acid deficiency and hence epigenetic mutations. Meiosis in females occurs very early in embryogenesis. The first meiotic division occurs between the 6th and 8th week of intrauterine life. The folic acid status of the grandmother at the time of oogenesis could, therefore, have a major effect on imprinting stability in the oocyte that will become the grandchild. This type of transgenerational effect needs to be considered when evaluating imprinting effects in human diseases.

8
Summary

In summary, there are a number of conditions where genomic imprinting effects are recognized to be associated clinical disorders of importance in humans. There may be many more. Genomic imprinting should be suspected in any disorder with overgrowth, undergrowth, or behavior abnormalities. Disorders with unusual pattern of inheritance should be studied for the possibility that genomically imprinted gene(s) are involved. Understanding the mechanisms of genomic imprinting has major ramifications in terms of recurrence risk, prediction of whether offspring will be affected, and risk of malignancy. Of particular concern is the potential for uniparental disomy when trisomy is found during prenatal diagnosis.

References

Arthur EI, Zlotogora J, Lerer I et al. (1997) Transient neonatal diabetes mellitus in a child with invdup(6)(q22q23) of paternal origin. Eur J Hum Genet 5:417–419

Beldjord C, Henry I, Bennani C, Vanhaeke D, Labie D (1992) Uniparental disomy: a novel mechanism for thalassemia major. Blood 80:287–289

Benlain P, Foubert L, Gagne E, Bernard L, Gennes JLD, Langlois S, Robinson W, Hayden M (1996) Complete paternal isodisomy for chromosome 8 unmasked by lipoprotein lipase deficiency. Am J Hum Genet 59:431–436

Brzustowicz LM, Allitto BA, Matseoane D, Theve R, Michaud L et al. (1994) Paternal isodisomy for chromosome 5 in a child with spinal muscular atrophy. Am J Hum Genet 54:482–488

Buchholz T, Jackson J, Smith A (1997) Methylation analysis at three different loci within the imprinted region of chromosome 15q11-13. Am J Med Genet 72:117–119

Bürger J, Buiting K, Dittrich B, Groß S et al. (1997) Different mechanisms and recurrence risks of imprinting defects in Angelman syndrome. Am J Hum Genet 61:88–93

Catchpoole D, Lam WWK, Valler D, Temple IK et al. (1997) Epigenetic modification and uniparental inheritance of H19 in Beckwith-Wiedemann syndrome. J Med Genet 34:353–359

Cavenee WK, Dryja TP, Phillips RA, Benedict WF, Godbout R, Gallie BL, Murphree AL, Strong LC, White RL (1983) Expression of recessive alleles by chromosomal mechanisms. Nature 305:779–784

Davies SJ, Hughes HE (1993) Imprinting in Albright's hereditary osteodystrophy. J Med Genet 30:101–103

Dimmick JE, Kalousek DK (1992) Developmental pathology of the embryo and fetus. JB Lippincott, Philadelphia, Pennsylvania

Donlon T (1997) Fishing out the Angelman syndrome gene. Nat Med 3:281

Engel E (1997) Uniparental disomy (UPD). Genomic imprinting and a case for new genetics. Ann Genet 40:24–34

Feil R, Kelsey G (1997) Insights from model systems. Genomic imprinting: a chromatin connection. Am J Hum Genet 61:1213–1219

Ferguson-Smith AC (1996) Imprinting moves to the centre. Nat Genet 14:119–121

Giddings SJ, King CD, Harman KW et al. (1994) Allele specific inactivation of insulin 1 and 2, in the mouse yolk sac, indicates imprinting. Nat Genet 6:310–313

Glenn CC, Driscoll DJ, Yang TP, Nicholls RD (1997) Genomic imprinting: potential function and mechanisms revealed by the Prader-Willi and Angelman syndromes. Mol Hum Reprod 3:321–332

Hall JG (1990) Genomic imprinting: review and relevance to human diseases. Am J Hum Genet 46:857–873

Hall JG (1997) Genomic imprinting: nature and clinical relevance. Annu Rev Med 48:35–44

Hall JG, Solehdin F (1997) Uniparental disomy: HK J Paediatr 2:3–8

Hoglund P, Holmberg C, de la Chapelle A, Kere J (1994) Paternal Isodisomy for chromosome 7 is compatible with normal growth and development in a patient with congenital chloride diarrhea. Am J Hum Genet 55:747–752

Jay P, Rougeule C, Massacrier A et al. (1997) The human *necdin* gene, *NDN*, is maternally imprinted and located in the Prader-Willi syndrome chromosomal region. Nat Genet 17:357–360

Jones KL (1988) Smith's Recognizable Patterns of Malformation (4th edn). WB Saunders, Philadelphia, Pennsylvania

Joyce JA, Lam WK, Catchpoole DJ et al. (1997) Imprinting of *IGF2* and *H19*: lack of reciprocity in sporadic Beckwith-Wiedemann syndrome. Hum Mol Genet 6:1543–1548

Kishino T, Lalande M, Wagstaff J (1997) *UBE3A/E6-AP* mutations cause Angelman syndrome. Nat Genet 15:70–73

Kobayashi S, Kohda T, Miyoshi N et al. (1997) Human *PEG1/MEST*, an imprinted gene on chromosome 7. Hum Mol Genet 6:781–786

Korenstein A, Ravia Y, Avivi L (1994) Uniparental disomy of chromosome 16 in offsprings of familial Mediterranean fever (FMF) patients treated with colchicine. ASHG abstract 616 (Suppl.): pp A109

Kotzot D, Schmitt S, Bernasconi F et al. (1995) Uniparental disomy 7 in Russell – Silver syndrome. Hum Mol Genet 4:583–587

Langlois S, Lopez-Rangel E, Hall JG (1995) New mechanisms for genetic disease and nontraditional modes of inheritance. Adv Pediatr 42:91–111

Ledbetter DH, Engel E (1995) Uniparental disomy in humans: development of an imprinting map and its implications for prenatal diagnosis. Hum Mol Genet 4:1757–1764

Lee MP, Hu R-J, Johnson LA, Feinberg AP (1997) Human *KVLQT1* gene shows tissue-specific imprinting and encompasses Beckwith-Wiedemann syndrome chromosomal rearrangements. Nat Genet 15:181–185

Lindsay S, Ireland M, O'Brien O et al. (1997) Large scale deletions in the *GPC3* gene may account for a minority of cases of Simpson-Golabi-Behmel syndrome. J Med Genet 34:480–483

MacDonald HR, Wevrick R (1997) The *necdin* gene is deleted in Prader-Willi syndrome and is imprinted in human and mouse. Hum Mol Genet 6:1873–1878

Mannens M, Wilde A (1997) *KVLQT1*, the rhythm of imprinting. Nat Genet 15:113–115

Matsuura T, Sutcliffe JS, Fang P et al. (1997) De novo truncating mutations in E6-AP ubiquitin-protein ligase gene (*UBE3A*) in Angelman syndrome. Nat Genet 15:74–77

McGuffin P, Scourfield J (1997) A father's imprint on his daughter's thinking. Nature 387:652–653

Ngo KY, Lee J, Dixon B, Liu D, Jones OW (1993) Paternal uniparental isodisomy, in a hydrops fetalis alpha thalassemia fetus. Am J Hum Genet 53 (Suppl):1207

Nishita Y, Yoshid I, Sado T, Takagi N (1996) Genomic imprinting and chromosomal localization of the human *MEST* gene. Genomics 36:539–542

O'Keefe D, Dao D, Zhao L et al. (1997) Coding mutations in $p57^{KIP2}$ are present in some cases of Beckwith-Wiedemann syndrome but are rare or absent in Wilms tumors. Am J Hum Genet 61:295–303

Pentao L, Lewis RA, Ledbetter DH, Patel PI, Lupski JR (1992) Maternal uniparental Isodisomy of chromosome 14: association with autosomal rod monochromacy. Am J Hum Genet 50:690 699

Quan F, Janas J, Toth-Fejel S, Johnson DB, Wolford JK, Popovich BW (1997) Uniparental disomy of the entire X chromosome in a female with Duchenne Muscular Dystrophy. Am J Med Genet 60:160–165

Reik W, Maher E (1997) Imprinting in clusters: lessons from Beckwith-Wiedemann syndrome. Trends Genet 13:330–334

Skuse DH, James RS, Bishop DVM et al. (1997) Evidence from Turner's syndrome of an imprinted X-linked locus affecting cognitive function. Nature 387:705–708

Spotila LD, Sereda L, Prockop DJ (1992) Partial isodisomy for maternal chromosome 7 and short stature in an individual with a mutation at the COL1A2 locus. Am J Hum Genet 51:1396–1405

Sulisalo T, de la Chapelle A, Kaitila I (1994) Uniparental disomy as an explanation of presumptive low penetrance (abstract). Am J Hum Genet 55 [Suppl]:A7

Temple IK, Gardner RJ, Robinson DO et al. (1996) Further evidence for an imprinted gene for neonatal diabetes localised to chromosome 6q22-q23. Hum Mol Genet 5:1117–1121

Tycko B, Trasler J, Bestor T (1997) Genomic imprinting: gametic mechanisms and somatic consequences. J Androl 18:480–486

Versteeg R (1997) Aberrant methylation in cancer. Am J Hum Genet 60:751–754

Vidaud D, Vidaud M, Plassa F, Gazengel C, Noel B, Goossens M (1989) Father-to-son transmission of hemophilia A due to uniparental disomy. AHSG, Abstr 1989, 889

Voss R, Ben-Simon E, Vital A, Zlotogora Y, Dogan J, Godfry S, Tikochinski Y (1989) Isodisomy of chromosome 7 in a patient with cystic fibrosis: could uniparental disomy be common in human? Am J Hum Genet 45:373–380

Vu TH, Hoffman AR (1997) Imprinting of the Angelman syndrome gene, *UBE3A*, is restricted to brain. Nat Genet 17:12–13

Ward A (1997) Beckwith-Wiedemann syndrome and Wilms' tumour. Mol Hum Reprod 3:157–168

Watrin F, Roeückel N, Lacroix L et al. (1997) The mouse *necdin* gene is expressed from the paternal allele only and lies in the 7C region of the mouse chromosome 7, a region of conserved synteny to the human Prader-Willi syndrome region. Eur J Hum Genet 5:324–332

Welch TR, Beischel LS, Choi E, Balakrischan K, Bishof NA (1990) Uniparental isodisomy 6 associated with deficiency of the fourth component of complement. J Clin Invest 86:675–678

Wevrick R, Francke U (1997) An imprinted mouse transcript homologous to the human imprinted in Prader-Willi syndrome (*IPW*) gene. Hum Mol Genet 6:325–332

Wiedemann HR (1983) Tumors and hemihypertrophy associated with Wiedemann-Beckwith syndrome. Eur J Pediatr 141:129

Woodage T, Prasad M, Dixon JW, Selby RE, Romain DR et al. (1994) Bloom syndrome and maternal uniparental disomy for chromosome 15. Am J Hum Genet 55:74–80

Zeschnigk M, Lich C, Buiting K et al. (1997) A single-tube PCR test for the diagnosis of Angelman and Prader-Willi syndrome based on allelic methylation differences at the SNRPN locus. Eur J Hum Genet 5:94–98

Genomic Imprinting and Cancer

Benjamin Tycko

1
Introduction

Fifteen years have gone by since the initial demonstrations of mammalian genomic imprinting (Barton et al. 1984; McGrath andSolter 1984; Surani et al. 1984), 7 years since the identification of specific imprinted genes in mice (DeChiara et al. 1991; Barlow et al. 1991; Bartolomei et al. 1991) and 6 years since the demonstration of conservation of imprinting in humans (Zhang and Tycko 1992; Giannoukakis et al. 1993; Ohlsson et al. 1993; Rainier et al. 1993; Zhang et al. 1993). During this time our understanding of the importance of this phenomenon for human diseases has advanced in some areas but remained incomplete in others. In this chapter we review data bearing on the role of genomic imprinting and imprinted genes in human cancer. We highlight the possible relationship of the regionality of imprinting and the role of imprinted genes in controlling fetal growth to the involvement of imprinting in human neoplasia. We describe data linking three imprinted genes on chromosome 11p15.5, *H19*, *IGF2*, and *p57KIP2*, to the pathogenesis of Wilms' tumor (WT) and other neoplasms, as well as to the fetal overgrowth associated with the Beckwith-Wiedemann Syndrome(BWS). We discuss evidence for epimutations at the *H19* locus, causing *H19* silencing and biallelic activation of *IGF2*, as site-specific molecular lesions with a crucial permissive role in Wilms' tumorigenesis. We also discuss the possible roles of imprinted genes in other pediatric tumors, in other proliferative and tumor syndromes, and in high grade adult malignancies. Lastly, we consider the feasibility of designing rational anticancer therapy targeting imprinted genes.

2
Imprinted Domains, Growth-Regulation and Cancer

As described in the other chapters, genomic imprinting is an epigenetic process which occurs prior to fertilization and results in relative transcriptional

Department of Pathology, Divisions of Oncology and Neuropathology, Columbia University College of Physicians and Surgeons, New York, NY 10032 USA

repression of particular genes on one parental allele in normal tissues of the offspring. The functional readout of imprinting is monoallelic or biased allelic expression – this has led to proposals that imprinted growth-regulating genes, that is tumor suppressor (TS) genes or dominant proto-oncogenes might be frequent targets of one-hit functional deletion (imprinted TS genes) or duplication (imprinted oncogenes) in human neoplasia (Wilkins 1988; Scrable et al. 1989; Zhang and Tycko 1992; Ogawa et al. 1993; Rainier et al. 1993). Important for understanding not only the mechanisms of imprinting, but also the involvement of imprinted genes in tumorigenesis, is the fact that many of the known imprinted genes are clustered together in particular chromosomal regions. The imprinted chromosomal domain which is most strongly implicated in neoplasia, chromosome 11p15.5, contains at least ten imprinted genes (see Beechey, this volume, and Fig. 1). Imprinted domains may correspond to extended regions of the chromosome which are differentially accessible to epigenetic modification in male vs. female gametogenesis. This idea is supported by low-resolution data indicating that the rates of recombination for markers within or near putative imprinted domains are markedly different in female vs. male meioses (Thomas and Rothstein 1991; Paldi et al. 1995; Robinson and Lalande 1995). As shown in Figure 1b, for chromosome 11p15.5 there is a higher rate of recombination in male meioses, suggesting a more accessible chromatin configuration in male spermatocytes than in developing oocytes. Some type of epigenetic modification (DNA methylation and/or binding of specific chromosomal proteins which later confer templating of allelic asymmetry) may then differentially affect the DNA in the two types of developing gametes. Most of the imprinted genes on chromosome 11p15.5 are, in fact, relatively repressed on the paternal allele, consistent with a single large differentially accessible chromatin domain in gametogenesis. The two exceptions, *IGF2* and, at least in mouse, *Ins2*, are more apparent than real, since the relative maternal repression of these two genes is adequately explained by *cis*-interaction with the downstream paternally repressed *H19* gene, both in mice and humans (Moulton et al. 1994; Steenman et al. 1994; Leighton et al. 1995; Taniguchi et al. 1995; Ripoche et al. 1997): in other words, the *IGF2/H19* couple, contained in about 100 kb of DNA, is a local circuit embedded in the more extended megabase-scale imprinted domain.

Imprinting and imprinted genes are particularly implicated in childhood tumors and the role of imprinting and imprinted genes in neoplasia may well be closely intertwined with the more general biological/evolutionary rationale for imprinting and the role of imprinted genes in regulating normal fetal and placental growth. The theory of imprinting which potentially ties these two strands together is the model proposed by Haig and his colleagues (Haig and Graham 1991; Moore and Haig 1991; see also Burt and Trivers, Chapter 1 and Messing and Grossniklaus, Chapter 2, this volume) in which imprinting is posited to be a mode of gene dosage regulation whereby males imprint an array of genes in a way which will promote large offspring: this may

A.

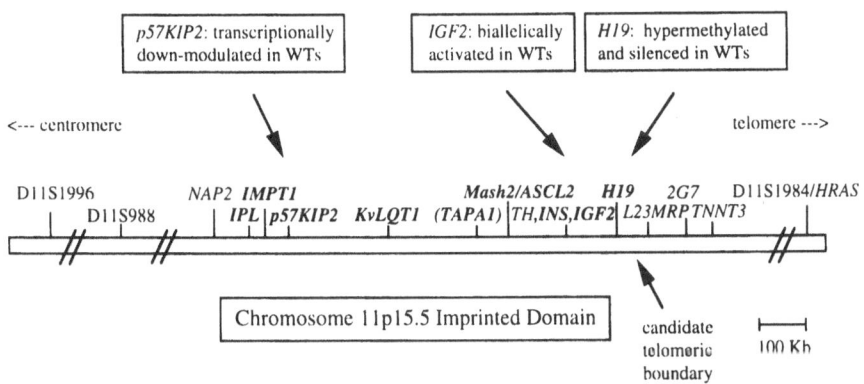

B.

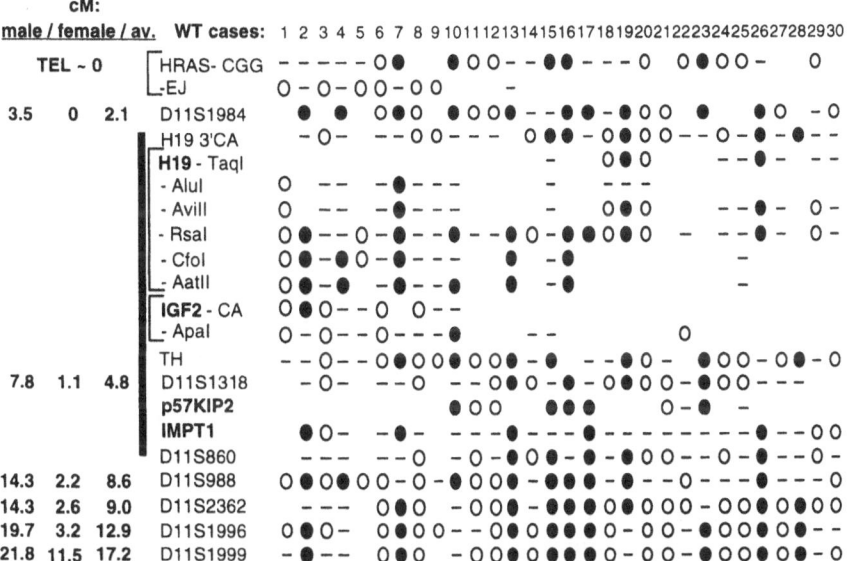

Fig. 1A, B. Nucleus of the chromosome 11p15.5 imprinted domain and overlap with the region of LOH in WTs. A Physical map of the imprinted region. Imprinted genes are in bold and genes which have shown biallelic expression without evidence of imprinting in studies to date are in plain type. Synonyms for *IPL* are *TSSC3* and *BWR1C*. Synonyms for *IMPT1* are *ORCTL2* and *BWR1A.*. For reference to syntenic mouse loci, *KVLQT1* is *Kcnq1*, *NAP2* is *Nap1l4*, *L23MRP* is *Rpl23l*, *TAPA1* is *Cd81*, and *p57KIP2* is *Cdkn1c* (Paulsen et al., 1998). B Genetic and LOH maps anchored to markers in the imprinted domain. The genetic recombination frequencies show a strong sex bias in the imprinted domain. All WTs in this series of 30 primary tumors show either coordinate 11p15.5 LOH at all markers, including those throughout the imprinted domain, or no LOH. Thus, the imprinted domain is entirely included in the minimal region of LOH defining the *WT2* locus. LOH mapping of an additional 25 cases has confirmed this (unpubl. data). The *black circles* indicate LOH; the *open circles* retention of heterozygosity, and the *dashes* noninformative markers. Genetic markers in the left column descend in order of increasing distance from the telomere. The imprinted domain is indicated by the *vertical bar*

lead to greater probabilities of future reproduction of their genome, while females imprint the genome in a way which will restrain fetal and placental growth, leading to greater probabilities that they will be able to bear multiple offspring fathered by multiple males. This parental competition model makes two predictions: first, the set of imprinted mammalian genes should be enriched in genes which affect fetal and postnatal growth and second, the direction of imprinting (paternal vs. maternal repression) should correlate systematically with the growth-related effects (negative vs. positive) of all or at least most imprinted genes. Ultimately, the hypothesis will be tested by accumulating a large list of *bona fide* imprinted genes and asking whether the functions and directions of imprinting of these genes are consistent with its predictions.

The number will have grown by the time this book appears, but at present there are still only about twenty entries in the list of known imprinted genes; those within the megabase-scale imprinted domain on human chromosome 11p15.5 (matching a region with conserved synteny on mouse distal chromosome 7) are shown in the map in Fig. 1 and are listed with summaries of their biochemical roles in the appendix (Beechey, this volume). There are both frustrating and intriguing aspects to this information. Even for this domain, which is the most densely annotated, there are only about ten imprinted genes characterized to date and of these only some can even provisionally be assigned an effect on cell and tissue growth. Since several of these genes (*IPL*, *IMPT1*, *2G3-8*) have only recently been identified, functional tests, which should include mouse knockout experiments, tests in cell transfections, and biochemical experiments, have not yet been reported. However, there are some trends which seem to support the Haig hypothesis: the limited available data suggest that for at least a majority of the genes in this domain which are imprinted in the paternal direction (relatively repressed on the paternal allele) there are negative effects on growth of fetal and placental tissues. The paternally repressed *H19* gene encodes an abundant fetal RNA lacking conserved translational reading frames (Brannan et al. 1990). An *H19* expression construct made from a genomic fragment including the entire gene caused inhibition of growth in RD rhabdomyosarcoma cells and suppression of anchorage-independent growth and tumorigenicity, without significant inhibition of anchorage-dependent growth, in G401 rhabdoid tumor cells (Hao et al. 1993), as well as suppression of tumorigenicity in syrian hamster sarcoma cells (Isfort et al. 1997), although no phenotypic effects were seen in other highly transformed lines such as SiHa cervical cancer cells (Hao et al. 1993) and choriocarcinoma cells (Lustig-Yariv et al. 1997). We have also recently observed inducible suppression of anchorage-dependent growth by an ecdysone promoter-*H19* expression vector in human 293 cells, a transformed line derived from fetal kidney (Luwa Yuan and B.T., unpubl. data). The mechanism of this growth suppressive and tumor suppressive activity of an untranslated RNA is not known, but other non-translated RNAs have also had TS activity (Rastinejad et al. 1993; Fan et al. 1996), and a reasonable hypothesis is that *H19* RNA interferes in *trans* with the translation of growth-promoting mRNAs in the cytoplasm. Examples of growth-promoting mRNAs with stringent translational regulation are ornithine decarboxylase (Manzella and Blackshear 1990; Pegg and Shantz 1994) and, interestingly, insulin-like growth factor-2 (*IGF2*) (Nielson et al. 1990, 1995). In two studies, deletion of the *H19* gene from the mouse germline caused generalized somatic overgrowth of the fetuses, which was attributed genetically to activation in *cis* of the nearby oppositely imprinted *Igf2* gene via alleviation of enhancer-competition (Leighton et al. 1995; Ripoche et al. 1997). As discussed in Section 3, silencing of the *H19* gene is a hallmark of WT, adrenocortical car-cinoma (ADCC), and at least a subset of cases of other human embryonal tumors including hepatoblastoma (HB) and embryonal rhabdomyosarcoma (ER).

The paternally repressed *p57KIP2* gene (Hatada and Mukai 1995; Chung et al. 1996; Hatada et al. 1996a; Matsuoka et al. 1996; Taniguchi et al. 1997) encodes a tissue-specific cyclin-cdk inhibitor (Matsuoka et al. 1995; Lee et al.1995) which, as predicted from its biochemical function in the cell cycle, is growth suppressive in some human tumor cells (O'Keefe et al. 1997). As dis-cussed in Section 5, *p57KIP2* is inactivated by mutations in a definite minority of cases of BWS and yields some, though not all, of the phenotypes of BWS when deleted from the mouse germline. The paternally repressed *Mash2* gene (human homologue *ASCL2*) encodes an achaete-scute family transcription factor which is essential for normal placental development (Guillemot et al. 1995), although its precise effects on growth vs. cell differentiation have not yet been sorted out. The paternally repressed *IPL* gene (Qian et al. 1997) encodes a member of a new gene family whose founding member, *TDAG51*, was shown to be essential for Fas expression and apoptosis in mouse T-lymphocytes (Park et al. 1996). *Ipl* mRNA is particularly abundant in placenta and yolk sac, as well as in mouse kidney (Qian et al. 1997), but the distribution of its protein product has not yet been fully characterized. The function of IPL protein is under investigation in biochemical and mouse germline deletion experiments. The paternally repressed *IMPT1* gene (Dao et al. 1998) encodes a member of the polyspecific membrane transporter family, and this gene, which is highly expressed in placenta and yolk sac as well as in adult organs such as kidney and liver, may influence fetal/placental metabolite exchange: information on the biological function of *Impt1* and whether it is growth suppressive also await germline deletion experiments. The *2G3-8* gene gives rise to multiple tran-scripts (Diem Dao and B.T., unpublished data): at least one of these is im-printed and another is partially anti-sense to *IMPT1*. The *Tapa1* gene (Oren et al. 1990; Andria et al. 1991; Reid et al. 1997) is paternally repressed, albeit weakly, in early mouse development (Luwa Yuan and B.T., unpublished data) and it encodes a widely-expressed membrane protein which is growth-inhibitory when cross-linked by antibodies on the surface of B-lymphocytes.

The *KvLQT1* gene (Wang et al. 1996) is paternally repressed (Lee et al. 1997) and encodes an ion channel. At this time it is not possible to assign it an effect on cell or tissue growth. Thus, while about half of the genes in this domain have not yet been characterized functionally, for all of the genes in the 11p15.5 imprinted domain which are paternally repressed and for which any functional data is available, there is either a negative or neutral effect on cell or tissue growth, and for others there are preliminary hints of growth-related functions from sequence analysis.

The maternally repressed *IGF2* gene (DeChiara et al. 1991; Ohlsson et al. 1993; Giannoukakis et al. 1993) encodes a major fetal mitogen, insulin-like growth factor-II. Its underexpression in mice causes stunting (DeChiara et al. 1991) and its overexpression contributes to overgrowth of many somatic tissues (Leighton et al. 1995; Ripoche et al. 1997; Sun et al. 1997). As discussed in later sections, pathological biallelic expression of *IGF2* is a hallmark of WT, ADCC, HB and RMS and biallelic expression of this gene is also seen in some cases of a somatic overgrowth disorder of humans, BWS. The only other gene in this domain which is maternally repressed, at least in yolk sac, is mouse *Ins2* (Giddings et al. 1994), encoding the insulin peptide, a growth-promoting and anabolic hormone. Thus, both of the genes in the 11p15.5 domain which are functionally imprinted (silenced) in the maternal direction are growth-promoting.

If imprinted genes are clustered together and if multiple genes with the same direction of imprinting have coordinate effects on cell and tissue growth, then this might have basic implications for the involvement of imprinted genes in neoplasia. Loss of heterozygosity (LOH) is a common mechanism for loss of function of tumor suppressor (TS) genes. In the classical two-hit model for the inactivation of *non-imprinted* TS genes (Knudson 1971), the process of LOH, mediated either by outright loss of one chromosome homologue or, frequently, by mitotic recombination and appropriate segregation of the recombined chromosomes to daughter cells, serves to unmask a germline or somatic mutation in the TS gene by eliminating the normal allele (Fig. 2a; model 1). LOH eliminates the alleles inherited from one type of parent over a broad region of DNA typically many megabases) encompassing many genes. In the classical pathway, all but one of these genes are irrelevant bystanders, since their net expression is not altered. For TS genes in an imprinted domain, there are two fundamental differences. First, for an imprinted TS gene LOH by itself can be sufficient to eliminate gene activity in one-hit – no mutation is required (Fig. 2a, model 2); and if an oppositely imprinted growth-promoting gene (dominant oncogene) is in the same region then its dose will be doubled by LOH (when the LOH is due to mitotic recombination). Second, if the imprinted TS gene is in an extended imprinted domain, then LOH involving this gene will also coordinately affect the net expression of other imprinted genes in its vicinity. If imprinted genes are disproportionately involved in growth regulation and if a common direction of imprinting implies similar effects on growth, then LOH might cause a true multi-gene effect, either by coordinate loss of function of multiple growth-inhibiting imprinted genes or by gain of function of one or more growth-promoting genes, or both (Fig. 2a, model 3).

In view of this last possibility, it is important to define the full gene content and boundaries of imprinted domains implicated in neoplasia. We and others have taken a systematic approach to this problem for chromosome 11p15.5. In addition to characterizing imprinted genes centromeric to *H19*, we carried out a chromosome walk in the downstream (telomeric) direction to determine whether the imprinted domain extends in this direction. This yielded three genes, *L23MRP*, *2G7* and *TNNT3* (Tsang et al. 1995; Yuan et al. 1996). The first two genes encode a mitochondrial ribosomal protein L23-related sequence (most likely encoding the human mitochondrial L23 protein) and a ubiquitously expressed spliced and polyadenylated RNA without open reading frames, respectively; the *TNNT3* gene encodes the fast isoform of skeletal muscletroponin-T. All three are biallelically expressed, without any reproducible allelic bias, in panels of human fetal and adult tissues. This suggests that the DNA immediately downstream of *H19* may define the telomeric boundary of the 11p15.5 imprinted domain (Fig. 1a). Mouse *L23Mrp* (*Rpl23*) is also positioned downstream of *H19* and recent DNA replication timing data, which show a boundary of allelic replication asynchrony between *H19* and *Rpl23* (Greally et al. 1998), as well as recent allelic expression data for this gene (Zubair et al. 1997), support the *H19-L23MRP* boundary hypothesis. Since non-imprinted genes can be interspersed within the imprinted domain (human *TH* has been the single example on chromosome 11p15.5) it will be necessary to check for the presence or absence of imprinting of several additional genes farther downstream of *H19*, such as the *M3/6* phosphatase gene (Nesbit et al. 1997), before this boundary can be regarded as definite.

It will also be important to establish the position of the centromeric boundary. The *NAP2* gene, encoding a putative nucleosome assembly protein, is biallelically expressed in some human tissues (Hu et al. 1996); genes centromeric of *NAP2* are currently under investigation for imprinting. The *GOK* gene, encoding a transmembrane protein of unknown function which was proposed as a candidate tumor suppressor, and the *RRM1* gene, encoding a ribonucleotide reductase subunit, are located centromeric of *NAP2* but still within chromosome 11p15.5, and based on preliminary studies both genes have been suggested to lack functional imprinting (Byrne and Smith 1993; Parker et al. 1996; Sabbioni et al. 1997).

Since mutations are not required for loss of imprinted TS genes or gain of function of imprinted dominant oncogenes, and since multiple genes can be functionally altered by LOH affecting an imprinted region of the genome, the question arises as to how to test for the relative importance of specific genes in the tumorigenic pathway. In other words, in the absence of the "smoking gun", which has been point mutations for classical TS genes, how can one test candidate genes? Attempts can be made to answer this question by functional

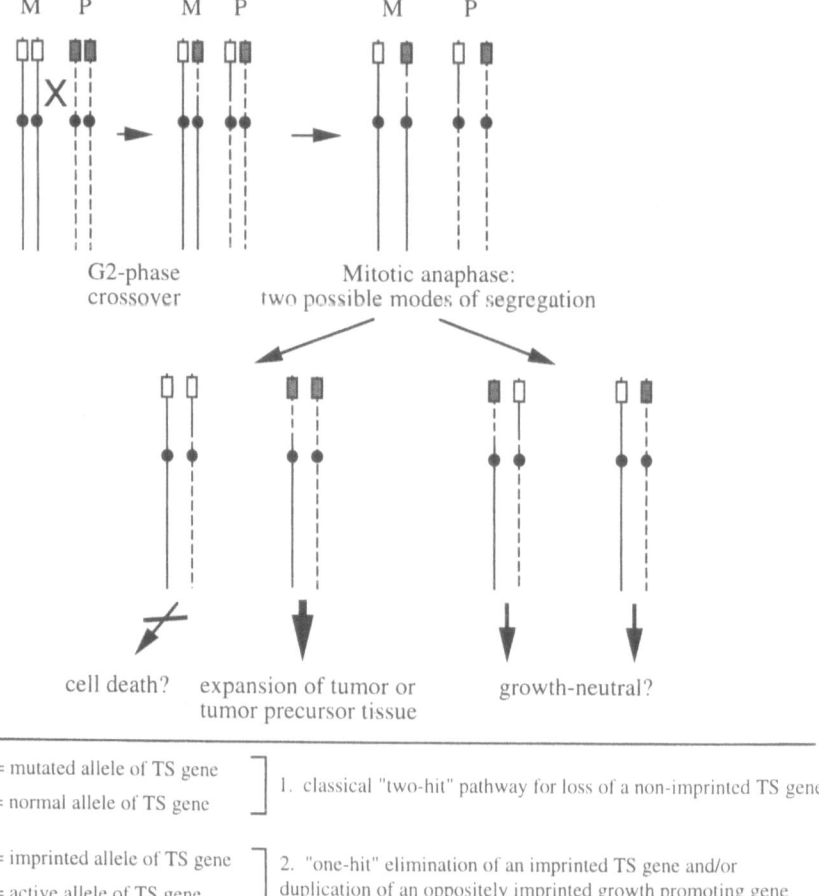

Fig. 2A, B. LOH and non-LOH (epigenetic) pathways for tumorigenesis involving imprinted genes. A One-hit functional elimination of an imprinted TS gene or multiple imprinted TS genes in an imprinted domain by LOH. Mitotic recombination can simultaneously result in duplication of growth-promoting genes. B A hypothetical model for imprint transfer *via* the action of DNA methyltransferase on transient recombination intermediates. The *black squares in the upper panel* represent methylated CpG dinucleotides. The *ovals and rectangles in the lower panel* represent the promoters and transcribed regions of the genes; *black shading* indicates hypermethylation. This mechanism can explain the endpoint of gene-specific biallelic *H19* hypermethylation and silencing and biallelic *IGF2* expression observed in many WTs which retain 11p15.5 heterozygosity

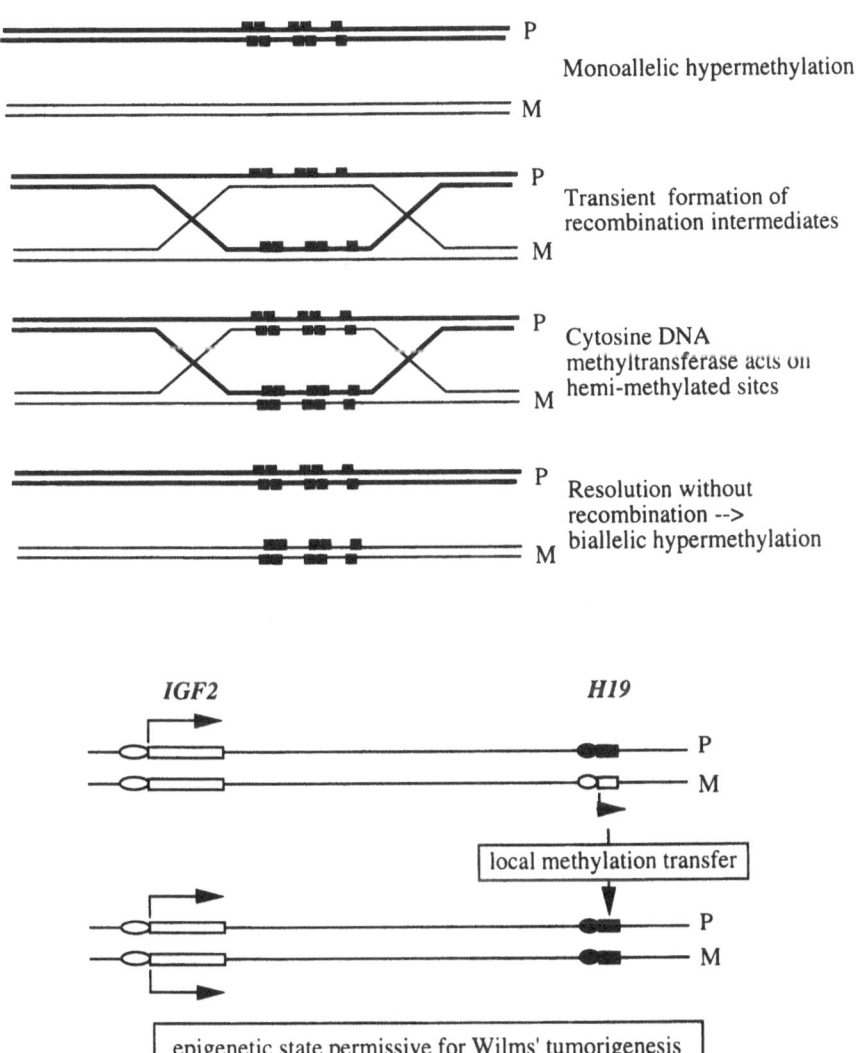

P Monoallelic hypermethylation

M

P Transient formation of
   recombination intermediates

M

P Cytosine DNA
   methyltransferase acts on
   hemi-methylated sites

M

P Resolution without
   recombination -->
   biallelic hypermethylation

M

IGF2   *H19*

local methylation transfer

epigenetic state permissive for Wilms' tumorigenesis

B

Fig. 2. *Continued*

genetics in animal models; but in terms of data from the actual human tumors it has been informative to examine candidate genes for abnormalities of RNA expression and for epigenetic lesions, specifically DNA methylation, in tumors of the same histopathology which have arisen *via* an alternative, non-LOH, pathway. Results of this analysis applied to WTs and other pediatric tumors are described in Section 4.

3
General Evidence for Involvement of Imprinted Genes in Neoplasia

In early studies, about half of sporadic WTs and ERs showed LOH for chromosome 11p15.5, importantly with an absolute bias toward loss of maternal rather than paternal alleles (Schroeder et al. 1987; Scrable et al. 1989; Williams et al. 1989; Pal et al. 1990). Subsequent studies have confirmed this in WTs (Rainier et al. 1993; Moulton et al. 1994; Chung et al. 1996), as well as in HBs (Albrecht et al. 1994) and, by an indirect assay examining parent-of-origin-specific DNA methylation patterns, in a large series of ADCCs (Gicquel et al. 1997). Based on the selective loss of maternal 11p15.5 alleles, one criterion for candidate TS genes in this region, often referred to as *WT2* gene(s) to distinguish them from the previously described *WT1* gene in band 11p13, is that they be expressed only from the maternal allele. However, in the first published model, it was argued that a paternally expressed growth-promoting gene(s) might be the key factor in Wilms' tumorigenesis (Wilkins 1988). This was reasonable since it is invariably observed on Southern blotting that in spite of the LOH for chromosome 11p15.5, there is no net loss of band intensity; that is, the WT precursor cells which have lost maternal alleles (by mitotic crossing-over) have in the process duplicated the paternal alleles (Fig. 2a).

A recent addition to the list of proliferative disorders with a parent-of-origin effect is persistent hyperinsulinemic hypoglycemia of infancy of the focal and sporadic type (FoPHHI). In this endocrine disorder, in which there is focal adenomatous hyperplasia of the pancreas, it was found, analogous to the data in WTs, ERs, HBs and ADCCs, that the hyperproliferative but benign pancreatic nodules can frequently have LOH for chromosome 11p15.5, again with a complete bias towards loss of maternal alleles (de Lonlay et al. 1997).

Selective loss of one type of parental allele was also reported in series of cases of neuroblastoma. Here, the lost alleles of markers on chromosome 1p35-36 appeared to be predominantly, though by no means exclusively, of maternal origin (Caron et al. 1993; Cheng et al. 1993), and there have been subtleties to the trend in a larger series of cases, such that only a particular subclass of neuroblastoma (those without N-*myc* amplification) is now considered to show the parent-of-origin effect (Caron et al. 1995). This work is apparently still in progress, so the conclusions could change with more data. The *P73* gene was suggested as a candidate imprinted TS gene in this region (Kaghad et al. 1997), and further studies of the imprinting and function of this gene in mice and humans are in progress. Even if *p73* proves not to be a TS gene, confirmation of its imprinting would strongly suggest the presence of a chromosome 1p35-36 imprinted domain.

Another type of evidence for imprinted genes in tumorigenesis came from pedigree and linkage analysis in the inherited glomus tumor syndrome. Here it has been observed that the phenotype, usually bilateral carotid body tumors, is manifested only after transmission of the disease gene from fathers (van der Mey et al. 1989). Since the high frequency of affected individuals in the pedigrees is otherwise consistent with autosomal dominant transmission, it has been predicted that the chromosome 11q23-qter gene accounting for this syndrome will turn out to be a maternally imprinted/paternally expressed dominant oncogene (Heutink et al. 1992). Follow-up studies with defined genetic loci in these families and in the associated tumors will be of great interest. Lastly, uniparental disomy (UPD; see Hall, this volume) associated with a phenotype can also be evidence for involvement of imprinted gene(s) and this type of evidence was produced for DNA markers on chromosome 11p15 in non-neoplastic tissues of a small but significant minority of cases of BWS, an overgrowth syndrome which is associated with an increased risk of childhood cancers (Grundy et al. 1991; Henry et al. 1991). In these cases, the duplicated alleles were always paternal and the lost alleles maternal in origin. This strongly implied that one or more imprinted genes on chromosome 11p15.5 would turn out to be BWS genes. Additional genetic and epigenetic findings in BWS are discussed in Section 5.

4
Imprinted Genes in WT and Other Pediatric Tumors: the *WT2* Locus

Since imprinted genes are monoallelically expressed, if there are imprinted TS genes then they might be expected to show frequent one-hit deletion in tumors. Objections to this idea with statements such as "if there were imprinted TS genes then everybody would get cancer" are not valid if the loss of imprinted TS genes is not *rate-limiting*, but rather *permissive* for tumorigenesis. That is, if inactivation of imprinted TS genes facilitates tumor formation or progression, but only in the presence of rare rate-limiting mutations at non-imprinted TS loci, then imprinted TS genes might be very common targets of one-hit inactivation in tumors and yet the overall rate of tumorigenesis would remain low. Altered expression and functional imprinting of a discreet set of imprinted genes on chromosome 11p15.5 is, in fact, implicated in WTs, ERs, HBs, and ADCCs, and there is good evidence that the altered expression of these genes plays a permissive role in the multistep pathway of tumor formation. All of these malignant embryonal tumors have a high frequency of LOH for distal chromosome 11, with loss of markers spanning a large genetic and physical distance but often restricted to band 11p15.5. Data from informative LOH cases taken from our series of primary untreated WTs are shown in Fig. 1b. In this large series of intensively genotyped cases, the minimal region of LOH encompasses the entire known extent of the chromosome 11p15.5 imprinted domain and all of the tumors with LOH have arisen after single mitotic crossover events well centromeric of this region (for a study which suggested a similar but slightly smaller minimal region of LOH in ERs, see Besnard-Guerin et al. 1996).

Studies from several laboratories, examining well over 100 cases in aggregate, have shown that WTs can be grouped into three molecular classes with respect to chromosome 11p15.5. The first group (~45%) show 11p15.5 LOH, with loss of maternal alleles and reduplication of paternal alleles. What is the effect of LOH on the known genes in the imprinted domain? Since *H19* is efficiently silenced on the paternal allele and since the *H19* imprint is stable in Wilms' tumorigenesis, these cases uniformly show loss of *H19* RNA, that is, a reduction to less than 5% of fetal kidney levels (Moulton et al. 1994, 1996; Steenman et al. 1994; O'Keefe et al. 1997). Since the imprinted paternal allele of *H19* is normally methylated at virtually every CpG dinucleotide within both the promoter and body of the gene (Zhang et al. 1993), these genetically bipaternal tumors show biallelic hypermethylation of *H19* DNA. Since *IGF2* is paternally expressed and since the paternal allele is duplicated, these cases tend to overexpress *IGF2* mRNA relative to fetal kidney, although the levels can be variable, probably depending on the physiological and differentiation state of the cells (Moulton et al. 1994; Steenman et al. 1994; Taniguchi et al. 1995b; Wang et al. 1996). Finally, since the *p57KIP2* gene is subject to "leaky" functional imprinting in humans, with from 10–30% of mRNA derived from the imprinted paternal allele (Chung et al. 1996), tumors with 11p15.5 LOH show variable reductions in the major *p57KIP2* mRNA relative to fetal kidney, as indicated by standard Northern analysis, RNAse protection and RT-PCR (Chung et al. 1996; Hatada et al. 1996a; Overall et al. 1996; O'Keefe et al. 1997; Taniguchi et al. 1997).

Since tumors with 11p15.5 LOH coordinately lose the maternal alleles of all genes in a large span of DNA, it is the group of *non-LOH* tumors which has been more informative for testing the importance of specific genes in tumorigenesis. Among the ~55% of cases of WT which retain both parental alleles, more than half nonetheless show specific inactivation of the previously active maternal *H19* allele, correlating with extensive biallelic CpG-hypermethylation of both the body and the promoter of this gene (Moulton et al. 1994, 1996; Steenman et al. 1994; Taniguchi et al. 1995b). These tumors, constituting the second molecular class, invariably show activation of the normally silent maternal *IGF2* allele. The biallelic expression of *IGF2* is indeed associated with high total mRNA levels from this gene in most WTs, although as in the LOH cases there is quantitative variation (Moulton et al. 1994; Steenman et al. 1994; Taniguchi et al. 1995b; Wang et al. 1996). That the fully developed WTs with biallelic *IGF2* expression have amounts of *IGF2* mRNA which are statistically greater than those with monoallelic expression of this gene was supported by one study (Steenman et al. 1994), but questioned by another (Wang et al. 1996). Since the second study used an RT-PCR approach instead of northern blots, the different conclusions are most likely due to differences in methodology.

In fact, since *IGF2* expression is responsive to the metabolic state of the tumor cells it is not clear whether measurements in the late stage tumors are biologically relevant: some minimal amount of IGF-II peptide is probably essential for growth of the transformed tumor cells (Qing et al. 1996), but it may be that there is a window early in tumorigenesis during which the amounts of *IGF2* mRNA and protein and *H19* RNA are particularly critical in determining whether the *preneoplastic* precursor cell responds to the second, rate-limiting and as yet uncharacterized, genetic event by becoming neoplastic. We currently favor this model for *WT2* as a permissive TS locus, and it is diagrammed in Fig. 3. Consistent with this, in two cases in our original series in which the epigenetic lesion (biallelic *H19* hypermethylation) was widespread in the non-neoplastic kidney parenchyma adjacent to the tumors, there was indeed reduced *H19* RNA and increased *IGF2* mRNA relative to normal juvenile control kidneys (Moulton et al. 1994).

The third molecular class of WTs are the remaining 25% of cases in which there is retention of heterozygosity, *H19* RNA is persistently expressed, and *H19* DNA normally methylated. In some of these cases, *in situ* hybridization shows focal loss of *H19* RNA in areas of the tumor, with persistent monoallelic expression of *IGF2* (Cui et al. 1997). This may indicate an alternative mechanism for *H19* silencing, perhaps through loss of a required transcription factor rather than by pathological DNA methylation.

Two lines of evidence suggest that the inactivation of *H19* via biallelic hypermethylation in WTs is a locus-specific epigenetic lesion, reflecting neither a global derangement of DNA methylation nor a domain-wide disruption of imprinting: First, the abnormal DNA methylation is highly localized. Considering the three genes in the 100 kb of DNA immediately downstream of *H19* (*L23MRP*, *2G7*, and *TNNT3*), none of these is functionally imprinted in normal tissues, the gene closest to *H19* (*L23MRP*) is not abnormally methylated in WTs and neither *L23MRP* nor *2G7* is transcriptionally dysregulated in WTs (Tsang et al. 1995; Yuan et al. 1996). Similarly, the *TAPA1* gene, which is situated squarely within the imprinted domain (Reid et al. 1997; Fig. 1a) and which shows weak imprinting in extra-embryonic tissues, though not in kidney, produces abundant mRNA at equivalent levels in normal fetal kidney and WTs (L. Yuan and B. Tycko, unpubl. data). As another example, the *p57KIP2* gene is transcriptionally less active in WTs than in normal fetal kidney, but it does not show detectable CpG hypermethylation in the WTs and its transcriptional inactivation does not track precisely with inactivation of *H19* (Chung et al. 1996; O'Keefe et al. 1997; Taniguchi et al. 1997). Our analyses by Southern blotting have also not revealed gross DNA methylation abnormalities in WTs in the CpG islands associated with two other genes, *IPL* and *IMPT1*, in the imprinted domain, and similarly the *ZNF195* gene, located near the centromeric border of the known 11p15.5 imprinted domain (Hussey et al. 1997) is neither hypermethylated nor transcriptionally silenced in WTs (L.

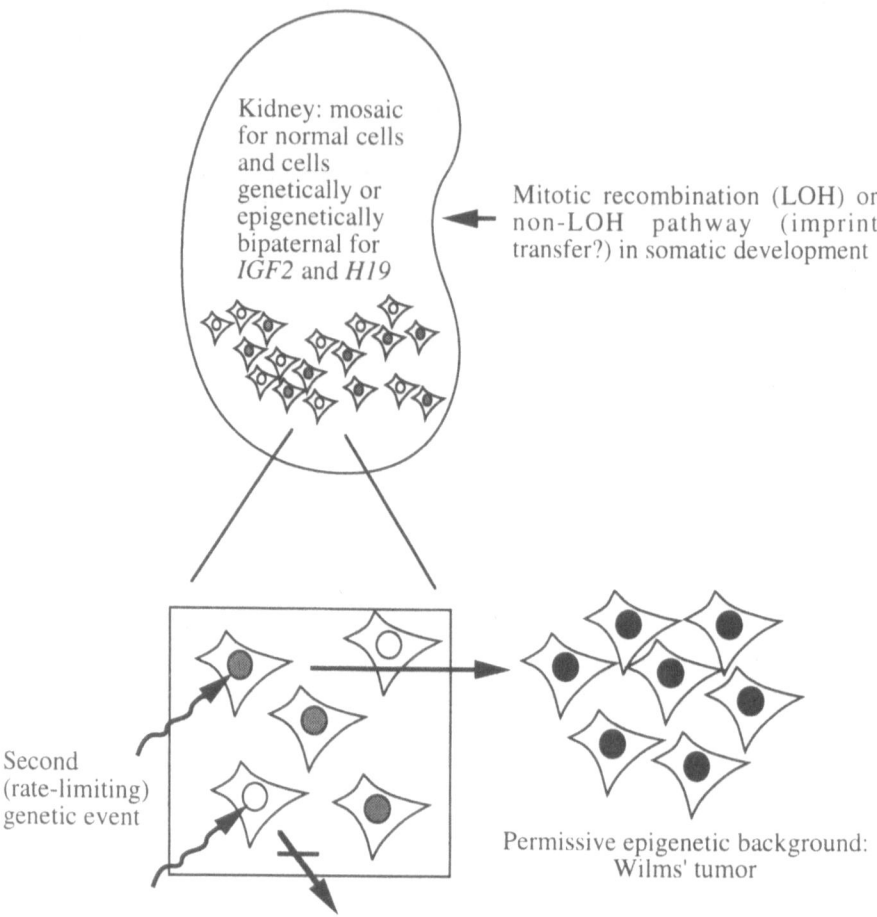

Fig. 3. Permissive model for the imprinted *WT2* locus. The kinetics of tumorigenesis indicate that *WT2* must be a permissive rather than rate-limiting TS locus, and the *IGF2* and *H19* genes most likely correspond to this locus. The experimental evidence for the illustrated epigenetic mosaicism in kidneys of WT patients is described in the text. Altered expression of other imprinted genes in the same domain, including *p57KIP2*, may also play a role in WT progression.

Yuan and B. Tycko, unpubl. data). Moreover, Taniguchi et al. (1995b) showed lack of methylation abnormalities in WTs with two other 11p15.5 probes, *INS* and *HRAS*.

In terms of DNA methylation in the mouse and human *IGF2* genes, there are subtle methylation differences between the maternal and paternal alleles in normal tissues, but most of the CpG dinucleotides in the promoter region and body of this gene are hypomethylated on both alleles (Koide et al. 1994) and in humans the inactive maternal allele is paradoxically less methylated than the active paternal allele at a CpG site in exon 9 (Schneid et al. 1993). The lack of a strong methylation imprint on this gene is consistent with functional evidence from the *H19* knockout mice showing that, for most tissues, the allelic expression and DNA methylation of *IGF2* is controlled in *cis* by the transcriptional activity of the *H19* gene (Leighton et al. 1995; Ripoche et al. 1997), as well as with functional data showing that demethylation of the *H19* gene by azaC correlates with a reciprocal down-modulation of *IGF2* mRNA in tumor cells (Chung et al. 1996). These data suggest that the decreases in methylation of a few CpG dinucleotides in *IGF2* which are present systematically in WTs with a bipaternal epigenotype (Taniguchi et al. 1995), occur as a consequence of the massive pathological hypermethylation of *H19* and alleviation of enhancer-competition (Fig. 2b). Parallel data indicating that the conversion to a bipaternal epigenotype at *IGF2/H19* is a specific pathological event came from analysis of *IGF2* promoter-specific imprinting by Reeve's laboratory (Taniguchi et al. 1995a). Consistent with other studies (Vu and Hoffman 1994; Ekstrom et al. 1995), they found that in normal kidney, expression from the P1 promoter became biallelic as the tissue matured in the postnatal period while expression from the fetal promoters P2, P3 and P4 remained imprinted; in contrast, in WTs with the epigenetic lesion there was a specific molecular pathology – all four *IGF2* promoters were active on both alleles.

Second, the epigenetic inactivation of *H19* can precede overt tumorigenesis – in an early study (Moulton et al. 1994) two patients were found with biallelic *H19* hypermethylation (and biallelic *IGF2* expression) not only in their tumors but also in nonneoplastic kidney parenchyma adjacent to the tumors, and within this nonneoplastic tissue the abnormal methylation was confined to the body and promoter of the *H19* gene. Similar observations of epigenetic inactivation of *H19* in nonneoplastic kidney in the setting of WT were made subsequently in four children with a somatic overgrowth syndrome, probably distinct from classical BWS, who had biallelic *IGF2* expression in their nonneoplastic tissues, nephromegaly and, in two cases, WTs (Morison et al. 1996). Since the tumors in these patients were not multifocal, these observations are consistent with a permissive, rather than rate-limiting, effect of this lesion in the tumorigenic pathway. Also consistent with this is the striking observation that frank 11p15.5 LOH can also be seen in non-neoplastic tissues (blood and kidney) of some WT patients who present with nonsyndromic unilateral unifocal tumors (Chao et al. 1993; Moulton et al. 1996). Recently, a study using quantitative Southern blot analysis has suggested that lower-level mosaicism for biallelic *H19* hypermethylation (anywhere from 15–80% of cells affected) may in fact be extremely frequent in nonneoplastic kidney cells from WT patients (Okamoto et al. 1997). Lastly, a recent study using in situ hybridization has confirmed that *H19* inactivation is an early event in WT formation, since loss of expression of this gene was observed in nephrogenic rests, as well as in non-neoplastic kidney adjacent to WTs (Cui et al. 1997).

Thus, inactivation of *H19* in the tumor precursor cells, which, as in the tumors themselves, is associated with biallelic *IGF2* expression (Moulton et al. 1994; Okamoto et al. 1997), is an early and permissive step in the multi-stage tumorigenic pathway of most WTs (Fig. 3). The necessary accumulation of genetic and epigenetic lesions is the key to this model. For example, a variation on the theme are the rare cases of WT which occur in the setting of germline *WT1* (chromosome 11p13) deletions or mutations. In these cases, there can also be chromosome 11p15.5 LOH in the tumor, which has obviously occurred as a second event (Henry et al. 1989a). Either the combination of *WT1* deletion and 11p15.5 LOH is sufficient for tumorigenesis, or these tumors have also undergone a third hit at an as yet undefined rate-limiting locus.

Many, though not all, WTs which retain 11p15.5 heterozygosity also show marked down-regulation of *p57KIP2* mRNA relative to whole fetal kidney (Chung et al. 1996; O'Keefe et al. 1997). There is overlap of the *p57KIP2*-inactivated cases and the *H19*-inactivated cases, but there is no absolute correlation. In contrast to *H19*, *p57KIP2* inactivation in WTs is not associated with detectable DNA hypermethylation of this gene and, also in contrast to *H19*, *p57KIP2* expression cannot be induced in embryonal rhabdomyosarcoma cells in culture by exposure to 5-azaC (Chung et al 1996). So in spite of the proximity of these two genes on the chromosome and their shared paternal imprinting, there may not be a single concerted mechanism for their inactivation in WTs. Although germline mutations in *p57KIP2* are present in some people with BWS (Section 5), mutations in this gene are rare or absent in WTs (Orlow et al. 1996; Overall et al. 1996; O'Keefe et al. 1997). Since low resolution in situ hybridization pictures have suggested that *p57Kip2* mRNA levels may depend on the differentiation status of cells in the normal developing kidney of mice, with lower levels in blastema than in more differentiated cells (Matsuoka et al. 1995), it is even possible that the down-modulation of *p57KIP2* mRNA in WTs (which can have varying amounts of blastemal component) may be a passive consequence of the block to cell differentiation. This is not true for *H19* inactivation, since hypermethylation of the maternal allele is not seen in normal fetal kidney, and since loss of *H19* RNA in WTs is independent of the amount of blastemal component (Moulton et al. 1994).

Like WTs, ADCCs frequently (46% of cases) show 11p15.5 LOH (Henry et al. 1989a; Gicquel et al. 1997). This rare but difficult-to-treat tumor can occur in children or adults and it appears to have an increased incidence in the setting of BWS (Henry et al. 1989b). Recently, Gicquel et al. (1997) reported information on chromosome 11p15.5 LOH, *IGF2* mRNA expression, *H19* RNA expression and *IGF2* allelic usage in a substantial series of adrenal tumors (82 tumors of various histological grades of malignancy). The data are reminiscent of those from WTs. All 29 unequivocally malignant ADCCs showed complete loss of *H19* RNA; *H19* RNA could be either present or absent in the tumors with borderline histological grade and it was usually detectable in the benign adrenal adenomas. Similar findings were made by Liu et al. (1997), who additionally observed reduced *p57KIP2* mRNA expression tracking with *H19* silencing in ADCCs. In a reciprocal trend, *IGF2* mRNA was substantially higher in a large majority of the malignant ADCCs than in the normal adrenal control tissue; intermediate grade tumors had less markedly elevated levels on average and benign tumors had normal levels. *IGF2* expression was pathologically biallelic in the tumors with retention of 11p15.5 heterozygosity and *H19* inactivation. No DNA methylation analyses were shown, but presumably the *H19* silencing was due to methylation of the maternal allele. There would seem to be two possibilities to explain these data: either the malignant ADCCs arise in benign adenomas and inactivation of *H19* is a late event in tumor progression, or alternatively the malignant tumors are biologically distinct from the benign ones and arise de novo, and *H19* is inactivated early in their formation. In any case, *H19* inactivation is the best molecular marker for malignancy in adrenal tumors identified to date, and since these tumors can be difficult to evaluate for malignancy by histology (hence the frequent borderline diagnosis; Weiss et al. 1989), this could have practical application in molecular diagnosis.

Both ERs and HBs are rare but highly malignant embryonal tumors which are known to undergo LOH for chromosome 11p15.5 and, although the data are based on collections of case reports, they appear to be more common in association with BWS and/or hemihypertrophy than in the general population. A study of ERs was among the first to report selective loss of maternal 11p15.5 alleles and to propose a model for tumorigenesis involving imprinting (Scrable et al. 1989). There is also a convincing report of exclusively maternal 11p15.5 LOH in HBs (Albrecht et al. 1994), and since then there have been a few studies of the *IGF2* and *H19* genes in small series of these types of tumors. Li et al. (1995) described significant down-regulation of *H19* RNA in 3/3 Hbs, associated with biallelic *IGF2* expression in one of them. Rainier et al. (1995) did a similar study and found biallelic *IGF2* expression in one of five HBs. Three of four HBs examined by Northern blotting in that study showed detectable *H19* RNA while in one case there was no expression; a fetal liver control lane was not shown so it is not clear whether the three expressing cases had down-regulated *H19* expression relative to the normal precursor tissue. Davies (1993) did not find biallelic *IGF2* expression in any of three HBs, even though she observed biallelic expression in control postnatal livers (a normal phenomenon which is due to use of the nonimprinted P1 promoter of the *IGF2* gene in this tissue and which does not reflect a disruption of imprinting); *H19* was not examined. Montagna et al. (1994) also found a low but detectable incidence of 11p15.5 LOH (3 of 13 cases) and reported one case of HB with retention of heterozygosity but with biallelic *IGF2* expression; *H19* DNA methylation and the level of its RNA were not quantitated. In an in situ hybridization study of 11 cases, expression of *IGF2* mRNA in HBs correlated with the degree of cellular differentiation (Akmal et al. 1995); *H19* RNA was not assessed. Consistent with the permissive model for the effect of the epigenetic chromosome

11p15.5 lesions in these tumors, HBs frequently have missense mutations (a second genetic hit) in a classical rate-limiting TS gene, the adenomatous polyposis coli gene, *APC* (Kurahashi et al. 1995; Oda et al. 1996). Also consistent with a permissive model similar to that which we have proposed for WTs, Smith and colleagues have observed somatic mosaicism for chromosome 11p15.5 LOH in non-neoplastic liver of HB patients (Simms et al. 1995).

In terms of ERs, one study described biallelic *IGF2* expression in six rhabdomyosarcomas, at least one of which had embryonal histology (Zhan et al. 1994) and another laboratory reported biallelic expression of this gene in 5/7 rhabdomyosarcomas with retention of 11p15.5 heterozygosity, several of which had ER histology (Pedone et al. 1994). Neither of these studies examined expression or DNA methylation of the *H19* gene. An ER cell, RD, which expresses abundant *IGF2* mRNA and which has silenced the *H19* gene, has been used to show that DNA hypermethylation is causative of the gene silencing – this was done by treating these cells with the DNAmethyltransferase inhibitor 5-aza-2'-deoxy-cytidine (azaC) and observing prominent reexpression of *H19* RNA, which correlated with an approximately two-fold reduction in the expression of *IGF2* mRNA, exactly as predicted from reciprocal regulation in *cis* of these two oppositely imprinted genes (Chung et al. 1996). After 2 weeks of treatment with azaC, the cells entered a terminal growth arrest, probably due to altered expression of multiple genes regulated by DNA methylation, including *H19* and *IGF2*.

Thus, there are clear similarities between the epigenetic lesions present at the *IGF2* and *H19* loci in WTs and ADCCs and, while the data in HBs and ERs are incomplete, these tumors may show both similarities and differences with the WTs. Six years after we initially proposed it (Zhang and Tycko 1992), the data reviewed above indicate that *H19* remains the strongest candidate for the *WT2* TS gene. Its anti-WT activity may be mediated either by its negative regulation of *IGF2* in *cis*, its negative regulation of tumor cell growth by its RNA product in *trans*, or most likely by both mechanisms. The downmodulation of *p57KIP2* mRNA in WTs is also most likely significant for tumor progression, but the evidence for participation of this gene in the *WT2* locus would be more convincing if tumor-specific epigenetic lesions could be identified in *p57KIP2* DNA. Chromosome 11p15.5 probably contains other as yet unidentified TS genes, as suggested from studies of adult malignancies such as lung carcinomas, in which there is 11p15.5 LOH but in which *H19* is often not silenced (Sect. 9), and for some time there has been speculation that there may be another *WT* locus (a *WT3* gene) centromeric to the known imprinted domain, in a region of 11p15.3 defined by certain BWS-associated chromosomal translocation breakpoints (Redeker et al. 1995).

5
Possible Mechanisms for *H19* Epimutation in WTs

How does the inactivation of *H19* by DNA methylation come about in the precursor cells of the non-LOH tumors? Since the abnormal methylation can be present in non-neoplastic kidney cells of WT patients, but not in controls, the pathway must correspond to a rare event which can occur early in somatic development. Since the methylation abnormality is highly localized, and since other genes in the region are either unaffected or, in the case of *p57KIP2*, affected by a non-methylation-dependent mechanism which does not precisely track with *H19* inactivation, any candidate mechanism must be specific to the *IGF2/H19* locus.

A mechanism which involves transfer of the methylation imprint from the paternal to the maternal *H19* allele *via* a transient recombination intermediate (double Holliday junction) is shown in Figure 2b. We previously described this model in a review of the creation and propagation of methylation patterns (Bestor and Tycko 1996). Supporting the feasibility of this mechanism, experimental data from Rossignol's laboratory have shown that inter-allelic methylation transfer occurs prior to meiosis in the ascomycete *Ascobolus immersus*, and that it most likely depends on the formation of recombination intermediates (Colot et al. 1996). In this system, a visible phenotype(spore color) and tetrad analysis were used to investigate the characteristics of transfer of methylation from an inactive hypermethylated and genetically marked allele of the *b2* marker gene to an active hypomethylated allele. It was found that transfer of methylation correlated with gene conversion and that the two phenomena showed the same "polarity", with greater frequencies of both gene conversion and methylation transfer at the 5' end of the b2 marker locus. The finding of physical proximity of homologous alleles of imprinted domains in interphase nuclei of human somatic cells provides some support for this model (LaSalle and Lalande 1996; see also chapters by Paldi and Marshall/Sedat, this vol.).

Are there other possible mechanisms? There is good evidence that the disruption of normal imprinting on chromosome 15q11–q13 which occurs in some rare cases of Prader-Willi syndrome (PWS) and Angelman syndrome (AS) is caused by submicroscopic DNA deletions in a putative imprinting center upstream of the *SNRPN* gene (Sutcliffe et al. 1994; Buiting et al. 1995). These patients showa regional shift in epigenotype (allelic DNA methylation patterns and allelic gene expression) over a region encompassing more than a megabase of DNA. This model must be considered in WTs, but in fact the lack of evidence for a regional shift in epigenotype and the inability to detect DNA deletions in studies to date argue against it. Interestingly, recent data from other cases of AS which have abnormal functional and methylation imprinting but which lack detectable deletions have led to the proposal that inter-allelic imprint transfer might be an alternative mechanism in that disease (Burger et al. 1997). Since one case in this study shared the same maternal alleles with her unaffected sibling, the putative imprint transfer might have occurred early in somatic development. Thus, there may be a closer than expected parallel between the mechanisms of biallelic inactivation (gain-of-imprinting) of *H19* in Wilms' tumorigenesis and the inactivation of *UBE3A* or other relevant genes in a nonneoplastic epigenetic disease, AS.

6
Genomic Imprinting and Imprinted Genes in BWS

BWS is diagnosed by exomphalos (omphalocele), macroglossia, visceromegaly (including organomegaly affecting tongue, kidney, liver and adrenal), and sometimes hemihyperthrophy and gigantism, all of which reflect overgrowth of developing tissues. While ascertainment bias may be present, from combined case reports it is thought that about 7.5% of affected individuals will develop Wilms' tumor (WT) or other embryonal neoplasms (Wiedemann 1983, described this rate of tumors in combined reports of 388 cases; see also Pettenati et al. 1986, for an unselected sequential series of cases, with a low frequency of tumors, 1/20 cases). In discussing the genetics and epigenetics of this syndrome, it is important to remember that it is a *syndrome*, not a homogeneous disease. The evidence for involvement of imprinted genes has taken several forms. Early studies showed that BWS or a BWS-like condition with widespread somatic overgrowth can occur with partial trisomies involving extra material from chromosome 11p15.5 (Turleau et al. 1984; Journel et al. 1985; Turleau and deGrouchy 1985; Brown et al. 1992). In rare families, the trait is associated with constitutional chromosomal inversions or translocations involving 11p15.3–p15.4 or, more usually 11p15.5, but the phenotype is only expressed after passage of the abnormal chromosome through the maternal germline (Redeker et al. 1995; Lee et al. 1997b; refs. therein). The syndrome can be transmitted with no cytogenetic abnormalities but with genetic linkage to chromosome 11p15.5, and here too the phenotype is usually seen after passage of the disease gene through the maternal germline (Koufos et al. 1989; Ping et al. 1989). In a third group of cases, the syndrome occurs de novo in association with paternal isodisomy for 11p15.5 (Grundy et al. 1991; Henry et al. 1991), which is often present as somatic mosaicism, suggesting a mitotic recombination event early in somatic development (Henry et al. 1993). Most cases are seemingly sporadic, without uniparental disomy or detectable cytogenetic abnormalities.

BWS is most likely caused in some cases by abnormally high expression of the *IGF2* gene. Paternal disomies or duplications of chromosome 11p15.5 are predicted to lead to an increase in IGF-II protein production and increased growth of IGF-II-responsive tissues, accounting for the organomegaly. In fact, early reports described BWS patients with increased circulating somatomedin (bioactive IGF protein) levels (Ashton and Aynsley-Green 1978; Spencer et al.

1980). Findings of biallelic *IGF2* mRNA expression, i.e. erasure of funtional imprinting, in fibroblasts and tongue tissue of a minority of BWS patients who lacked paternal 11p15 disomies provided critical support for the *IGF2* hypothesis (Weksberg et al. 1993). Since DNA lesions in the vicinity of the *IGF2* gene were not found in this or in another study of a large series of cases (Schneid et al. 1993), the disruption of imprinting was postulated to be a long range chromosomal effect. The existence of cases of BWS or a BWS-like overgrowth syndrome due to partial trisomies of chromosome 11p (Turleau and de Grouchy 1985; Brown et al. 1992) suggested that overexpression of *IGF2* (and perhaps other paternally expressed growth-promoting genes on chromosome 11p15.5) may be sufficient to cause some features of the syndrome (but not severe omphalocele?), and that loss of a TS gene may not be necessary for many aspects of the growth phenotype. The pattern of *IGF2* mRNA expression in the human fetus correlates well with the organs most affected by BWS (Hedborg et al. 1994).

The gene encoding the cyclin-cdk inhibitor *p57KIP2* was originally isolated and mapped to human chromosome 11p15.5 by the Elledge and Massague laboratories (Lee et al. 1995; Matsuoka et al. 1995). This gene was then localized centromeric to *IGF2*, near several BWS translocation breakpoints, by the Feinberg/Mannens groups (Hoovers et al. 1995), and the mouse version was shown to be imprinted by Hatada and Mukai (1995), who proposed that *p57KIP2* would turn out to be equivalent to *WT2*. Inactivating coding mutations in this gene were then identified, first by the Hatada laboratory and then by ourselves and others in a small but definite subset of individuals, amounting to ~10 percent of cases with BWS (Hatada et al. 1996b; Hatada et al. 1997; Lee et al. 1997a; O'Keefe et al. 1997) and in at least some of the rare cases with 11p15.5 chromosomal rearrangements the translocation breakpoints occur near, though not in, this gene – interrupting the large *KvLQT1* locus immediately downstream (Lee et al. 1997b). Whether these interesting translocation cases are associated with a position effect causing biallelic *IGF2* expression or silencing of *p57KIP2*, or both, has not yet been reported, but this information will obviously be of interest. As noted above, in our comprehensive sequencing analysis which uncovered one out of five cases of BWS with a *p57KIP2* mutation, we did not find any *p57KIP2* mutations in a large series of WTs enriched for cases which expressed *p57KIP2* mRNA (O'Keefe et al. 1997). All of the BWS patients reported with *p57KIP2* mutations have had obvious omphaloceles, and the patient in our series had a severe omphalocele requiring surgical correction.

7
H19, IGF2 and the Epigenetic Pathway for BWS

Since evidence to date suggests that most cases of BWS are not explained by mutations or by chromosomal translocations, BWS may in many cases be a true epigenetic disease. Consistent with this possibility, there have been a number of documented cases of this disease occuring as a discordant trait in twins, many of whom have been dizygotic (reviewed by Pettenati et al. 1986; see also Hall, Chap. 6, this vol.), but some of whom have been monozygotic (Clayton-Smith et al. 1992; refs. therein). In these cases, an attractive possibility is that there has been a shift to a bipaternal epigenotype, either by methylation transfer at *H19*, resulting in biallelic *IGF2* expression, or by some other *trans*-sensing mechanism, in an early somatic cell of one of the twins (see also Paldi, Chap. 13, this vol.). This would be the epigenetic counterpart of the paternal isodisomy cases which result from the genetic pathway of mitotic recombination (Henry et al. 1993). Interestingly, the distinct benign proliferative and endocrine disorder FoPHHI, which also maps to chromosome 11p15 and is associated withexclusively maternal LOH in the pancreatic nodules (de Lonlay et al. 1997), has also been reported as a discordant trait in two pairs of monozygotic twins (Santer et al. 1995; Thornton et al. 1995; de Lonlay et al. 1997).

Some BWS cases who are non-UPD do indeed show biallelic hypermethylation and silencing of *H19*, associated with biallelic expression of *IGF2* (Reik et al. 1995; Joyce et al. 1997; Catchpoole et al. 1997), and in the last study, which involved the largest number of subjects, one of the five BWS patients with this bipaternal *IGF2/H19* epigenotype developed a WT. In another study using *in situ* hybridization, *H19* RNA expression was absent in a pathologically relevant tissue, kidney, in two other BWS cases, one of whom developed a WT and the other diffuse blastemal overgrowth (Cui et al. 1997). A study of fibroblasts from other BWS cases suggested that *IGF2* can sometimes be expressed biallelically without *H19* silencing (Joyce et al. 1997). However, in that report no DNA methylation analyses were shown, *H19* RNA was assessed only by RT-PCR, and pathologically relevant tissues such as tongue and kidney were not available for analysis. Cases of BWS with epigenetic inactivation of *H19* are not uncommon – Catchpoole et al. (1997) reported that five of 63 non-UPD cases (8%) had *H19* hypermethylation in the available tissue samples, which were peripheral blood, and since blood is not a pathologically involved tissue in BWS this may even be a low estimate of the true frequency.

8
BWS and WT2

The data cited above are relevant to the important question of the relationship of BWS and BWS genes to embryonal tumors and the *WT2* locus. Since the *syndrome* BWS can be caused by lesions in at least two different genes, including *p57KIP2* and *IGF2*, and since there are at least five different general pathways: autosomal dominant with maternal transmission (including the cases with *p57KIP2* mutations), "sporadic" de novo (sometimes discordant between twins and in at least some cases reflecting the epigenetic pathway including biallelic *IGF2* activation and biallelic *H19* hypermethylation), uniparental paternal disomy for a large region of chromosome 11 including band 11p15, trisomy with duplicated paternal 11p15 alleles, and apparently balanced chromosomal translocations in at least two different breakpoint cluster regions on chromosome 11p15, and since less than 10 percent of BWS cases are associated with any type of neoplasm, it seems reasonable to ask whether the cases with and without tumors are molecularly distinct. That is, it is technically accurate but also potentially misleading to state generically that "BWS conveys an increased predisposition to tumors" – it may be that tumor predisposition is in fact restricted to a specific etiopathological type of BWS.

This is a difficult question because the literature is limited and because it is difficult to judge the effects of ascertainment bias (the cases which end up at referral centers may on average have more severe phenotypes and a higher frequency of tumors). However, several presumably independent studies of reasonable large numbers of cases have suggested that those with uniparental paternal isodisomy (UPD) for chromosome 11p15.5 develop tumors more frequently than those without UPD (Grundy et al. 1991; Chao et al. 1993; Henry et al. 1993; Schneid et al. 1993; Reik et al. 1994; Steenman et al. 1994; Weksberg 1994; Catchpoole et al. 1997; Cui et al. 1997). Compilation of 41 cases of 11p15.5 UPD, most of which showed features of complete BWS, revealed 13 cases with a tumor. This ratio of 32% can be compared to the classical 7.5%. The absolute rate is probably an overestimate due to ascertainment bias, but this bias should also apply to non-UPD cases. In one study (Schneid et al. 1993) the rate of tumor formation was 66% (2/3) in UPD patients. Two out of three children with UPD in this series developed tumors and one other tumor occurred in a patient carrying both parental alleles but with a uniparental *IGF2* methylation pattern in the hypertrophied tongue but not in the blood (both alleles were methylated at the exon 9 site; *H19* methylation was not examined); the remaining individual with a tumor was homozygous for the markers used and therefore not informative. In another study (Henry et al., 1993) among 16 cases with complete BWS and UPD there were five with tumors. In the series of Catchpoole et al. (1997), WT occurred in 1/5 non-UPD patients with *H19* hypermethylation, in 0/19 non-UPD patients without *H19* hypermethylation and in 1/12 UPD patients. Similarly, in the 4 non-syndromic overgrowth patients with epigenetic inactivation of *H19* and biallelic activation of *IGF2* described by Morison et al. (1996), 2/5 had WTs. Clearly, larger studies are needed, but the trend is intriguing.

At least 17 cases with 11p15 trisomy with paternal duplication (as opposed to UPD) have been reported (Slavotinek et al. 1997) and of these the only case with a paternal duplication who may have had a neoplasm was one in the study by Journel et al. (1985) with an adrenal lesion found at autopsy . This was thought to be an ADCC initially, but the pathology was later felt to be more consistent with adrenocortical cytomegaly (C. Junien, pers. commun.). More

11p15 trisomy cases need to be evaluated to test this preliminary suggestion of low tumor incidence, but if the trend persists then it will support the hypothesis that tumor predisposition in BWS is dependent not only on *IGF2* overexpression, but also on the elimination of one or more imprinted 11p15.5 TS genes.

In contrast to this apparently high rate of WTs in the UPD patients and probably in the epigenetic pathway patients, none of the 7 BWS patients with germline mutations of *p57KIP2* described to date has been reported to have WTs or other neoplasms, with the exception of one tumor, a neuroblastoma (a type of tumor which has been reported in BWS, but much less frequently than the other neoplasms), in one patient described by Lee et al. (1997). Note: the clinical descriptions in some of the papers are brief; when tumors are discussed in general and *p57KIP2* mutations are then described in specific cases, and no tumors are specifically assigned to the mutation-bearing cases, we conclude that these cases were not the ones that had the tumors. Consistent with the *p57KIP2* mutation cases and non-UPD cases not being strongly predisposed to tumors, in our series of five cases of BWS one had a *p57KIP2* mutation, all had a nondisomic pattern of 11p15.5 markers, and none developed tumors (O'Keefe et al. 1997). Earlier, Grundy et al. (1994) noted that "although the incidence of WT in BWS patients, 85% of whom are sporadic, has been reported to be 7.5%, the incidence of tumors in *familial* BWS patients, as assessed from reports in the literature in which there is sufficient clinical description, is 1/61 from 14 families. Since the inherited defect would be more likely to involve only one gene (*read p57KIP2, others?*), the greater incidence of tumors observed in sporadic cases of BWS, some of whom will prove to have isodisomy, would be more consistent with there being two separate genes for BWS and WT." With the knowledge that we have accumulated since, we can suggest a variation on this: BWS can be caused by *p57KIP2* mutations or *IGF2* biallelic expression, but the critical permissive event for WT formation is a conversion to a bipaternal genotype or epigenotype at the *IGF2* and *H19* loci. This is exactly the situation which is achieved by UPD or by the epigenetic inactivation of *H19*, linked to biallelic activation of *IGF2*. Tumor progression at later stages is probably enhanced by down-modulation of *p57KIP2* mRNA and possibly influenced by other as yet uncharacterized imprinted genes in the 11p15.5 domain.

9
Relevance of Mouse Models for Understanding BWS

Developing a mouse model for BWS has been a long-term goal of several laboratories. Reviewing this literature one has the impression of steady progress over time, with a series of mouse models coming closer to the actual phenotype. First, in mice made mosaic for paternal disomy of distal chromosome 7 (by blastocyst injections of cells carrying the disomy) there was definite somatic overgrowth *in utero* (Ferguson-Smith et al. 1991), and moderate generalized overgrowth was the major phenotype of mice in which *Igf2* was overexpressed due to *H19* deletions (Leighton et al. 1995; Ripoche et al. 1997). Germline deletion of the *p57Kip2* gene produced mice which had phenotypes similar to some of the specific lesions of BWS, including omphalocele, adrenal cortical cytomegaly and adrenal medullary dysplasia (Yan et al. 1997; Zhang et al., 1997). In these experiments, there was fetal lethality without net overgrowth, and the presence of increased programmed cell death suggested that this was due to activation of a cell cycle checkpoint response. This may represent a species-specific aspect, as a similar low-threshold apoptotic response can explain other divergent phenotypes in mice and humans, including the lack of retinoblastomas in *Rb1*-deletion mice (Hamel et al. 1993).

The two most recent "BWS mice" both involved high-level overexpression of IGF-II protein. In one strategy, an *H19* deletion strain, which overexpresses *Igf2* mRNA and protein, was crossed with another strain in which the IGF-II clearance receptor, *Igf2r*, was deleted, leading to a further marked increase in circulating IGF-II peptide (Eggenschwiler et al. 1997). In another strategy, *Igf2* transgenes were introduced and these somehow caused transactivation of the endogenous *Igf2* gene, leading to elevated circulating levels of peptide, but probably not as high as in the Eggenschwiler et al. mice (Sun et al. 1997). In both of these models, there was pronounced overgrowth of the mice (which were often viable to birth), nephromegaly and macroglossia and, in the first system more prominently than in the second, omphaloceles. The differences between the *Igf2* overexpression and *p57Kip2* deletion phenotypes in mice are catalogued in detail in the paper by Eggenschwiler et al. (1997). There may be a similarity with the findings in humans, since the BWS cases with *p57KIP2* mutations have uniformly manifested omphaloceles, while at least some cases with the bipaternal *IGF2/H19* epigenotype and at least some cases with the incomplete syndrome due to partial trisomies of 11p15.5 have not. No mouse model of BWS has so far manifested tumors, and in fact mice have never been observed to develop WT-like neoplasms under any circumstances, including hemizygous *Wt1* deletions (Krieder et al. 1993).

10
Imprinted Genes in Adult Cancers

The evaluation of regional LOH for significance can be difficult in high-grade tumors with unstable genomes (virtually all chromosomal regions show detectable LOH in these tumors), but some investigators feel that adult carcinomas of the lung and breast have a significant site of loss of alleles, above the background of genomic instability, on chromosome 11p15.5. In breast cancers in one large study 12.5% of cases showed LOH at the telomeric marker *HRAS*, 26.8% showed LOH at *TH*, in the middle of band 11p15.5, and 33.3% showed

LOH at *D11S860* and *HBB*, closer to the centromeric border of band 11p15.5 (Winqvist et al. 1993). Similarly, based on extensive allelotyping in lung cancers, it has been proposed that there are several distinct subregions of significant rates of LOH on chromosome 11p15, but the allelotypes are very complex (Bepler and Garcia-Blanco 1994). The identification of specific candidate 11p15.5 TS genes relevant to these cancers is awaited. Evidence for involvement of genomic imprinting and imprinted genes in these types of cancers is also preliminary. In primary breast cancers, one study using RT-PCR concluded that there was no evidence of systematic alterations in *IGF2* or *H19* expression or imprinting (Yballe et al. 1996), and another study using in situ hybridization found *H19* overexpression in focal areas of the tumor in a subset of breast cancers, but suggested that many cases of apparent overexpression might actually represent signal coming from the nonneoplastic reactive stromal cells, with *H19* silent in the carcinoma cells (Dugimont et al. 1995). We are not aware of studies of parent-of-origin of the lost 11p15.5 alleles in these cancers.

Using an indirect assay for parent-of-origin, Takahashi and coworkers reported random loss of 11p15.5 alleles independent of parental origin in their first series of 36 cases of lung cancer (Kondo et al. 1994), but they found an incomplete bias towards loss of maternal alleles when they used a *p57KIP2* gene polymorphism as a marker (Kondo et al. 1996). Lung cancers can show biallelic expression of *IGF2* (Suzuki et al. 1994) and this does not correlate with *H19* silencing; indeed, in contrast to the situation in the BWS-associated tumors WTs and ADCCS, *H19* can be highly expressed in some of these high-grade tumors (Suzuki et al. 1994), as well as in some bladder cancers, cervical cancers and uterine leiomyosarcomas (Ariel et al. 1995; Vu et al. 1995; Douc-Rasy et al. 1996). Whether the aberrant expression of *IGF2* and *H19* in these high-grade tumors is related to their pathogenesis or instead reflects only a global disruption of chromosomal structure and gene expression remains to be seen – if evidence could be found for alterations of functional imprinting in the appropriate *preneoplastic* cells, then the argument for biological significance would be more convincing. The 11p15.5 LOH in some of the tumor types which do not lose *H19* RNA suggests that there is at least one other TS gene relevant to these tumors in this chromosomal region. This gene could be *p57KIP2* (Kondo et al. 1996) or an as yet unidentified imprinted or non-imprinted gene. LOH studies in a mouse model of SV40-drivenhepatocellular carcinoma, in which there is chromosome 7 LOH but high expression of *H19* RNA in some of tumors, have also suggested this (Haddad and Held 1997).

11
Imprinted Genes in Germ Cell Tumors (GCTs) and Trophoblastic Neoplasia

Since parental imprinting is switched by passage of a gene through the germline of the opposite sex, we know that the imprint must be erased at some point in formation of the germ cell lineages and then re-established as the germ cells mature (Tycko et al. 1997). The precise stages of germ cell development during which these events occur are being investigated (see Horsthemke et al., Chap. 5, this vol.), but in considering the expression of imprinted genes in GCTs it is important to keep in mind that these tumors derive from a cell lineage which is normally capable of erasing the imprint. Strong suggestion of a dominant process for erasure, which is probably dependent on DNA replication, has come from a study by Tada et al. (1997), who fused embryonic germ cells with thymic lymphocytes and observed widespread demethylation of both imprinted and non-imprinted genes after clonal expansion of the hybrid cells.

Imprinted genes, including *H19*, *IGF2* and others, can often be abnormally expressed in male and female GCTs (Rachmilewitz et al. 1996, refs. therein), and part of this "aberrant" expression may reflect cell lineage. One study, which found frequent 11p15.5 LOH in male germ cell tumors but no parent-of-origin bias, concluded that there is as yet no specific evidence for involvement of imprinted genes in germ cell tumor formation (Mishina et al. 1996). On the other hand, there may be additional complex changes in the epigenetic states of these genes during tumor progression (Arima et al. 1997). This dialogue is likely to continue until specific GCT TS genes are found. Special classes of human tumors, benign ovarian teratomas (a GCT) and hydatidiform moles (not a GCT, but often discussed under this rubric), can be either completely gynogenetic (teratomas) or androgenetic (moles), and the phenotypes of these benign or premalignant tumors are roughly consistent with the phenotypes of nonneoplastic gynogenetic and androgenetic conceptuses in mice; so in the special case of these tumors genomic imprinting does have an obvious role in determining their biological characteristics (reviewed in Roberts and Mutter 1994). Consistent with this, immunoreactive p57[kip2] protein was recently found to be reduced or absent in all cases in a large series of androgenetic moles (Chilosi et al., 1998).

12
Rational Cancer Therapy Targeted to Epigenetic Lesions

WTs can usually be cured by current clinical protocols, but some are treatment failures and the other tumors in this class, ADCC, ER, and HB, are often resistant to current therapies. In the long term, a better understanding of normal imprint maintenance may allow rational design of cancer treatments to revert the 11p15.5 epigenotype to normal. In the preceding sections we have made the argument that inactivation of *H19* by pathological DNA methylation on the maternal allele creates a permissive state for Wilms' tumorigenesis and that this pathway is also important in ADCCs and probably in some ERs and HBs. Although this is a permissive rather than rate-limiting step in tumor formation, it is still likely that its reversal will inhibit tumor cell growth. Supporting this are both direct experiments in which overexpression of *H19*

RNA can inhibit tumorigenicity in some cell systems (Hao et al. 1993; Isfort et al. 1997) and, in terms of the mechanistically linked phenomenon of overexpression of *IGF2*, that addition of inhibitors of its signaling receptor can inhibit tumor cell growth (Qing et al. 1996). It is well established that azaC can erase DNA methylation and the mechanism of its activity is covalent binding to the DNA methyltransferase enzyme (Bender et al. 1998, refs. therein). What is perhaps less well appreciated is that there is some specificity to the activity of this drug. In at least two studies (Chung et al. 1996; Barletta et al. 1997), exposure of human tumor cells to azaC reverted the *IGF2/H19* epigenotype to normal by erasing *H19* DNA hypermethylation specifically on one allele (presumably the pathologically methylated maternal one) and restoring the methylation pattern to monoallelism, *not* by fully erasing *H19* methylation on both alleles, and in another tumor cell line maximally effective dosages of the drug only partially activated the imprinted (presumably paternal) *H19* allele (Tycko 1997). Since *p57KIP2* inactivation was not reversed by azaC in the RD cell line, derived from an ER, and since this gene is not strongly methylated in WTs (Chung et al. 1996), its down-modulation may not be methylation-dependent and demethylating drugs may not be able to reactivate it in tumors. However, the possibility of using azaC to reactivate TS genes is obviously not restricted to imprinted ones like *H19* – a number of classical TS genes, including the Von-Hippel Lindau gene *VHL*, the gene encoding the tumor suppressive cdk inhibitor p16, and even the most classical of all TS genes, *RB1*, are inactivated by a non-mutational DNA hypermethylation pathway in some tumors (Ohtani-Fujita et al. 1997; Baylin et al. 1998; Bender et al. 1998, refs. therein).

Can azaC or other demethylating drugs be applied in human diseases? Of course the answer is yes, since azaC has been used previously for treatment of a variety of disorders, including sickle cell anemia and thallasemia (reactivation of fetal hemoglobin) and myelodysplastic syndrome and chronic myelogenous leukemia (anti-cancer therapy) (Lowrey and Nienhuis 1993; Kantarjian et al. 1997). All of these disorders have responded to treatment, though none have been cured by the drug. Obviously, this will depend on decisions by clinicians, not molecular biologists, but the embryonal tumors with altered functional imprinting may turn out to be one of the most appropriate settings for therapy with DNA demethylating agents.

Acknowledgments. This work was supported by grant R01CA60765 from the N.I.H. We wish to thank Dale Frank and Naifeng Qian for comments on the manuscript and we are grateful to Claudine Junien both for comments on the manuscript and for essential observations concerning the risk of neoplasia in different molecular categories of BWS. We also thank Rolf Ohlsson for his helpful suggestions. The LOH mapping shown in Fig. 1 was done by Thomas Moulton, Diem Dao, Lin Feng and Dmitri Gorelov.

References

Akmal SN, Yun K, MacLay J, Higami Y, Ikeda T (1995) Insulin-like growth factor 2 and insulin-like growth factor binding protein 2 expression in hepatoblastoma. Hum Pathol 26:846-851

Albrecht S, von Schweinitz D, Waha A, Kraus JA, von Deimling A, Pietsch T (1994) Loss of maternal alleles on chromosome arm 11p in hepatoblastoma. Cancer Res 54:5041-5044

Andria ML, Hsieh, C-H, Oren R, Franke U, Levy S (1991) Genomic organization and chromosomal localization of the TAPA-1 gene. J Immunol 147:1030-1036

Ariel I, Lustig O, Schneider T, Pizov G, Sappir M, DeGroot N, Hochberg A (1995) The imprinted *H19* gene as a tumor marker in bladder carcinoma. Urology 45:335-338

Arima T, Matsuda T, Takagi N, Wake N (1997) Association of *IGF2* and *H19* imprinting with choriocarcinoma development. Cancer Genet Cytogenet 93:39-47

Ashton IK, Aynsley-Green A (1978) Plasma somatomedin activity in an infant with Beckwith Wiedemann syndrome. Early Hum Dev 1:357-362

Barletta JM, Rainier S, Feinberg AP (1997) Reversal of loss of imprinting in tumor cells by 5-aza-2'-deoxycytidine. Cancer Res 57:48-50

Barlow DP, Stoger R, Hermann BG, Saito K, Schweifer N (1991) The mouse insulin-like growth factor type-2 receptor is imprinted and closely linked to the *Tme* locus. Nature 349:84-87

Bartolomei MS, Zemel S, Tilghman SM (1991) Parental imprinting of the mouse *H19* gene. Nature 351:153-155

Barton SC, Surani MAH, Norris ML (1984) Role of paternal and maternal genomes in mouse development. Nature 311:374-376

Baylin SB, Herman JG, Graff JR, Vertino PM, Issa JP (1998) Alterations in DNA methylation: a fundamental aspect of neoplasia. Adv Cancer Res 72:141-196

Bender CM, Pao MM, Jones PA (1998) Inhibition of DNA methylation by 5-aza-2'-deoxycytidine suppresses the growth of human tumor cell lines. Cancer Res 58:95-101

Bepler G, Garcia-Blanco M (1994) Three tumor-suppressor regions on chromosome 11p identified by high-resolution deletion mapping in human non-small-cell lung cancer. Proc Natl Acad Sci USA 91:5513-5517

Besnard-Guerin C, Newsham I, Winqvist R, Cavanee WK (1996) A common region of loss of heterozygosity in Wilms' tumor and embryonal rhabdomyosarcoma distal to the D11S988 locus on chromosome 11p15.5. Hum Genet 97:163-170

Bestor TH, Tycko B (1996) Creation of genomic methylation patterns. Nat Genet 12:363-367

Brannan CI, Dees EC, Ingram RS, Tilghman SM.(1990) The product of the *H19* gene may function as an RNA. Mol Cell Biol 10:28-36

Brown KW, Gardner A, Williams JC, Mott MG, McDermott A, Maitland NJ (1992) Paternal origin of 11p15 duplications in the Beckwith-Wiedemann syndrome. A new case and review of the literature.Cancer Genet Cytogenet 58:66-70

Buiting K, Saitoh S, Gross S, Dittrich B, Schwartz S, Nicholls RD, Horsthemke B (1995) Inherited microdeletions in the Angelman and Prader-Willi syndromes define an imprinting centre on human chromosome15. Nat Genet 9:395-400

Burger J, Buiting K, Dittrich B, Gross S, Lich C, Sperling K, Horsthemke B, Reis A (1997) Different mechanisms and recurrence risks of imprinting defects in Angelman syndrome. Am J Hum Genet 61:88-93

Byrne JA, Smith PJ (1993) The 11p15.5 ribonucleotidereductase M1 subunit locus is not imprinted in Wilms' tumour and hepatoblastoma. Hum Genet 91:275-277

Caron H, van Sluis P, van Hoeve M, de Kraker J, Bras J, Slater R, Mannens M, Voute PA, Westerveld A, Versteeg R (1993) Allelic loss of chromosome 1q36 in neuroblastoma is of preferential maternal origin and correlates with N-*myc* amplification. Nat Genet 4:187-190

Caron H, Peter M, van Sluis P, Speleman F, de Kraker J, Laureys G, Michon J, Brugieres L, Voute PA, Westerveld A, Slater R, Delattre O, Versteeg R (1995) Evidence for two tumour suppressor loci on chromosomal bands 1p35–36 involved in neuroblastoma: one probably imprinted, another associated with N-*myc* amplification. Hum Mol Genet 4:535–539

Catchpoole D, Lam WW, Valler D, Temple IK, Joyce JA, Reik W, Schofield PN, Maher ER (1997) Epigenetic modification and uniparental inheritence of *H19* in Beckwith-Weidemann syndrome. J Med Genet 34:353–359

Chao L-Y, Huff V, Tomlinson G, Riccardi VM, Strong LC, Saunders GF (1993) Genetic mosaicism in normal tissues of Wilms' tumor patients. Nat Genet 3:127–131

Cheng JM, Hiemstra JL, Schneider SS, Naumova A, Cheung N-KV, Cohn SL, Diller L, Sapienza C, Brodeur G (1993) Preferential amplification of the paternal allele of the N-*myc* gene in human neuroblastomas. Nat Genet 4:191–193

Chilosi M, Piazzola E, Lestani M, Benedetti A, Guasparri I, Granchelli G, Aldovini D, Leonard E, Pizzolo G, Doglioni C, Menestrina F, Mariuzzi GM (1998) Differential expression of p57kip2, a maternally imprinted (sic) cdk inhibitor, in normal human placenta and gestational trophoblastic disease. Lab Invest 78:269–276

Chung W-Y, Yuan L, Feng L, Hensle T, Tycko B (1996) Chromosome 11p15.5 regional imprinting: comparative analysis of *KIP2* and *H19* in human tissues and Wilms' tumors. Hum Mol Genet 8:1101–1108

Clayton-Smith J, Read AP, Donnai D (1992) Monozygotic twinning and Wiedemann-Beckwith syndrome. Am J Med Genet 42:633–637

Colot V, Maloisel L, Rossignol JL (1996) Interchromosomal transfer of epigenetic states in Ascobolus: transfer of DNA methylation is mechanistically related to homologous recombination. Cell 86:855–864

Cui H, Hedborg F, He L, Nordenskjold A, Sandstedt B, Pfeifer-Ohlsson S, Ohlsson R (1997) Inactivation of *H19*, an imprinted and putative tumor repressor gene, is a preneoplastic event during Wilms' tumorigenesis. Cancer Res 57:4469–4473

Dao D, Frank D, Qian N, Vosatka R, Walsh CP, Tycko B (1998) *IMPT1*, An imprinted gene similar to polyspecific transporter and multi-drug resistance genes. Hum Mol Genet 7:597–608

Davies SM (1993) Maintenance of genomic imprinting at the *IGF2* locus inhepatoblastoma. Cancer Res 53:4781–4783

Davies SM (1994) Developmental regulation of genomicimprinting of the *IGF2* gene in human liver. Cancer Res 54:2560–2562

DeChiara TM, Robertson EJ, Efstratiadis A (1991) Paternal imprinting of the mouse insulin-like growth factor II gene. Cell 64:849–859

de Lonlay P, Fournet J-C, Rahler J, Gross-Morand M-S, Poggi-Travert F, Foussier V, Bonnefont J-P, Brusset M-C, Brunelle F, Robert J-J, Nihoul-Fekete C, Saudubray J-M, Junien C (1997) Somatic deletion of theimprinted 11p15 region in sporadic persistent hyperinsulinemic hypoglycemia of infancy is specific of focal adenomatous hyperplasia and endorses partial pancreatectomy. J Clin Invest 100:802–807

Douc-Rasy S, Barrois M, Fogel S, Ahomadegbe JC, Stehelin D, Coll J, Riou G (1996) High incidence of loss of heterozygosity and abnormal imprinting of *H19* and *IGF2* genes in invasive cervical carcinomas. Uncoupling of *H19* and *IGF2* expression and biallelic hypomethylation of *H19*. Oncogene 12:423–430

Dugimont T, Curgy JJ, Wernert N, Delobelle A, Raes MB, Joubel A, Stehelin D, Coll J (1995) The *H19* gene is expressed within both epithelial and stromal components of human invasive adenocarcinomas. Biol Cell 85:117–124

Eggenschwiler J, Ludwig T, Fisher P, Leighton PA, Tilghman SM, Efstratiadis A (1997) Mouse mutant embryos overexpressing IGF-II exhibit phenotypic features of the Beckwith-Wiedemann and Simpson-Golabi-Behmel syndromes. Genes Dev 11:3128–3142

Ekstrom TJ, Cui H, Li X, Ohlsson R (1995) Promoter-specific *IGF2* imprinting status and its plasticity during human liver development. Development 121:309–316

Fan H, Villegas C, Huang A, Wright JA (1996) Suppression of malignancy by the 3′ untranslated regions of ribonucleotide reductase R1 and R2 messenger RNAs. Cancer Res 56:4366–4369

Ferguson-Smith AC, Cattanach BM, Barton SC, Beechey CV, Surani MA (1991) Embryological and molecular investigations of parental imprinting on mouse chromosome 7. Nature 351:667-670

Giannoukakis N, Deal C, Paquette J, Goodyer CG, Polychronakos C (1993) Parental genomic imprinting of the human *IGF2* gene. Nat Genet 4:98-101

Gicquel C, Raffin-Sanson ML, Gaston V, BertagnaX, Plouin PR, Schlumberger M, Louvel A, Luton JP, Le Bouc Y (1997) Structural and functional abnormalities at 11p15 are associated with the malignant phenotype in sporadic adrenocortical tumors:study on a series of 82 tumors. J Clin Endocrinol Metab 82:2559-2565

Giddings SJ, King CD, Harman KW, Flood JF, Carnaghi LR (1994) Allele-specific inactivation of insulin 1 and 2, in the mouse yolk sac, indicates imprinting. Nat Genet 6:310-313

Greally JM, Starr DJ, Hwang S, Song L, Jaarola M, Zemel S (1998) The mouse *H19* locus mediates a transition between imprinted and non-imprinted DNA replication patterns. Hum Mol Genet 7:91-96

Grundy P, Telzerow P, Paterson MC, Haber D, Herman B, Li F, Garber J (1991) Chromosome 11 uniparental isodisomy predisposing to embryonal neoplasms. Lancet 338:1079-1080

Grundy P, Wilson B, Telzerow P, Zhou W, Paterson MC (1994) Uniparental disomy occurs infrequently in Wilms tumor patients. Am J Hum Genet 54:282-289

Guillemot F, Caspary T, Tilghman SM, Copeland NG, Gilbert DJ, Jenkins NA, Anderson DJ, Joyner AL, Rossant J, Nagy A (1995) Genomic imprinting of *Mash2*, a mouse gene required for trophoblast development. Nat Genet 9:235-241

Haddad R, Held WA (1997) Genomic imprinting and *Igf2* influence liver tumorigenesis and loss of heterozygosity in SV40 T antigen transgenic mice.Cancer Res 57:4615-4623

Haig D, Graham C (1991) Genomic imprinting and the strange case of the insulin-like growth factor II receptor. Cell 64:1045-1046

Hamel PA, Phillips RA, Muncaster M, Gallie BL (1993) Speculations on the roles of RB1 in tissue-specific differentiation, tumor initiation, and tumor progression. FASEB J 7:846-854

Hao Y, Crenshaw T, Moulton T, Newcomb E, Tycko B (1993) Tumour-suppressor activity of *H19* RNA. Nature 365:764-767

Hatada I, Mukai T (1995) Genomic imprinting of p57KIP2, a cyclin-dependent kinase inhibitor, in mouse. Nat Genet 11:204-206

Hatada I, Inazawa J, Abe T, Nakayama M, Yasuhiko K, Jinno Y, Niikawa N, Ohashi H, Yoshimitsu F, Iida K, Yutani C, Takahashi S, Chiba Y, Ohishi S, Mukai T (1996a) Genomic imprinting of human *p57KIP2* and its reduced expression in Wilms' tumors. Hum Mol Genet 5:783-788

Hatada I, Ohashi H, Fukushima Y, Kaneko Y, Inoue M, Komoto Y, Okada A, Ohishi S, Nabetani A, Morisaki H, Nakayama M, Niikawa N, Mukai T (1996b) An imprinted gene, *p57KIP2* is mutated in Beckwith-Wiedemann syndrome. Nat Genet 13:171-173

Hatada I, Nabetani A, Morisaki H , Xin Z, Ohishi S, Tonoki H, Niikawa N, Inoue M, Komoto Y, Okada A, Steichen E, Ohashi H, Fukushima Y, Nakayama M, Mukai T (1997) New *p57KIP2* mutations in Beckwith-Wiedemann syndrome. Hum Genet 100:681-683

Hedborg F, Holmgren L, Sandstedt B, Ohlsson R (1994) The cell-type specific *IGF2* expression during early human development correlates to the pattern of overgrowth and neoplasia in the Beckwith-Wiedemann syndrome. Am J Pathol 145:802-817

Henry I, Grandjouan S, Couillin P, Barichard F, Huerre-Jeanpierre C, Glaser T, Philip T, Lenoir G, Chaussain JL, Junien C (1989a) Tumor-specific loss of 11p15.5 alleles in del11p13 Wilms' tumor and in familial adrenocortical carcinoma. Proc Natl Acad Sci USA 86:3247-3251

Henry I, Jeanpierre M, Couillin P, Barichard F, Serre JL, Journel H, Lamouroux A, Turleau C, de Grouchy J, Junien C (1989b) Molecular definition of the 11p15.5 region involved in Beckwith-Wiedemann syndrome and probably in predisposition to adrenocorticalcarcinoma. Hum Genet 81:273-277

Henry I, Bonaiti-Pellié C, Chehensse V, Beldjord C, Schwartz C, Utermann G, Junien C (1991) Uniparental disomy in a genetic cancer-predisposing syndrome. Nature 351:665-667

Henry I, Puech A, Riesewijk A, Ahnine L, Mannens M, Beldjord C, Bitoun P, Tournade MF, Landrieu P, Junien C (1993) Somatic mosaicism for partial paternal isodisomy in Wiedemann-Beckwith syndrome: a post-fertilization event. Eur J Hum Genet 1:19–29

Heutink P, van der Mey AGL, Sandkuijl A, van Gils APG, Bardoel A, Breedveld GJ, van Vliet M, van Ommen G-JB, Cornelisse CJ, Oostra BA, Weber JL, Devilee P (1992) A gene subject to genomic imprinting and responsible for hereditary paragangliomas maps to chromosome 11q23-qter. Hum Mol Genet 1:7–10

Hoovers JMN, Kalikin LM, Johnson LA, Alders M, Redeker B, Law DJ, Bliek J, Steenman M, Benedict M, Wiegant J, Lengauer C, Taillon-Miller P, Schlessinger D, Edwards MC, Elledge SJ, Ivens A, Westerveld A, Little P, Mannens M, Feinberg A (1995) Multiple genetic loci within 11p15 defined by Beckwith-Wiedemann syndrome rearrangement break-points and subchromosomal transferable fragments. Proc Natl Acad Sci USA 92:12456–12460

Hu R-J, Lee MP, Johnson LA, Feinberg AP (1996) A novel human homologue of yeast nucleosome assembly protein, 65 kb centromeric to the *p57KIP2* gene, is biallelically expressed in fetal and adult tissues. Hum Mol Genet 5:1743–1749

Hussey DJ, Parker NJ, Hussey ND, Little PFR, Dobrovic A (1997) Characterization of a KRAB family zinc finger gene, ZNF195, mapping to chromosome band 11p15.5. Genomics 45:451–455

Isfort RJ, Cody DB, Kerckaert GA, Tycko B, LeBoeuf RA (1997) Role of the H19 gene in Syrian hamster embryo cell tumorigenicity. Mol Carcinogenesis 20:189–193

Journel H, Lucas J, Allaire C, Le Mee F, Defawe G, Lecornu M, Jouan H, Roussey M, Le Marec B (1985) Trisomy 11p15 and Beckwith-Wiedemann syndrome. Report of two new cases. Ann Genet 28:97–101

Joyce JA, Lam WK, Catchpoole DJ, Jenks P, Reik W, Maher ER, Schofield PN (1997) Imprinting of *IGF2* and *H19*: lack of reciprocity in sporadic Beckwith-Wiedemann syndrome. Hum Mol Genet 6:1543–1548

Kaghad M, Bonnet H, Yang A, Creancier L, Biscan JC, Valent A, Minty A, Chalon P, Lelias JM, Dumont X, Ferrara P, McKeon F, Caput D (1997) Monoallelically expressed gene related to p53 at 1p36, a region frequently deleted in neuroblastoma and other human cancers. Cell 90:809–819

Kantarjian HM, O'Brien SM, Keating M, Beran M, Estey E, Giralt S, Kornblau S, Rios MB, de Vos D, Talpaz M (1997) Results of decitabine therapy in the accelerated and blastic phases of chronic myelogenous leukemia. Leukemia 11:1617–1620

Knudson A (1971) Mutation and cancer: statistical study of retinoblastoma. Proc Natl Acad Sci USA 68:820–823

Koide T, Ainscough J, Wijgerde M, Surani MA (1994) Comparative analysis of Igf-2/H19 imprinted domain: identification of a highly conserved intergenic DNase I hypersensitive region. Genomics 24:1–8

Kondo M, Suzuki H, Ueda R, Takagi K, Takahashi T, Takahashi T (1994) Parental origin of 11p15 deletions in human lung cancer. Oncogene 9:3063–3065

Kondo M, Matsuoka S, Uchida K, Osada H, Nagatake M, Takagi K, Harper JW, Takahashi T, Elledge SJ, Takahashi T (1996) Selective maternal-allele loss in human lung cancers of the maternally expressed *p57KIP2* gene at 11p15.5. Oncogene 12:1365–1368

Koufos A, Grundy P, Morgan K, Aleck KA, Hadro T, Lampkin BC, Kalbakji A, Cavenee WK (1989) Familial Wiedemann-Beckwith syndrome and a second Wilms tumor locus both map to 11p15.5. Am J Hum Genet 44:711–719

Krieder JA, Sariola H, Loring JM, Maeda M, Pelletier J, Housman DE, Jaenisch R (1993) WT-1 is required for early kidney development. Cell 74:670–691

Kurahashi H, Takami K, Oue T, Kusafuka T, Okada A, Tawa A, Okada S, Nishisho I (1995) Biallelic inactivation of the APC gene in hepatoblastoma. Cancer Res 55:5007–5011

LaSalle JM, Lalande M (1996) Homologous association of oppositely imprinted chromosomal domains. Science 272:725–728

Lee MH, Reynisdottir I, Massague J (1995) Cloning of p57KIP2, acyclin-dependent kinase inhibitor with unique domain structure and tissue distribution. Genes Dev 9:639–649.

Lee MP, DeBaun M, Randhawa G, Reichard BA, Elledge SJ, Feinberg AP (1997a) Low frequency of *p57KIP2* mutation in Beckwith-Wiedemann syndrome. Am J Hum Genet 61:304–309

Lee MP, Hu R-J, Johnson LA, Feinberg AP (1997b) Human *KVLQT1* gene shows tissue specific imprinting and encompasses Beckwith-Wiedemann syndrome chromosomal rearrangements. Nat Genet 15:181–185

Leighton PA, Ingram RS, Eggenschwiler J, Efstratiatis A, Tilghman SM (1995) Disruption of imprinting caused by deletion of the *H19* gene region in mice. Nature 375:34–39

Li X, Adam G, Cui H, Sandstedt B, Ohlsson R, Ekstrom TJ (1995) Expression, promoter usage and parental imprinting status of insulin-like growth factor II (*IGF2*) in human hepatoblastoma: uncoupling of *IGF2* and *H19* imprinting. Oncogene 11:221–229

Liu J, Kahri AI, Heikkila P, Voutilainen R (1997) Ribonucleic acid expression of the clustered imprinted genes, p57KIP2, insulin-like growth factor II, and H19, in adrenal tumors and cultured adrenal cells. J Clin Endocrinol Metab 82:1766–1771

Lowrey CH, Nienhuis AW (1993) Treatment with azacitidine of patients with end stage beta-thalassemia. N Engl J Med 329:845–848

Lustig-Yariv O, Schulze E, Komitowski D, Erdmann V,Schneider T, deGroot N, Hochberg A (1997) The expression of the imprinted genes *H19* and *IGF2* in choriocarcinoma cell lines. Is *H19* a tumor suppressor gene? Oncogene 15:169–177

Manzella JM, Blackshear PJ (1990) Regulation of rat ornithine decarboxylase mRNA translation by its 5′-untranslated region. J Biol Chem 265:11817–11822

Matsuoka S, Edwards MC, Bai C, Parker S, Zhang P, Baldini A, Harper JW, Elledge SJ (1995) p57KIP2, a structurally distinct member of the p21CIP1Cdk inhibitor family, is a candidate tumor suppressor gene. Genes Dev 9:650–662

Matsuoka S, Thompson JS, Edwards MC, Barletta JM, Grundy P, Kalikin LM, Harper JW, Elledge SJ, Feinberg AP (1996) Imprinting of the gene encoding a human cyclin-dependent kinase inhibitor, p57*KIP2*, on chromosome 11p15.5. Proc Natl Acad Sci USA 93:3026–3030

McGrath J, Solter D (1984) Completion of mouse embryogenesis requires both the maternal and paternal genomes. Cell 37:179–183

Mishina M, Ogawa O, Kinoshita H, Oka H, Okumura K, Mitsumori K, Kakehi Y, Reeve AE, Yoshida O (1996) Equivalent parental distribution of frequently lost alleles and biallelic expression of the H19 gene in human testicular germ cell tumors.Jpn J Cancer Res 8, 816–823

Montagna M, Menin C, Chieco-Bianchi L, D'Andrea E (1994) Occasional loss of constitutive heterozygosity at 11p15.5 and imprinting relaxation of the *IGF2* maternal allele in hepatoblastoma. J Cancer Res Clin Oncol 120:732–736

Moore T, Haig D (1991) Genomic imprinting in mammalian development: a parental tug of war. Trends Genet 7:45–49

Morison IM, Becroft DM, Taniguchi T, Woods CG, Reeve AE (1996) Somatic overgrowth associated with overexpression of insulin-like growth factor II. Nature Med 2:311–316

Moulton T, Crenshaw T, Hao Y, Moosikasuwan J, Lin N, Dembitzer F, Hensle T, Weiss L, McMorrow L, Loew T, Kraus W, Gerald W, Tycko B (1994) Epigenetic lesions at the *H19* locus in Wilms' tumor patients. Nat Genet 7:440–447

Moulton T, Chung W-Y, Yuan L, Hensle T, Waber P, Nisen P, Tycko B (1996) Genomic imprinting and Wilms' tumor. Med Pediatr Oncol 27:476–483

Nesbit MA, Hodges MD, Campbell L, de Meulemeester TMAMO, Alders M, Rodrigues NR, Talbot K, Theodosiou AM, Mannens MA, Nakamura Y, Little PFR, Davies KE (1997) Genomic organization andchromosomal localization of a member of the MAP kinase phosphatase gene family to human chromosome 11p15.5 and a pseudogene to 10q11.2. Genomics 42:284–294

Nielsen FC, Gammeltoft S, Christiansen J (1990) Translational discrimination of mRNAs coding for human insulin-like growth factor II. J Biol Chem 265:13431–13434

Nielsen FC, Ostergaard L, Nielsen J, Christiansen J (1995) Growth-dependent translation of IGF-II mRNA by a rapamycin-sensitive pathway. Nature 377:358–62.

Oda H, Imai Y, Nakatsuru Y, Hata J, Ishikawa T (1996) Somatic mutations of the APC gene in sporadic hepatoblastomas. Cancer Res 56:3320–3323

Ogawa O, Eccles MR, Szeto J, McNoe LA, Yun K,Maw MA, Smith PJ, Reeve AE (1993) Relaxation of insulin-like growth factor II gene imprinting implicated in Wilms' tumour. Nature 362:749–751

Ohlsson R, Nystrom A, Pfeifer-Ohlsson S, Tohonen V, Hedbourg F, SchofieldP, Flam F, Ekstrom TJ (1993) IGF2 is parentally imprinted during human embryogenesis and in the Beckwith-Wiedemann syndrome. Nat Genet 4:94–97

Ohtani-Fujita N, Dryja TP, Rapaport JM, Fujita T, Matsumura S, Ozasa K, Watanabe Y, Hayashi K, Maeda K, Kinoshita S, Matsumura T, Ohnishi Y, Hotta Y, Takahashi R, Kato MV, Ishizaki K, Sasaki MS, Horsthemke B, Minoda K, Sakai T (1997) Hypermethylation in the retinoblastoma gene is associated with unilateral sporadic retinoblastoma. Cancer Genet Cytogenet 98:43–49

Okamoto K, Morison IM, Taniguchi T, Reeve AE (1997) Epigenetic changes at the insulin-like growth factor/H19 locus indeveloping kidney is an early event in Wilms' tumorigenesis. Proc Natl Acad Sci USA 94:5367–5371

O'Keefe D, Dao D, Sanderson R, Weiss L, Warburton D, Yeboa K, Tycko B(1997) Coding mutations in p57KIP2 occur in some cases of Beckwith-Wiedemann syndrome but are rare or absent in Wilms' tumors. Am J Hum Genet 61:295–303

Oren R, Takahashi S, Doss C, Levy R, Levy S (1990) TAPA-1, the target of an antiproliferative antibody, defines a new family of transmembrane proteins. Mol Cell Biol 10:4007–4015.

Orlow I, Iavarone A, Crider-Miller SJ, Bonilla F, Latres E, Lee M-H, Gerald WL, Massague J, Weissman BE, Cordon-Cardo C (1996) Cyclin-dependent kinase inhibitor p57KIP2 in soft tissue sarcomas and Wilms' tumors. Cancer Res 56:1219–1221

Overall ML, Spencer J, Bakker M, Dziadek M, Smith PJ (1996) p57KIP2 isexpressed in Wilms' tumor with LOH of 11p15.5. Genes Chromosomes Cancer 17:56–59

Pal N, Wadey RB, Buckle B, Yeomans E, Pritchard J, Cowell JK (1990) Preferential loss of maternal alleles in sporadic Wilms' tumor. Oncogene 5:1665–1668

Paldi A, Gyapay G, Jami J (1995) Imprinted chromosomal regions of the human genome display sex-specific meiotic recombination frequencies. Curr Biol 5:1030–1035

Park CG, Lee SY, Kandala G, Lee SY, Choi, Y (1996) A novel gene product that couples TCR signaling to Fas (CD95) expression in activation-induced cell death. Immunity 4:583–591

Parker NJ, Begley CG, Smith PJ, Fox RM (1996) Molecular cloning of a novel human gene (D11S4896E) at chromosomal region 11p15.5. Genomics 37:253–256

Paulsen M, Davies KR, Bowden LM, Villar A, Franck O, Fuermann M, Dean WL, Moore TF, Rodrigues N, Davies KE, Hu R-J, Feinberg AP, Maher ER, Reik W, Walter J (1998) Syntenic organization of the mouse distal chromosome 7 imprinting cluster and the Beckwith-Wiedemann syndrome region in chromosome 11p15.5. Hum Mol Genet 7:1149–1159

Pedone PV, Tirabosco R, Cavazzana AO, Ungaro P, Basso G, Luksch R, Carli M, Bruni CB, Frunzio R, Riccio A (1994) Mono- and bi-allelic expression of insulin-like growth factor II gene in human muscle tumors. Hum Mol Genet 3:1117–1121

Pegg AE, Shantz LM (1994) Overproduction of ornithine decarboxylase caused by relief of translational repression is associated with neoplastic transformation. Cancer Res 54:2313–2316

Pettenati MJ, Haines JL, Higgins RR, Wappner RS, Palmer CG, Weaver DD (1986) Wiedemann-Beckwith syndrome: presentation of clinical and cytogenetic data on 22 new cases and review of the literature. Hum Genet 74:143–154

Ping AJ, Reeve AE, Law DJ, Young MR, Boehnke M, Feinberg AP (1989) Genetic linkage of Beckwith-Wiedemann syndrome to 11p15. Am J Hum Genet 44:720–723

Qian N, Frank D, O'Keefe D, Dao D, Zhao L, Yuan L, Wang Q, Keating M, Walsh CP, Tycko B (1997) The IPL gene on chromosome 11p15.5 is imprinted in humans and mice and is similar to TDAG51, implicated in Fas expression and apoptosis. Hum Mol Genet 6:2021–2029

Qing RQ, Schmitt S, Ruelicke T, Stallmach T, Schoenle EJ (1996) Autocrine regulation of growth by insulin-like growth factor (IGF)-II mediated by type I IGF-receptor in Wilms tumor cells. Pediatr Res 39:160–165

Rachmilewitz J, Elkin M, Looijenga LH, Verkerk AJ, Gonik B, Lustig O,Werner D, deGroot N, Hochberg A (1996) Characterization of the imprinted IPW gene: allelic expression in normal and tumorigenic human tissues. Oncogene 13:1687–1692

Rainier S, Johnson L, Dobry CJ, Ping AJ, Grundy PE, Feinberg AP (1993) Relaxation of imprinted genes in human cancer. Nature 362:747–749

Rainier S, Dobry CJ, Feinberg AP (1995) Loss of imprinting in hepatoblastoma. Cancer Res 55:1836–1838

Rastinejad F, Conboy MJ, Rando TA, Blau HM (1993) Tumor suppression by the 3' untranslated region of alpha-tropomyosin. Cell 75:1107–1117

Redeker E, Alders M, Hoovers JMN, Richard CW III, Westerveld A, Mannens M (1995) Physical mapping of 3 candidate tumor suppressor genes relative to Beckwith-Wiedemann syndrome associated chromosomal breakpoints at 11p15.3. Cytogenet Cell Genet 68:222–225

Reid LH, Davies C, Cooper PR, Crider-Miller SJ, Sait SNJ, Nowak NJ, Evans G, Stanbridge EJ, deJong P, Shows TB, Weissman BE, Higgins MJ (1997) A 1-Mb physical map and PAC contig of the imprinted domain in 11p15.5 that contains TAPA1 and the BWSCR1/WT2 region. Genomics 43:366–375

Reik W, Brown KW, Slatter RE, Sartori P, Elliott M, Maher ER (1994) Allelic methylation of H19 and IGF2 in the Beckwith-Wiedemann syndrome. Hum Mol Genet 3:1297–1301

Reik W, Brown KW, Schneid H, Le Bouc Y, Bickmore W, Maher E (1995) Imprinting mutations in the Beckwith-Wiedemann syndrome suggested by an altered imprinting pattern in the IGF2-H19 domain. Hum Mol Genet 4:2379–2385

Ripoche M-A, Kress C, Poirier F, Dandolo L (1997) Deletion of the H19 transcription unit reveals the existence of a putative imprinting control element. Genes Dev 11:1596–1604

Roberts DJ, Mutter GL (1994) Advances in the molecular biology of gestational trophoblastic disease. J Reprod Med 39:201–208

Robinson WP, Lalande M (1995) Sex-specific meiotic recombination in the Prader-Willi/ Angelman syndrome imprinted region. Hum Mol Genet 4:801–806

Sabbioni S, Barbanti-Brodano G, Croce CM, Negrini M (1997) GOK: a gene at 11p15 involved in rhabdomyosarcoma and rhabdoid tumor development. Cancer Res 57:4493–4497

Santer R, Hoffman H, Suttorp M, Simeoni E, Schaub J (1995) Discordance for hyperinsulinemic hypoglycemia in monozygotic twins. J Pediatr 126:1017

Schneid H, Seurin D, Vazquez M-P, Gourmelen M, Cabrol S, Le Bouc Y (1993) Parental allele specific methylation of the human insulin-like growth factor II gene and Beckwith-Wiedemann syndrome. J Med Genet 30:353–362

Schroeder WT, Chao L-Y, Dao DT, Strong LC, Pathak S, Riccardi VM, Lewis WK, Saunders GF (1987) Nonrandom loss of maternal chormosome 11 alleles in Wilms' tumors. Am J Hum Genet 40:413–420

Scrable H, Cavenne W, Ghavimi F, Lovell M, Morgan K, Sapienza C (1989) A model for embryonal rhabdomyosarcoma tumorigenesis that involves genome imprinting. Proc Natl Acad Sci USA 86:7480–7484

Simms LA, Reeve AE, Smith PJ (1995) Genetic mosaicism at the insulin locus in liver associated with childhood hepatoblastoma. Genes Chromosomes Cancer 13:72–73

Slavotinek A, Gaunt L, Donnai D (1997) Paternally inherited duplications of 11p15.5 and Beckwith-Wiedemann syndrome. J Med Genet 34:819–826

Spencer GS, Schabel F, Frisch H (1980) Raised somatomedin in associated with normal growth hormone. A cause of Beckwith-Wiedemann syndrome? Arch Diseases Childhood 55:151–153

Steenman MJC, Rainier S, Dobry CJ, Grundy P, Horon IL, Feinberg AP (1994) Loss of imprinting of IGF2 is linked to reduced expression and abnormal methylation of H19 in Wilms' tumor. Nat Genet 7:433–439

Sun F-L, Dean WL, Kelsey G, Allen N, Reik W (1997) Transactivation of *Igf2* in a mouse model of Beckwith-Wiedemann syndrome. Nature 389:809–815

Surani MAH, Barton SC, Norrris ML (1984) Development of reconstituted mouse eggs suggests imprinting of the genome during gametogenesis. Nature 308:548–550

Sutcliffe JS, Nakao M, Christian S, Orstavik KH, Tommerup N, Ledbetter DH, Beaudet AL (1994) Deletions of a differentially methylated CpG island at the *SNRPN* gene define a putative imprinting control region. Nat Genet 8:52–58

Suzuki H, Ueda R, Takahashi T, Takahashi T (1994) Altered imprinting in lung cancer. Nat Genet 6:332–333

Tada M, Tada T, Lefebvre L, Barton SC, Surani MA (1997) Embryonic germ cells induce epigenetic reprogramming of somatic nucleus in hybrid cells. EMBO J 16:6510–6520

Taniguchi T, Schofield AE, Scarlett JL, Morison IM, Sullivan MJ, Reeve AE (1995a) Altered specificity of *IGF2* promoter imprinting during fetal development and onset of Wilms tumour. Oncogene 11:751–775

Taniguchi T, Sullivan MJ, Osamu O, Reeve A (1995b) Epigenetic changes encompassing the *IGF2/ H19* locus associated with relaxation of *IGF2* imprinting and silencing of *H19* in Wilms tumor. Proc Natl Acad Sci (USA) 92:2159–2163

Taniguchi T, Okamoto K, Reeve A (1997) Human *p57KIP2* defines a new imprinted domain on chromosome 11p, but is not a tumor suppressor gene in Wilms' tumor. Oncogene 14:1201–1206

Thomas BJ, Rothstein R (1991) Sex, maps, and imprinting. Cell 64:1–3

Thornton PS, Baker L, Stanley CA (1995) Discordance for hyperinsulinemic hypoglycemia in monozygotic twins (Reply). J Pediatr 126:1017

Tsang P, Gilles F, Yuan L, Kuo Y-H, Lupu F, Samara G, Moosikasuwan J, Goye A, Zelenetz AD, Selleri L, Tycko B (1995) A novel L23-related gene 40 kb downstream of the imprinted *H19* gene is biallelically expressed in mid-fetal and adult human tissues. Hum Mol Genet 4:1499–1507

Turleau C, de Grouchy J (1985) Beckwith-Wiedemann syndrome. Clinical comparison between patients with and without 11p15 trisomy. Ann Genet 28:93–96

Turleau C, de Grouchy J, Chavin-Colin CF, Martelli H, Voyer M, Charlas R (1984) Trisomy 11p15 and Beckwith-Wiedemann syndrome: a report of two cases. Hum Genet 67:219–221

Tycko B (1997) DNA methylation in genomic imprinting. Mut Res 386:131–140

Tycko B, Trasler J, Bestor TH (1997) Genomic imprinting: gametic mechanisms and somatic consequences. J Androl 18:480–486

van der Mey AGL, Maaswinkel-Mooy PD, Cornellisse CJ, Schmidt PH, van de Kamp JJP (1989) Genomic imprinting in hereditary glomus tumors: evidence for a new genetic theory. Lancet 2:1291–1294

Vu TH, Hoffman AR (1994) Promoter-specific imprinting of the human insulin-like growth factor-II gene. Nature 371:714–717

Vu TH, Yballe C, Boonyanit S, Hoffman AR (1995) Insulin-like growth factor II in uterine smooth-muscle tumors: maintenance of genomic imprinting in leiomyomata and loss of imprinting in leiomyosarcomata. J Clin Endocrinol Metab 80:1670–1676

Walsh C, Miller SJ, Flam F, Fisher RA, Ohlsson R (1995) Paternally derived *H19* is differentially expressed in malignant and nonmalignant trophoblast. Cancer Res 55:1111–1116

Wang Q, Curran ME, Splawski I, Burn TC, Millholland JM, VanRaay TJ, Shen J, Timothy KW, Vincent GM, de Jager T, Schwartz PJ, Toubin JA, Moss AJ, Atkinson DL, Landes GM, Connors TD, Keating MT (1996) Positional cloning of a novel potassium channel gene: KVLQT1 mutations cause cardiac arrhythmias. Nat Genet 12:17–23

Wang W-H, Duan J-X, Vu TH, Hoffman AR (1996) Increased expression of the insulin-like growth factor-II gene in Wilms' tumor is not dependent on loss of genomic imprinting or loss of heterozygosity. J Biol Chem 271:27863–27870

Weiss L, Medeiros J, Vickey A (1989) Pathologic features of prognostic significance in adrenocortical carcinomas. Am J Surg Pathol 13:202–206

Weksberg R (1994) Wiedemann-Beckwith syndrome: genomic imprinting revisited. Am J Med Genet 52:235–236

Weksberg R, Shen DR, Fei YL, Song QL, Squire J (1993) Disruption of insulin-like growth factor 2 imprinting in Beckwith-Wiedemannsyndrome. Nat Genet 5:143–150

Wiedemann HR (1983) Tumours and hemihypertrophy associated with Wiedemann-Beckwith syndrome. Eur J Pediatr 141:129

Wilkins RJ (1988) Genomic imprinting and carcinogenesis. Lancet 1:329–331

Williams JC, Brown KW, Mott MG, Maitland NJ (1989) Maternal allele loss in Wilms' tumor. Lancet 1:283–284

Winqvist R, Mannermaa A, Alavaikko M, Blanco G, Taskinen PJ, Kiviniemi H, Newsham I, Cavenee W (1993) Refinement of regional loss of heterozygosity for chromosome 11p15.5 in human breast tumors. Cancer Res 53:4486–4488

Yan Y, Frisen J, Lee MH, Massague J, Barbacid M (1997) Ablation of the CDKinhibitor p57Kip2 results in increased apoptosis and delayed differentiation during mouse development. Genes Dev 11:973–983

Yballe CM, Vu TH, Hoffman AR (1996) Imprinting and expression of insulin-like growth factor-II and H19 in normal breast tissue and breast tumor. J Clin Endocrinol Metab 81:1607–1612

Yuan L, Qian N, Feng L, Tycko B (1996) An extended region of biallelic gene expression and rodent-human synteny downstream of the imprinted H19 gene on chromosome 11p15.5. Hum Mol Genet 5:1931–1937

Zhan S, Shapiro DN, Helman LJ (1994) Activation of an imprinted allele of the insulin-like growth factor II gene implicated in rhabdomyosarcoma. J Clin Invest 94:445–448

Zhang P, Liegeois NJ, Wong C, Finegold M, Hou H, Thompson JC, Silverman A, Harper JW, DePinho RA, Elledge SJ (1997) Altered cell differentiation and proliferation in mice lacking p57KIP2 indicates a role in Beckwith-Wiedemann syndrome. Nature 387:151–158

Zhang Y, Tycko B (1992) Monoallelic expression of the human H19 gene. Nat Genet 1:40–44

Zhang Y, Shields T, Crenshaw T, Hao Y, Moulton T, Tycko B (1993) Imprinting of human H19: allele-specific CpG methylation, loss of the active allele in Wilms' tumor and potential for somatic allele switching. Am J Hum Genet 53:113–124

Zubair M, Hilton K, Saam JR, Surani MA, Tilghman SM, Sasaki H (1997) Structure and expression of the mouse l23mrp gene downstream of the imprinted H19 gene: biallelic expression and lack of interaction with the H19 enhancers. Genomics 45:290–296

Mechanisms of Transcriptional Regulation

Gary C. Franklin

1
Introduction

The phenomenon of genomic imprinting, whatever its biological functions and evolutionary origins may be, ultimately manifests itself in the monoallelic expression of certain genes. The assumption has been, therefore, that the transcriptional regulatory systems of imprinted genes would directly or indirectly represent the principal focus of the imprinting process. As this chapter will discuss, however, our perceptions of gene regulation mechanisms are currently undergoing a radical re-evaluation and some of the emerging concepts may prove to be of importance to our understanding of how genomic imprinting functions.

2
Eukaryotic Gene Transcription: The Traditional View

Two decades or more of intense, painstaking work enabled a fairly detailed picture of the transcription of eukaryotic class II genes to be drawn, which until recently, was almost universally accepted in terms of its basic paradigms. Transcription is the result of the association of a multi-component general transcription complex with widely utilised core promoter sequences (Roeder 1996). The efficiency of this process is regulated mainly by the effect of sequence-specific DNA binding proteins (regulatory transcription factors) which bind to regulatory *cis*-elements that are arranged in unique or highly specific combinations within the promoter and/or enhancer elements of individual genes (Maniatis et al. 1987). This system has long been viewed in terms of diffusible protein factors interacting with an essentially static component; i.e., the DNA sequences comprising the genes and their control elements.

Department of Animal Development & Genetics, Uppsala University, Norbyvägen 18A, S-752 36, Uppsala, Sweden

2.1
Basal and Regulated Transcription

An enormous effort has been put into the dissection and characterisation of the numerous general transcription factors (GTFs) which comprise the basal transcription machinery and defined roles for the individual GTFs have been well-established; including enhancement of DNA binding, protein phosphorylation and DNA helicase activities (Roeder 1996; Svejstrup et al. 1996). Only one of these proteins (TATA box binding protein; TBP) actually possesses any sequence-specific requirements for DNA binding (Wang et al. 1996) and the remainder of the so-called preinitation complex (PIC) is built-up via protein-protein interactions. Until recently, the picture was apparently clear: RNA polymerase II (RNAP II) and several associated GTFs were shown to form the PIC on promoter sequences via a functionally obligatory, DNA dependent sequential mechanism of assembly of the individual protein components (Buratowski 1994). It should be pointed out that the basal transcription process just described is not sufficient for physiological levels of gene activity, but merely defines the start point and direction of transcriptional initiation from specific promoter sequences.

A considerable volume of data has also been accumulated with respect to the mechanisms of the regulation of transcription. The overwhelming conclusion from this work has been that the activation domains of regulatory transcription factors modulate the efficiency of the basal transcriptional machinery (as described above), probably by direct protein contacts with the GTFs (Roberts et al. 1993), or via co-activator molecules that are closely associated with some of the GTFs (Kaiser and Meisterernst 1996). Very intricate in vitro transcription studies using recombinantly generated sub-units of the general transcription machinery have been employed to investigate these issues; showing that transcriptional activators can play a role only after a defined point in the sequential assembly process of the PIC, for example (Choy and Green 1993). Quite how the DNA-bound regulatory factors make physical contact with the PIC components is another question entirely. In the case of promoter-proximal elements it is likely that local distortions in DNA bending are involved (Parvin et al. 1995), but in the case of enhancer elements located at kilobase distances from their target promoters, both DNA loop out and scanning models have been proposed. The looping alternative has long been the most favoured and a recent study of the transcriptional regulation of the globin locus has provided strong evidence for this model (Dillon et al. 1997). One important paradigm to be established was that the regulatory transcription factors are totally promiscuous with respect to promoter activation (Kermekchiev et al. 1991); presumably because the highly conserved activation domains interact with the general transcription factors associated with most, or all, promoter sequences. This means that enhancer elements should be able to activate promoters other than their natural targets, as has indeed been very commonly observed in experimental systems. This has long been considered to be an important phenomenon for the evolution of gene regulation in eukaryotes, but poses interesting practical problems for cellular gene regulation, especially in the case of gene clusters, as will be discussed later.

The actual mechanism of transcriptional activation also remains open to question. The rate of transcriptional initiation has long represented the obvious target for activation processes, but it is also clear that this is not the whole truth of the matter. Initiation obviously represents the first step in the transcription process and there are numerous studies to show that binding of TBP to the promoter can represent the rate-limiting step of transcription in some cases (Colgan and Manley 1992). Detailed studies have shown, however, that the transcription cycle involves a number of quite distinct steps; including initiation, promoter clearance, processive transcription and elongation (Goodrich and Tijan 1994). Initiation occurs when there is localised, ATP-independent melting of the DNA duplex at the transcriptional initiation site of the promoter and RNAP II catalyses the incorporation of the first few (around three) ribonucleotides into the RNA chain. A kinase sub-unit of the TFIIH complex then phosphorylates the carboxyterminal domain of RNA polymerase II, inducing a more extensive duplex melting and the extension of the RNA chain to around 16 nucleotides (the process called promoter clearance). After this point, most of the general transcription factors are thought to dissociate from the RNAP II, elongation factors are recruited and transcription enters the elongation phase. It has now been established that different regulatory transcription factors affect different stages of the transcription cycle. While some are known to directly increase the rate of initiation, others have no such effect, but greatly increase the number of initiated transcripts which progress into processive transcription and subsequently into elongation (Krumm et al. 1993). It is thought, therefore, that some promoters may undergo high-level, abortive transcriptional initiation cycling which never progresses into productive transcription unless triggered to do so by transcription factors specific for processive transcription.

2.2
Transcription in the Chromosomal Context

Whereas many of the investigations on the transcription process have been carried out using in vitro transcription analysis methods; the transcriptional machinery in vivo has to operate in a chromatin environment. Intuitively, it seems obvious that chromatin structure would present a barrier to the RNA polymerase and its associated factors and scanning the literature often leaves the impression that the transcription machinery is essentially at the mercy of epigenetic repression mechanisms. This topic will be covered in some depth by Pirotta (Chap. 10) and Razin and Shemer (Chap. 9, this volume), but suffice it to say that an ever-growing number of chromatin-associated proteins and modification mechanisms (e.g., histone acetylation) have been identified which function to modulate the level of gene transcription (Davie and Hendzel 1994; Schlossherr et al. 1994). There is also a plethora of published data concerning the role of DNA methylation (particularly at CpG dinucleotides) in the control of gene transcription; again giving the general impression that the transcriptional machinery is inhibited by this epigenetic modification of chromosomal DNA (Selker 1990).

It is important to point out here that the transcription machinery is not, in fact, helpless in the face of epigenetic obstacles (also, see Sect. 3.2). It has been known for a number of years that some regulatory transcription factors, such as GAL4, are able to bind their recognition sites even in a nucleosomal environment and to induce an ATP-dependent chromatin remodelling at the promoter region (Pazin et al. 1994). Such factors may recruit auxiliary proteins which induce the local chromatin remodelling, such as the NURF factor (Tsukiyama and Wu 1995). Another protein (FACT) has now been identified which specifically functions to alleviate the nucleosomal block of transcriptional elongation seen with some promoters (Orphanides et al. 1998).

2.3
Implications for Genomic Imprinting

Within the context of the classical view of gene transcription discussed above, certain implications are readily apparent for the parent-of-origin-dependent, monoallelic transcription of imprinted genes. It is clear that functional imprinting must result from an original epigenetic discrimination between the two parental alleles and several attractive possibilities for how this might influence transcription present themselves. Firstly, allele-specific epigenetic modifications around the core promoter region could inhibit the basal transcription machinery and thereby silence one allele selectively. This is unlikely to involve a direct effect of methylation on DNA-binding, since neither of the only two sequence motifs known to be involved in direct DNA binding of pre-initiation complex proteins (the TATA box, TATAAA; and the initiator sequence, YYAN(T/A)YY) contain a CpG dinucleotide. In fact, there is evidence to show that methylation-dependent repression of transcription does not necessarily involve methylation of the promoter region per se, but may depend more upon the CpG methylation density around the promoter (Kass et al. 1997). Any effects of methylation may, therefore, be more likely to be involved in less precisely localised, chromatin-mediated repression. The latter type of modification could easily be envisaged to generate monoallelic expression of imprinted genes, by the selective repression of one of the parentally derived promoter regions.

The above report by Kass and colleagues did not definitively show that the effects of the methylation were at the level of transcriptional initiation, which is interesting given that methylation has been shown to specifically block transcriptional elongation and to have no effect on initiation itself in other systems (Rountree and Selker 1997). It is interesting to speculate as to whether the effects of epigenetic modifications such as methylation could exert their effects beyond the transcription level altogether. In the case of the tissue-specifically imprinted *Xist* gene, for example, the initially biallelic RNA products are stabilised in an allele-of-origin-specific manner during early development (Panning et al. 1997; Sheardown et al. 1997) and remains closely associated with the inactive X chromosome.

An alternative strategy could be to selectively modulate the activated transcription of one allele, by interfering with the DNA binding and/or transactivation potential of regulatory transcription factors, either within up-stream promoter regions or enhancer elements. A scenario in which repressive chromatin would preclude the binding of regulatory transcription factors from their target DNA sites is not difficult to envisage and indeed, the DNA-binding of many proteins has been shown to be excluded by the presence of nucleosomes for example (Venter et al. 1994). It is also likely that methylation within or adjacent to core recognition sequences can strongly influence the DNA binding of regulatory transcription factors. Data seems to point towards a generally negative effect of methylation on DNA binding. Such modifications could, therefore, lead to the selective silencing or activation of the methylated allele, depending upon whether the protein involved is a transactivator or repressor factor. It should also be pointed out that the DNA binding of some nuclear factors actually requires methylation of the target sites (Reith et al. 1994; Huntriss et al. 1997) and that the effect of methylation on transcription cannot be simply correlated to a role for activating or repressing factors in the absence of more detailed information.

The above considerations raise a number of issues, including the question of whether the monoallelic expression of imprinted genes primarily involves the selective activation of one of two inactive alleles; or the selective repression of one of two active alleles. If the basal transcription machinery represents the "imprinting target", this would imply that repression is the primary function, since there is no reason to believe that any epigenetic modifications exist which could directly potentiate formation of the preinitiation complex itself. A targeting of the regulatory transcription systems, however, could involve either activation and/or repression of one allele. Epigenetic-dependent blocking of transactivating factors, or potentiation of repressive factor binding would lead to silencing of that allele, whereas inhibition of repressive factors or potentiation of transactivators by epigenetic modifications would result in the selective activation of one allele. In either case, this would lead to a situation in which there could be an enormous difference in transcription levels between the two parental alleles, but in which the basal transcription activity of the two promoters would not be effected. On the other hand, it can be viewed that basal transcription is a phenomenon of the in vitro transcription technique and that any levels of gene activity depending on this process alone in vivo have no physiological significance. Such issues may have some relevance, however, in terms of experiments designed to detect extremely low levels of transcriptional activity, such as the use of RT-PCR analysis.

The plasticity of imprinted gene expression that has been observed, both in terms of cell type-specific functional imprinting and the abnormal allelic expression patterns frequently observed in tumours (Ekstrom et al. 1995; Ohlsson and Franklin 1995), might also be explained in terms of the issues discussed here. Many of the theories put forward to explain the existence of imprinting, for example, are based upon a role in early developmental stages (Moore and Haig 1991). It may be, therefore, that if specific activation pathways do represent the imprinting target for some genes, the utilisation of alternative gene activation pathways in different cell-types and/or at different development stages will result in changes in allelic expression. Such changes may not reflect any changes in the imprint itself, but merely result from the use of *cis*-elements and their associated factors which are outside the sphere of influence of the imprinting mechanisms. If the basal transcription process was the target of allele-specific imprinting repression, however, the plasticity of imprinting observed for some genes might be harder to explain.

3
Eukaryotic Gene Transcription: Recent Revelations

The issues discussed in the preceding section pertain to the classical view of gene transcription: i.e., that supported by the vast bulk of earlier transcription research and the one still described in most text books and frequently alluded to by many prominent researchers in the field. In the past 2 or 3 years, however, a number of astonishing developments have taken place, providing sometimes still controversial data which nevertheless entails a radical re-think into the whole mechanism of eukaryotic transcription. These issues are of vital importance for the wider field of gene transcription, but also raise a number of interesting possibilities with regard to genomic imprinting, as we will now consider.

3.1
Basal Transcription and the Holoenzyme Debate

One of the most controversial debates currently raging within the transcription field relates to the formation of the preinitiation complex itself and the much vaunted sequential assembly model described in Section 2.1. There have been so many data in support of this model (both from in vitro transcription studies and from the isolation of RNAP II complexes in vivo) that it is generally considered to represent one of the key differences between eukaryotic transcription and its prokaryotic counterpart; where in the latter case, the RNA polymerase and its associated factors bind to promoter sequences as a preassembled holoenzyme which forms in the absence of DNA (Eick et al. 1994). Recent data has cast severe doubts over this paradigm of eukaryotic transcription, with RNAP II-containing holoenzyme complexes described in both yeast (Koleske and Young 1994) and higher eukaryotes (Ossipow et al. 1995). The precise composition of the holoenzyme remains controversial, but one group have reported that an entity containing all the components of the PIC can be immunoprecipitated from human cells, using an antibody directed against the general transcription factor TFIIH (Ossipow et al. 1995). The transcription field remains divided on this subject: the sequential assembly camp maintains that the holoenzymes described could represent partial break-down products of previously utilised transcription complexes; whereas those who believe in the holoenzyme model assert that the sequential assembly model is based upon in vitro transcription artefacts and sub-optimal purification techniques of nuclear protein complexes. This debate is probably most readily assimilated by consulting a review article covering this controversy (Ranish and Hahn 1996). A simplified visual representation of the two basic models of PIC formation can also be found in Fig. 1.

A compromise, two-step model has recently been proposed, in which the DNA-binding component of the preinitiation complex, TBP (or its complex form, TFIID) is recruited to the promoter first and then the remaining proteins of the general transcription machinery associate as an RNA polymerase-containing holo-complex (Stargell and Struhl 1996). This would be in keeping with the vast majority of holoenzyme complexes described, which do not include TFIID, and presents a scenario in which the principle function of regulatory transcription factors is to aid the recruitment of TFIID and/or the holoenzyme complex to the promoter. Even if the association of the general transcription machinery with the core promoter is the main target of the imprint in some cases, therefore, plasticity in allelic expression patterns could reflect the differential use of regulatory proteins with very different efficiencies with respect to complex recruitment. Aside from this, the idea of a holoenzyme type of mechanism has contributed to an even more revolutionary concept of gene regulation which has profound implications for the mechanisms of imprinting; as will be discussed below.

3.2
Transcriptosomes: Factories for Gene Expression in the Nucleus?

Further investigations into the nature of the general transcription machinery have indicated that its size and complexity are far greater than has been considered until now. In addition to the fifty or more proteins involved in transcription per se, additional complexes with quite distinct biological roles have been identified within eukaryotic RNAP II holoenzymes. For example, TFIIH is now known to serve a dual role: it is involved in the phosphorylation of the carboxy terminal domain of the RNAP II immediately prior to

1) Sequential Assembly Model

1) RNA pol II Holoenzyme Model

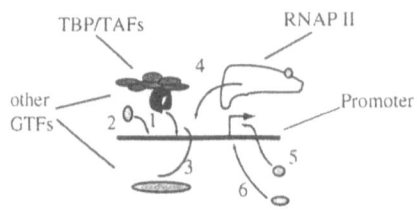

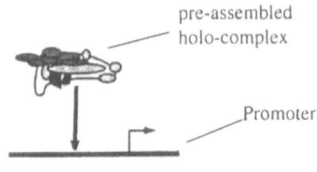

2) Promiscuous enhancer action

2) Promoter-specific enhancer action

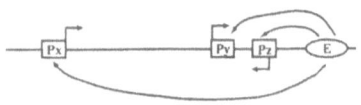

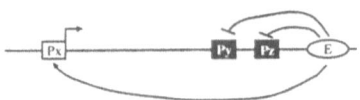

3) Diffusible transcription components

3) Transcription components in fixed factories

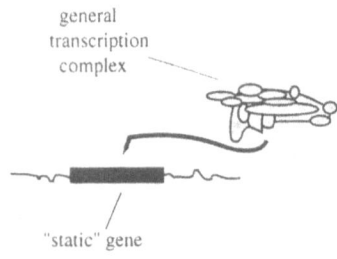

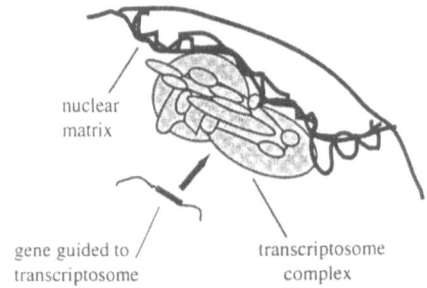

The sequential assembly model of the basal transcription complex would predict multiple steps at which regulatory processes, including imprinting, could exert their effects.

Promiscuous enhancer action within clusters of genes could provide severe regulatory problems for imprinting mechanisms: the fundamental accessibility of promoters to distal *cis*-elements would provide the most logical control point.

The "classical" view would suggest that imprints may function simply to block or enhance the access of diffusible proteins to promoter and/or enhancer sequences.

The recruitment of the holoenzyme complex could provide a focal target for imprinting mechanisms.

The association of chromatin-modifying complexes with the holoenzyme needs consideration in terms of the nature of the epigentic imprint.

Promoter-specific enhancer systems could provide one target for imprinting mehchanisms within some of the clusters of imprinted genes.

The transcription factory view of the nucleus suggests that imprinting elements could "target" genes to specific nuclear locations and that there might be a spatial separation of the two alleles of imprinted genes to permissive and non-permissive regions of the nucleus.

Fig. 1. Changing views of the mechanisms of transcriptional regulation and the potential implications for genomic imprinting

transcriptional initiation, but is also part of a multi-protein complex which is responsible for transcription-coupled DNA repair (Svejstrup et al. 1996). It is also now clear that the general transcription machinery includes large protein complexes, such as SWI/SNF (Wilson et al. 1996), which are capable of disrupting chromatin (Peterson 1996). Taken together, these findings indicate that the size of a eukaryotic holoenzyme complex could be as large as 3 MDa: this multi-functional super complex (capable of supporting transcription, chromatin disruption and DNA repair) has been termed the transcriptosome (Halle and Meisterernst 1996).

The physical size alone of this entity makes the traditional view of transcription as a system involving the interaction of diffusible protein factors with static genes seem less intuitively logical. Additional evidence has revealed that the factors which comprise the general transcription machinery are not distributed randomly throughout the nucleus, but can be localised in a relatively small number of discrete foci (Jackson et al. 1993). Calculations on the cellular concentrations of some of the key proteins of the PIC have also produced the rather surprising result that they are present in very low abundance. TBP, for example, which is essential for all transcription by RNA polymerases I, II and III, may be present in the nucleus at around only one molecule per active gene (Schmidt and Schibler 1995). These data taken together suggest that the components of the transcription machinery may well be localised in concentrated areas within the nucleus and that in fact, gene regulatory sequences need to find their way to immobilised transcription factories within a highly organised nuclear architecture (Jackson et al. 1993; see also Marshall and Sedat, Chap. 14, this vol.). This presents entirely new possibilities for the mechanisms of imprinting; where epigenetic modifications could control the localisation of active and inactive alleles to discrete nuclear compartments with very different transcriptional potentials. In such a scenario, regulatory transcriptional elements could be the target of the imprint, in that epigenetic modifications could modulate the binding of protein factors which in turn function to localise the gene to a defined nuclear location. With this in mind, it is worth noting that matrix attachment regions (MARs) appear to be commonly found within enhancer elements (Yu et al. 1994). While the evidence in favour of the transcriptosome is extremely interesting, it should be noted that this remains a highly controversial issue and that the interpretation of this data has been challenged from some quarters (Singer and Green 1997). The alternative interpretation of data suggesting that active genes must find their way to defined sub-nuclear locations enriched in transcription machinery components, for example, is that the foci of such proteins that are observed are actually the result of high-level gene activity. The resolution of these issues will represent a crucial step forward in our understanding of the gene regulation process in the coming years.

3.3
Promoter-Specificity of Enhancer Action

The thinking towards how enhancer elements operate has largely been domi-
nated by the earliest studies, showing that they were highly promiscuous
towards promoters (Kermekchiev et al. 1991). It has become clear in more
recent years, however, that at least some eukaryotic enhancers exhibit a strong
preference, or even exclusive specificity towards their natural promoters. Exactly what proportion of enhancers do show promoter specificity remains to be
seen, but there are now a significant number of published reports; including
examples in humans (Franklin et al. 1991), rodents (Goping et al. 1995) and
insects (Merli et al. 1996). In addition to the general transcription complex
components, the mechanisms involved must employ the binding of proteins to
specialised sequence elements; has have been demonstrated in some promoter
and enhancer elements (Franklin et al. 1995). On reflection, the operation of
enhancers which have no preference for particular promoters makes little
sense in the chromosomal/nuclear context, where the potential for unsched-
uled promoter activation by enhancers could lead to regulatory chaos. This
problem would be of particular importance in the case of tightly clustered
genes. Although the presence of chromatin-based boundary and insulator
systems undoubtedly plays an important role in this regard (Geramisova and
Corces, Chap. 11, this vol.), it has been convincingly demonstrated that it is
promoter-specific enhancer action and not chromatin domains which regulate
the expression of at least one gene cluster in *Drosophila* (Merli et al. 1996). The
authors of this report also suggested that promoter-specific enhancers could
represent the major mechanism for maintaining the regulatory autonomy of
genes in the cell and that chromatin boundaries may represent a specialised
mechanism which is particular for very highly expressed genes. A summary of
the traditional, classical view of transcription versus these recent, revolutionary new concepts is shown in Fig. 1, together with some of the major implications that these two views may have for our understanding of imprinting.

3.4
Implications for Genomic Imprinting

The impact of these new concepts in the transcription field on the quest to gain
insights into imprinting mechanisms remains to be seen. The sequential
assembly versus holoenzyme debate, for example, may not be so directly
pertinent. The predominant use of a holoenzyme type of mechanism for transcription would simplify the problem, however, since recruitment of the pre-
formed transcription complex to the promoter would present an obvious
target for any imprinting mechanisms, via interference with the positive/nega-
tive transcription factors involved in modulating the recruitment of the complex to the promoter of any given gene. The alternative, sequential assembly

model of transcription would provide an additional layer of complexity, since the build-up of the general transcription machinery on the promoter could conceivably be subject to allelic, epigenetic effects at any of a number of discrete steps. What may be of more importance is the demonstration that powerful chromatin disruption/remodelling complex SWI/SNF appears to intimately associated with the RNAP II holoenzyme complex (Wilson et al. 1996). Since the basal transcription machinery itself may be capable, therefore, of dealing with repressive chromatin, any observations of differential, allele-specific chromatin structure at the promoter regions of imprinted genes may not reflect a causative explanation for allele-specific expression. The regulation of transcriptional elongation may turn out to be interesting in this regard, although this will require a much clearer understanding of the function of elongation factors, such as the FACT protein (Orphanides et al. 1998)

One other intriguing aspect of the fundamental transcription process which has come to light is the highly atypical behaviour of the transcription machinery during the earliest stages of development. Elegant experiments have shown that in the earliest stages of mouse development, for example, both enhancers and the key core promoter element, the TATA box, are not required for gene activity (Majumder and DePamphilis 1994, 1995). The mechanisms underlying this highly atypical behaviour remain to be fully elucidated, but it has been reported that sperm contain extraordinarily high levels of some basal transcription factors, such as TBP (Schmidt and Schibler 1995) and a developmentally acquired coactivator activity has recently been described that correlates with the acquisition of enhancer function during development (Majumder et al. 1997). If imprinting mechanisms target the normal transcription systems of the cell, it may well be that the expression of imprinted genes could exhibit unexpected behaviour at the very earliest stages of development.

The operation of promoter-specific or promoter-selective enhancers within clusters of genes raises some interesting possibilities, however, with regard to imprinted gene regulation. The characteristic clustering of so many of the imprinted genes means that the issue of the maintenance of regulatory autonomy and the potential problems of promiscuous enhancer action are of particular relevance. Since it is now known that specialised elements within some promoters can greatly influence their interactions with distal enhancers and silencers (Franklin et al. 1995; Goping et al. 1995), epigenetically regulated versions of this kind of system could be particularly important within the imprinted gene clusters. One classic example of imprinted gene regulation is the *Igf2/H19* locus; where these genes are monoallelically expressed from the opposite parental chromosomes. A regulatory model has been suggested in which shared enhancer elements are competed for by the *Igf2* and *H19* promoter systems, with epigenetic modifications determining the choice of promoter on each allele (Bartolomei et al. 1993). These genes lie within an extended and quite tight cluster of several genes, however, only some of which are imprinted (Beechey, Appendix, this vol.). An enhancer with epigenetically switchable selectively for the *Igf2* and *H19* promoters could, therefore, provide the complex allele-specific expression of these two genes without perturbing the normal cell type-specific expression of the neighbouring genes within the cluster. Indeed, there is evidence suggesting that the *H19* endodermal enhancers exhibit a synergistic interaction with the *H19* promoter (Yoo-Warren et al. 1988).

The clustered nature of imprinted genes does present several potentially interesting problems with respect to their regulation. Data obtained from the study of non-imprinted gene clusters, such as the globin locus and surfeit gene cluster may have important consequences for our understanding of the regulation of the imprinted genes. The five *surf* genes within the surfeit locus have little or no structural or functional connection, but despite this, they are regulated in a highly co-ordinated and interdependent manner, even to the point of two of the genes sharing a partially overlapping, bi-directional promoter (Gaston et al. 1995). This raises the possibility that the co-ordinated regulation of genes within clusters may occur as a result of their physical co-localisation, rather than reflecting any functional overlap. One interesting possibility raised by this could be that at least some imprinted genes are imprinted as a by-product of their chromosomal location (Horsthemke et al., Chap. 5, this vol.).

Extensive studies of the globin locus have also provided clues as to how regulation of imprinted gene clusters might be controlled. The five genes within the globin locus are expressed in a sequential, developmental stage-specific pattern and all depend upon the presence of a distally located locus control region (LCR). Although a number of conflicting models have been proposed for the control of this locus, it appears that the LCR acts as a master enhancer for the globin genes and functions by making mutually exclusive direct contact with one of the five promoters, almost certainly via a looping-out mechanism (Dillon et al. 1997). It also appears that physical proximity to the LCR confers a formidable competitive advantage to any given promoter. This type of scenario would appear to present some problems when examined in the context of the imprinted *Igf2/H19* region, for example. If distal regulatory elements, such as the globin LCR and the *H19* endodermal enhancers, generally function in this way, then it becomes more difficult to envisage the enhancer competition model of *Igf2/H19* regulation. It is more difficult, for example, to understand how the distally located enhancers can simultaneously activate the multiple promoters of the *Igf2* gene. One could envisage a dynamic situation in which the enhancer would interact transiently with the different promoters in a stochastic fashion, but extrapolating from the globin scenario, it might be expected that the promoter closest to the *H19* gene would have a competitive advantage. This issue is also highly relevant with respect to the choice of *Igf2* or *H19* expression on a given allele: considering the distances involved, it could be difficult to uncouple the *H19* promoter/enhancer combination in favour of *Igf2* activation. As alluded to in the original enhancer competition model, this problem could be overcome if the promoter(s) of the silent gene was rendered inaccessible to enhancer activation by epigenetic means, on the appropriate allele. With this in mind, there are data to suggest that in the globin locus, it is the relative position of the five promoters themselves which controls the developmental stage-specific expression pattern: the function of the LCR is to enhance the level of transcription of which ever promoter is "on" at that stage (Martin et al. 1996). While this scenario could be extrapolated to the *Igf2/H19* region, there has been very recent transgenic data to demonstrate that the *Igf2* promoters maintain the potential for activation by the *H19* endodermal enhancers even on the maternal allele and may even be strongly favoured as a target by these enhancers (Webber et al. 1998), suggesting that some form of boundary mechanisms must be involved (Horsthemke et al, Chap. 5, this Vol. and Gerasimova and Corces, Chap. 11, this vol.).

Perhaps the most exciting of the new developments within the transcription field with regard to imprinting is the concept of the transcriptosome or transcription factory. It is conceivable that epigenetic modifications of specific gene regions could control the targeting of genes to nuclear locations, either permissive or repressive for transcription, in an allele-specific manner. Recent work has shown that specific chromosomal regions can occupy quite distinct spatial positions within the nucleus; being either closely associated with the nuclear envelope, or more internally localised within the nucleoplasm (Marshall et al. 1997; see also Marshall and Sedat, Chap. 14, this vol.). It is conceivable that imprinting might involve the differential epigenetic targeting of chromosomal regions containing the two parental alleles of a gene to regions of the nuclear periphery or interior; where the environment could be either permissive or restrictive for expression, on the basis of local concentrations of transcription factories. The problem with this concept comes when considering the tight clusters of imprinted genes. Current data suggests that the chromosomal loops that can be defined as occupying internal or peripheral regions in the nucleus are quite large, of the order of 1–2 Mb (Marshall et al. 1996). This makes it harder to visualise how such a mechanism would operate in cases where oppositely imprinted, as well as non-imprinted genes, are co-localised within regions that are considerably shorter than the predicted size of the chromosomal loops. An extremely complex and very highly ordered mechanism would have to be invoked in which transcription factories with different gene-specific compositions would have to be assembled in close association, at various sub-nuclear locations. At the current time, it is not possible to say whether or not the nucleus is so highly organised and if it could contain transcription "factories" with such a high degree of gene specificity.

One exciting aspect of the general concept of the highly ordered nucleus, however, could be that this might allow the visualisation of the imprinting mechanism in action, since the expressed and non-expressed allele of an imprinted gene could be situated in quite distinct locations within the nucleus. This sort of approach has already begun to reveal quite intricate details of gene expression in the nuclear context (Dernburg et al. 1996) and the technical developments in DNA and RNA FISH, as well as in microscopy and digital image processing should enable a thorough investigation of these possibilities in the near future.

One interesting recent hypothesis is that the whole process of homologous pairing of chromosomes during meiosis may be critically dependent on DNA-protein interactions specified by the unique combinations of transcription factors found associated with any given promoter/enhancer system (Cook 1997). This would present an interesting problem with regard to imprinted genes, where the expressed and silent alleles could have difficulty in pairing up, although there is evidence to suggest that process does occur by some means (LaSalle and Lalande 1996; and also see Paldi and Youvenot, Chap. 13, this Vol.). Perhaps this would dictate that imprinted gene clusters must either also contain non-imprinted genes, or include oppositely imprinted genes: this seems to be the case for the clusters of imprinted loci described to date. The transcription-dependent homologous pairing hypothesis, combined with the transcription factory hypothesis, suggests an intriguing possibility: that each individual gene (or perhaps a limited number of genes located strategically along each chromosome) has a unique nuclear target site for its transcription. It could be, for example, that allelic exclusion and genomic imprinting are merely variations on a common theme; i.e., that in the latter case, one of two alleles is fixed as the one which can occupy the only permissible transcription site in the nucleus. In the former case, the choice is random (perhaps fixed by the first association of one allele with the transcription factory), whereas in the latter case, the targeting/blocking of one allele is fixed epigenetically in association with the imprinting process itself. Such a similarity in the basic mechanism of these two processes might explain the apparent drive towards monoallelic expression exhibited by some genes in cases where the functional imprint has been lost (He et al. 1998).

It is also interesting to speculate whether the same basic conditions might apply to other, apparently biallelically expressed genes. Even where there is clearly a biallelic contribution to steady-state RNA levels, it would be difficult to establish whether or not there could be a mutually exclusive (but dynamic) occupation of a unique transcriptional site in the nucleus at any given time. In other words, one could consider the possibility that monoallelic expression could be a feature of many more genes than is currently realised: where allelic exclusion represents a modification in which a random allele is fixed for expression and genomic imprinting has added the refinement of achieving this fixing process on the basis of parental origin.

4
Concluding Remarks

The imprinting field, like the transcription field, is currently in a dynamic and exciting phase; full of new possibilities and open for radical new concepts. The recent controversies in the study of gene regulation, which have turned the whole field upside down, also present important new ideas for our understanding of genomic imprinting. It is clear, therefore, that these two vitally important areas of modern biological research are irrevocably intertwined and that new developments in one will continue to have profound significance for the other.

References

Bartolomei MM, Webber AL, Brunkow ME, Tilghman SM (1993) Epigenetic mechanisms underlying the imprinting of the mouse H19 gene. Genes Dev 7:1663–1673

Buratowski S (1994) The basics of basal transcription by RNA polymerase II. Cell 77:1–3

Choy B, Green MR (1993) Eukaryotic activators function during multiple steps of preinitiation complex assembly. Nature 366:531–536

Colgan J, Manley JL (1992) TFIID can be rate limiting in vivo for TATA-containing, but not TATA-lacking, RNA polymerase II promoters. Genes Dev 6:304–315

Cook PR (1997) The transcriptional basis of chromosome pairing. J Cell Sci 110:1033–1040

Davie JR, Hendzel MJ (1994) Multiple functions of dynamic histone acetylation. J Cell Biochem 55:98–105

Dernburg AF, Broman KW, Fung JC, Marshall WF, Philips J, Agard DA, Sedat JW (1996) Perturbation of nuclear architecture by long-distance chromosome interactions. Cell 85:745–759

Dillon N, Trimborn T, Strouboulis J, Fraser P, Grosveld F (1997) The effect of distance on long-range chromatin interactions. Mol Cell 1:131–139

Eick D, Wedel A, Heumann H (1994) From initiation to elongation: comparison of transcription by prokaryotic and eukaryotic RNA polymerases. TIG 10:292–296

Ekstrom TJ, Cui H, Li X, Ohlsson R (1995) Promoter-specific IGF2 imprinting status and its plasticity during human liver development. Development 121:309–316

Franklin G, Donovan M, Adam G, Holmgren L, Pfeifer-Ohlsson S, Ohlsson R (1991) Expression of the human PDGF-B gene is regulated by both positively and negatively acting cell type-specific regulatory elements located in the first intron. EMBO J 10:1365–1373

Franklin GC, Adam GIR, Miller SJ, Moncrieff CL, Ullerås E, Ohlsson R (1995) An Inr-containing sequence flanking the TATA box of the human c-sis (PDGF-B) proto-oncogene promoter functions in cis as a co-activator for its intronic enhancer. Oncogene 11:1873–1884

Gaston K, Duhig T, Armes N, Colombo P, Fried M (1995) Surf5: a gene in the tightly clustered mouse surfeit locus is highly conserved and transcribed divergently from the rpL7A (Surf3) gene. Genomics 30:163–170

Goodrich JA, Tijan R (1994) Transcription factors IIE and IIH and ATP hydrolysis direct promoter clearance by RNA polymerase II. Cell 77:145–156

Goping IS, Lamontagne S, Shore GC, Nguyen M (1995) A gene-type-specific enhancer regulates the carbamyl phosphate synthetase I promoter by cooperating with the proximal GAG activating element. Nucleic Acids Res 23:1717–1721

Halle JP, Meisterernst M (1996) Gene expression: increasing evidence for a transcriptosome. Trends Genet 12:161–163

He LM, Cui HM, Walsh C, Mattsson R, Lin W, Anneren G, Pfeifer-Ohlsson S, Ohlsson R (1998) Hypervariable allelic expression patterns of the imprinted IGF2 gene in tumor cells. Oncogene 16:113–119

Huntriss J, Lorenzi R, Purewal A, Monk M (1997) A methylation-dependent DNA-binding activity recognising the methylated promoter region of the mouse Xist gene. Biochem Biophys Res Commun 235:730–738

Jackson DA, Hassan AB, Errington RJ, Cook PR (1993) Visualization of focal sites of transcription within human nuclei. EMBO J 12:1059–1065

Kaiser K, Meisterernst M (1996) The human general co-factors. Trends Biochem Sci 21:342–345

Kass SU, Landsberger N, Wolffe AP (1997) DNA methylation directs a time-dependent repression of transcription initiation. Curr Biol 7:157–165

Kermekchiev M, Pettersson M, Matthias P, Schaffner W (1991) Every enhancer works with every promoter for all the combinations tested: could new regulatory pathways evolve by enhancer shuffling? Gene Expr 1:71–81

Koleske AJ, Young RA (1994) An RNA polymerase II holoenzyme responsive to activators. Nature 368:466–469

Krumm A, Meulia T, Groudine M (1993) Common mechanisms for the control of eukaryotic transcriptional elongation. Bio Essays 15:659–665

LaSalle JM, Lalande M (1996) Homologous association of oppositely imprinted chromosomal domains. Science 272:725–728

Majumder S, DePamphilis ML (1994) TATA-dependent enhancer stimulation of promoter activity in mice is developmentally acquired. Mol Cell Biol 14:4258–4268

Majumder S, DePamphilis ML (1995) A unique role for enhancers is revealed during early mouse development. Bio Essay 17:879–889

Majumder S, Zhao Z, Kaneko K, DePamphilis ML (1997) Developmental acquisition of enhancer function requires a unique coactivator activity. EMBO J 16:1721–1731

Maniatis T, Goodbourn S, Fischer JA (1987) Regulation of inducible and tissue-specific gene expression. Science 236:1237–1245

Marshall WF, Dernburg AF, Harmon B, Agard DA, Sedat JW (1996) Specific interactions of chromatin with the nuclear envelope: positional determination within the nucleus in Drosophila melanogaster. Mol Biol Cell 7:825–842

Marshall WF, Fung JC, Sedat JW (1997) Deconstructing the nucleus: global architecture from local interactions. Curr Opin Genet Dev 7:259–263

Martin DIK, Fiering S, Groudine M (1996) Regulation of -globin gene expression: straightning out the locus. Curr Opin Genet Dev 6:488–495

Merli C, Bergstrom DE, Cygan JA, Blackman RK (1996) Promoter specificity mediates the independent regulation of neighbouring genes. Genes Dev 10:1260–1270

Moore T, Haig D (1991) Genomic imprinting in mammalian development: a parental tug-of-war. Trends Genet 7:45–49

Ohlsson R, Franklin G (1995) Normal development and neoplasia: the imprinting connection. Int J Dev Biol 39:869–876

Orphanides G, LeRoy G, Chang CH, Luse DS, Reinberg D (1998) FACT, a factor that facilitates transcript elongation through nucleosomes. Cell 92:105–116

Ossipow V, Tassan JP, Nigg EA, Schibler U (1995) A mammalian RNA polymerase II holoenzyme containing all components required for promoter-specific transcription initiation. Cell 83:137–146

Panning B, Dausman J, Jaenisch R (1997) X chromosome inactivation is mediated by Xist RNA stabilization. Cell 90:907–916

Parvin JD, McCormack RJ, Sharp PJ, Fisher DE (1995) Pre-bending of a promoter sequence enhances affinity for the TATA-binding factor. Nature 373:724–727

Pazin MJ, Kamakaka RT, Kadonaga JT (1994) ATP-dependent nucleosome reconfiguration and transcriptional activation from preassembled chromatin templates. Science 266:2007–2011

Peterson CL (1996) Multiple SWItches to turn on chromatin? Curr Opin Genet Dev 6:171–175

Ranish JA, Hahn S (1996) Transcription: basal factors and activation. Curr Opin Genet Dev 6:151–158

Reith W, Ucla C, Barras E, Gaud A, Durand B, Herrero Sanchez C, Kobr M, Mach B (1994) RFX1, a transactivator of hepatitis B virus enhancer I, belongs to a novel family of homodimeric and heterodimeric DNA-binding proteins. Mol Cell Biol 14:1230–1244

Roberts SGE, Ha I, Maldonado E, Reinberg D, Green MR (1993) Interaction between an acidic activator and transcription factor TFIIB is required for transcriptional activation. Nature 363:741–744

Roeder RG (1996) The role of general initiation factors in transcription by RNA polymerase II. Trends Biochem Sci 21:327–335

Rountree MR, Selker EU (1997) DNA methylation inhibits elongation but not initiation of transcription in Neurospora crassa. Genes Dev 11:2383–95

Schlossherr J, Eggert H, Paro R, Cremer S, Jack RS (1994) Gene inactivation in *Drosophila* mediated by the Polycomb gene product or by position-effect variegation does not involve major changes in the accessibility of the chromatin fibre. Mol Gen Genet 243:453–62

Schmidt EE, Schibler U (1995) High accumulation of components of the RNA polymerase II transcription machinery in rodent spermatids. Development 121:2373–2383

Selker EU (1990) DNA methylation and chromatin structure: a view from below. TIBS 15:103–107

Sheardown SA, Duthie SM, Johnston CM, Newall AE, Formstone EJ, Arkell RM, Nesterova TB, Alghisi GC, Rastan S, Brockdorff N (1997) Stabilization of Xist RNA mediates initiation of X chromosome inactivation. Cell 91:99–107

Singer RH, Green MR (1997) Comparmentilization of eukaryotic gene expression: Causes and effects. Cell 91:291–294

Stargell LA, Struhl K (1996) Mechanisms of transcriptional activation in vivo: two steps forward. Trends Genet 12:311–315

Svejstrup JQ, Vichi P, Egly JM (1996) The mutiple roles of transcription/repair factor TFIIH. Trends Biochem Sci 21:346–350

Tsukiyama T, Wu C (1995) Purification and properties of an ATP-dependent nucleosome remodeling factor. Cell 83:1011–1020

Venter U, Svaren J, Schmitz J, Schmid A, Hörz W (1994) A nucleosome precludes binding of the transcription factor Pho4 in vivo to a critical target site in the PHO5 promoter. EMBO J 13:4848–4855

Wang Y, Jensen RC, Stumph WE (1996) Role of TATA box sequence and orientation in determining RNA polymeraseII/III transcription specificity. Nucleic Acids Res 15:3100–3106

Webber AL, Ingram RS, Levorse JM, Tilghman SM (1998) Location of enhancers is essential for the imprinting of H19 and Igf2 genes. Nature 391:711–715

Wilson CJ, Chao DM, Imbalzano AN, Schnitzler GR, Kingston RE, Young RA (1996) RNA polymerase II holoenzyme contains SWI/SNF regulators involved in chromatin remodeling. Cell 84:235–244

Yoo-Warren H, Pachnis V, Ingram R, Tilghman S (1988) Two regulatory domains flank the mouse H19 gene. Mol Cell Biol 8:4707–4715

Yu J, Bock JH, Slightom JL, Villeponteau B (1994) A 5' beta-globin matrix-attachment region and the polyoma enhancer together confer position-independent transcription. Gene 139:139–145

Epigenetic Control of Gene Expression

Aharon Razin and Ruth Shemer

Epigenetics refers to changes in gene expression without change in nucleotide sequence. Genes destined to be silenced should be marked by an epigenetic signal, leading to the establishment of a heritable but potentially reversible inactive conformation of the gene. A large body of experimental data, that has been accumulated in the last two decades or so, clearly indicates that epigenetic control of gene expression in mammals is achieved by DNA methylation combined with chromatin structure.

1
DNA Methylation and Transcription

1.1
Why DNA Methylation?

Mammalian DNA is exclusively methylated at the 5 position of cytosine residues in CpG containing sequences (Sinsheimer 1955). Being palindromic, CpGs are methylated symmetrically on both DNA strands. Thus, this methylation is faithfully inherited through many cell cycles by a simple maintenance DNA methytransferase (Dmtase) which is present in all cells (Pollack et al. 1980; Wigler et al. 1981; Gruenbaum et al. 1982). Methylation of CpGs can be created anew by de novo methylation (Monk et al. 1987; Kafri et al. 1992) and erased by an active mechanism of demethylation (Razin et al. 1986; Kafri et al. 1993; Weiss et al. 1996; see Fig. 1). These features of CpG methylation and its profound effect on protein-DNA interaction make it a most suitable epigenetic tool in regulating differential gene expression (Shemer and Razin 1996).

The distribution of CpGs in gene sequences across the mammalian genome and the dynamic processes involved in establishing gene specific methylation patterns in the mouse embryo (Razin and Kafri 1994) suggest how CpG methylation may affect gene expression. The mammalian genome is characterized by a bimodal pattern of methylation. Most of the genome (60–80% of

Department of Cellular Biochemistry, Hebrew University, Hadassah Medical School, Jerusalem 91120, Israel

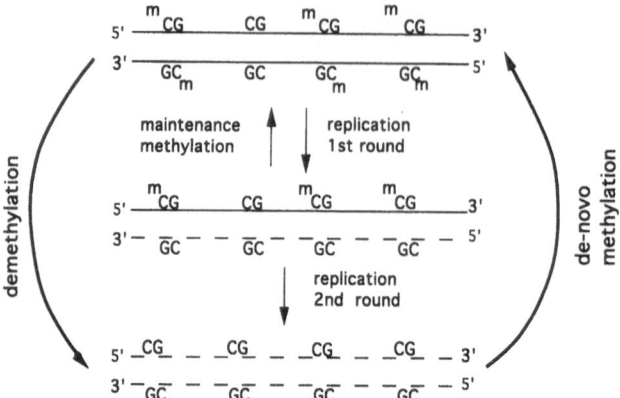

Fig. 1. Methylation metabolism. Methylation patterns in proliferating cells are stably maintained by a postreplication maintenance methylation using hemimethylated DNA as substrate. Although methylation patterns can be lost by two rounds of replication with no concomitant maintenance methylation, it is now believed that loss of methylation is exclusively achieved by an active demethylation mechanism. Methylation patterns can be established anew by de novo methylation followed by cell-specific demethylations. ^mC 5methylcytosine

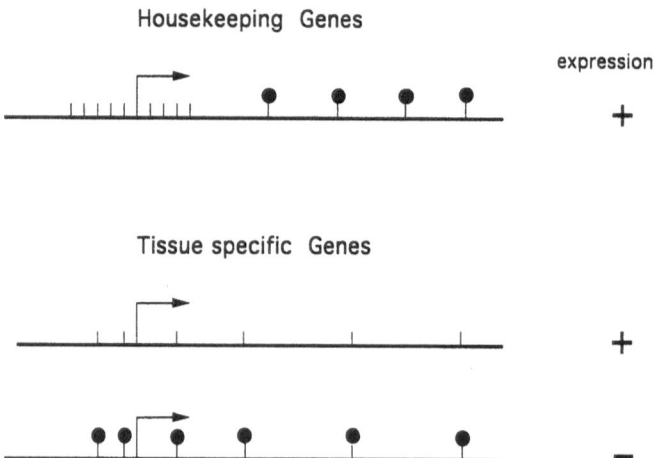

Fig. 2. Gene-specific methylation patterns. Housekeeping genes, as a rule, are characterized by an unmethylated CpG island. Tissue-specific genes are unmethylated in cell types that express the gene and methylated in cell types that do not express the same gene. (|) unmethylated CpG (●) methylated CpG

CpGs) is methylated. Islands of unmethylated CpGs are found in the promoter regions, extending to the first exon, of all housekeeping genes (Bird 1986). Tissue-specific genes are practically methylated in all tissues except for the tissue in which the gene is expressed (Fig. 2). The observed correlation between the methylation of CpGs and gene activity was the first indication that CpG

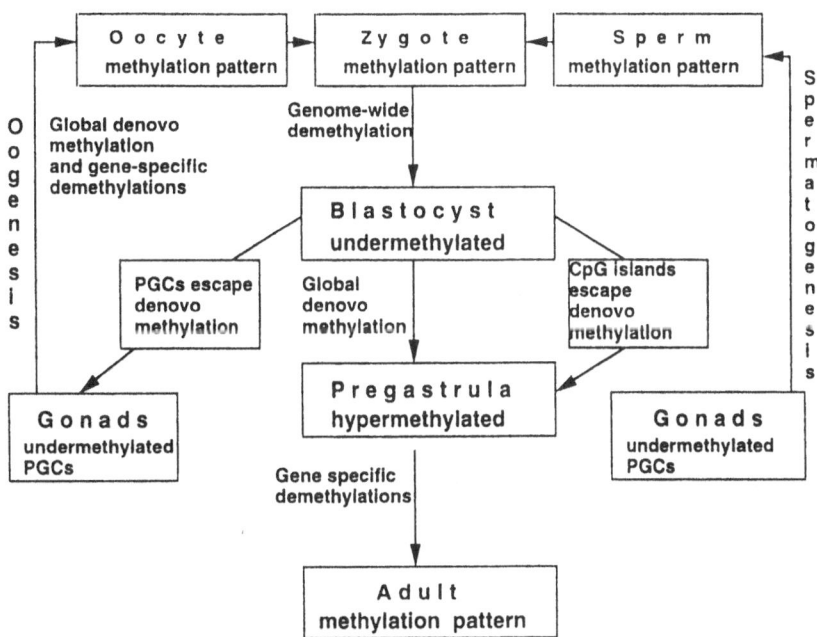

Fig. 3. Changes in gene-specific methylation patterns during development. The gene-specific methylation patterns in oocyte and sperm contribute to a combined methylation pattern in the zygote which is erased by an active demethylation mechanism during the first two to three cleavages. This undermethylated state persists through the blastula stage. Global de novo methylation takes place postimplantation, leaving CpG islands unmethylated. Primordial germ cells (PGCs) which emerge from the unmethylated epiblast remain unmethylated until later stages of gametogenesis (13.5–15.5 dpc in the mouse), when they undergo de novo methylation followed by gene specific demethylation (much like what happens in the embryo proper during gastrulation and further development)

methylation may play a role in the control of gene activity (Yeivin and Razin 1993).

Gene-specific methylation patterns which are observed in adult tissues are a result of dynamic changes in methylation in the early embryo (Razin and Kafri 1994). Gene sequences are heavily methylated in sperm and oocyte except for CpG islands in housekeeping genes. However, this heavy methylation is erased during the first two or three cell divisions post-fertilization. Only around the time of implantation a wave of de novo methylation leads to a bimodal pattern of methylation in which practically all CpGs, except those present in CpG islands, are methylated (Kafri et al. 1992). In the gastrula, a process of gene-specific demethylations starts, concomitant with cell differentiation (e.g., Benvenisty et al. 1985; Shemer et al. 1991). This process continues well into adult life and ends in the fully differentiated cell with the final gene-specific methylation patterns which are observed in adult tissues (Fig. 3).

Primordial germ cells emerging from the epiblast prior to implantation escape the global de novo methylation. Genes in primordial germ cells remain undermethylated until after the cells populate the gonads. De novo methylation and gene-specific demethylations take place during gametogenesis after differentiation to male and female gonads (Kafri et al. 1992; Fig. 3).

All imprinted genes studied so far possess differentially methylated regions (DMRs) (Razin and Cedar 1994). The ontogeny of these methylation patterns is different from the one described above for nonimprinted genes. Some of the CpGs in a DMR are methylated either in the oocyte or in sperm, or become methylated in the fertilized egg before syngamy. This parent-specific methylation is maintained throughout development and escapes the genome-wide demethylation at the precavitation stage (Brandeis et al. 1993; Shemer et al. 1996). Another striking difference between methylation in the imprinted genes as compared to that of the nonimprinted genes is that DMRs are often CpG islands which are normally unmethylated (Bird 1986). It is obvious that DMRs must undergo demethylation at early stages of gametogenesis and de novo methylation at later stages according to the gender of the individual embryo. This has indeed been shown for all imprinted genes which have been studied so far (Kafri et al. 1992; Shemer et al. 1997; Fig. 4).

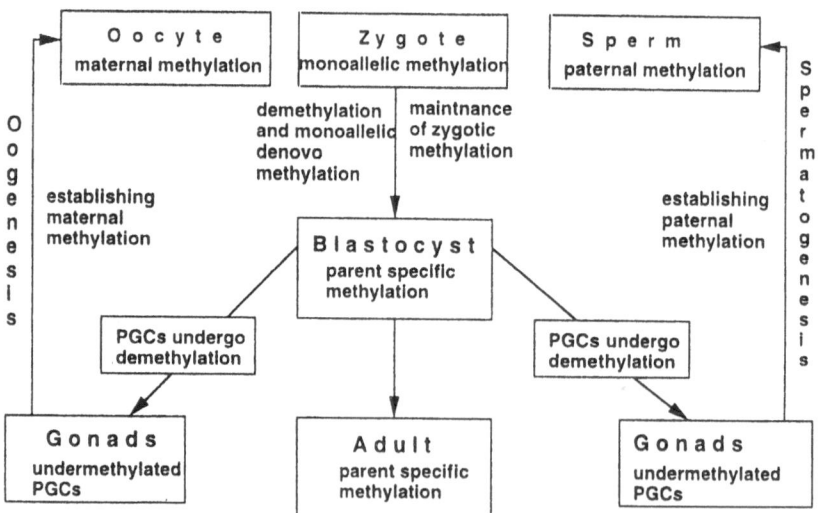

Fig. 4. Methylation changes in imprinted genes during development. Maternal and paternal methylation patterns established during gametogenesis are inherited and maintained by the embryo although some methylated sites lose their methylation and regain monoallelic methylation by the blastula stage. The inherited parent-specific methylation patterns are maintained in the soma throughout life. During gametogenesis the parent-specific methylation patterns are erased and reestablished according to the gender of the embryo

1.2
How Does Methylation Function in Gene Control?

Methyl groups on DNA are known to affect protein-DNA interactions. In most cases methylation prevents this interaction, as is the case in the classical example of restriction enzymes which recognize specific sequences but do not cut the DNA at those sequences when methylated. However, a few examples exist where restriction enzymes such as *Dpn* I cut only when the site is methylated. Methylations can therefore, in rare instances, facilitate protein binding to DNA (Smith and Kelly 1984).

The potential effect of DNA methylation on protein-DNA interactions, taken together with the observed gene-specific methylation patterns which are, in general, inversely correlated with gene activity (Yeivin and Razin 1993), strongly suggested that methylation acts to repress gene activity. In theory, however, methylation could also affect gene activity by preventing repressors from binding to promoter regions or facilitating the binding of activators, thereby assisting in maintaining an active state of the gene. Although such a mechanism has been suggested to operate in the regulation of monoallelic expression of some imprinted genes, such as *Ifg2* and *Igf2r* (Stoger et al. 1993; Li et al. 1993), it has not been corroborated by a direct experiment as yet.

CpG methylation at the regulatory sequences of a given gene can suppress transcription by one of three modes or a combination of two or three of the following:

1. Direct interference with binding of transcription factors, thereby preventing assembly of the transcription initiation complex or its interaction with activators. Although some transcription factors (Sp1, CTF, and YAY1) are not sensitive to methylation, the binding of other transcription factors, such as Ap2, cMyc/Myn, E1F, and NFkB is hindered when CpGs in their respective recognition site are methylated (Tate and Bird 1993). It is not certain, however, to what extent this weaker binding affects gene activity.

2. Protein-mediated inhibition, whereby a methyl binding protein renders the template inaccessible to transcription factors. Several lines of evidence support the possibility that the inhibition of transcription exerted by methylation is mediated by proteins that bind to the methylated CpGs which are positioned in the promoter region (Boyes and Bird 1991, 1992; Levine et al. 1991, 1992).

3. Methylation-directed formation of an inactive chromatin structure that sequesters transcription factors from the promoter region. It is a well-known fact that active genes are characterized by an open conformation of chromatin (Concklin and Groudine 1984). It has recently been shown that methylation plays a role in establishing the inactive chromatin conformation (Razin 1998). This process is maintained by the interaction of the methyl-binding protein MeCP2 (vide infra) with methylated CpG residues. MeCP2 in turn binds the

corepressor mSin3A. This corepressor protein constitutes the core of a multiprotein complex that includes histone deacetylases.

The first experimental evidence supporting the notion that methylation inactivates genes came from two types of experiments: (1) in vitro methylated genes introduced into fibroblasts growing in culture showed reduced or no activity (Stein et al. 1982; Vardimon et al. 1982). (2) Endogenous silenced genes could be activated by treating cells growing in culture with the potent methyltransferase inhibitor, 5-azacytidine (Jones et al. 1982). For many genes, methylation in the 5' regulatory region was sufficient for inhibition of gene activity (Busslinger et al. 1983; Keshet et al. 1985; Kruczek and Doerfler 1983; Yisraeli et al. 1988). Transfections with more accurate positioning of mCpG sites in the promoter region revealed that the extent of inhibition of promoter activity is correlated with the density of mCpG sites at the transcription preinitiation domain. Methylation at sequences flanking this region had little or no effect on the promoter activity.

These and other experiments led to the conclusion that methylation affects primarily transcription preinitiation (Levine et al. 1992; Boyes and Bird 1992). Thus, this inhibition of transcription can be achieved either by preventing binding of transcription factors (Watt and Molloy 1988) or facilitating interaction with proteins that exert their inhibitory effect by specifically binding to methylated CpGs (Boyes and Bird 1991; Levine et al. 1991). In fact, several methyl-binding proteins have been discovered over the past decade (Huang et al. 1984; Meehan et al. 1989; Jost and Hofsteenge 1992; Lewis et al. 1992; Meehan et al. 1992). While the methylated DNA-binding protein MBP-1 is methyl- and sequence-specific (Wang et al. 1986), MeCP1 and MeCP2 are methyl-dependent sequence-independent DNA-binding proteins (Meehan et al. 1989, 1992). So is the histone H1 like MDBP-2 (Jost and Hofsteenge 1992).

MeCP2, perhaps the most studied methyl-binding protein, is an abundant chromosomal protein that has been shown to be a methyl CpG-binding protein in vivo as well as in vitro (Nan et al. 1996). Disruption of the X-linked gene that codes for the MeCP2 protein resulted in death of the embryos at the gastrula stage (9.5 dpc) (Tate et al. 1996). The MeCP2 protein has an 80-amino-acid methyl binding domain and a region capable of long-range repression in vivo (Nan et al. 1997) consistent with the view that MeCP2 interacts with the transcriptional machinery or the initiation complex.

As mentioned above, active genes have to maintain an unmethylated state of the promoter region. Promoters of housekeeping genes being embedded in a CpG island region have been shown to involve an Sp1-mediated mechanism to maintain their unmethylated status (Brandeis et al. 1994; Macleod et al. 1994). It should be noted that cis-acting elements that direct demethylation of the promoter region have also been observed in genes expressed in a cell type-specific manner, such as α actin (Paroush et al. 1990) and IgG κ chain (Lichtenstein et al. 1994).

2
Chromatin Structure and Transcription

The nucleosomal structure of the genome is a basic element in the regulation of transcription in eukaryotic cells. This nucleoprotein structure, based on histones, provides the necessary infrastructure for allowing efficient transcription of one class of genes while all other genes are effectively silenced. The global chromatin structure of the genome must therefore be organized in two different conformations, an open conformation which accommodates genes in their active state, and a closed structure where genes are suppressed (Conklin and Gourdine 1984). Regulation of individual genes clearly involves basal transcription factors and regulatory factors, activators and suppressors. Nucleosomes provide the basis for assembling the DNA into a 30-nm fiber and higher order structures which are impermeable to certain transcription factors. Nucleosomes can also act individually to block transcription factor binding if positioned appropriately (Pina et al. 1990; Archer et al. 1992).

The concept of a regional organization of eukaryotic chromosomes into discrete functional domains is becoming widely accepted. Position effect variegation in *Drosophila* (Schaffer et al. 1993; Chap. ••, this vol.), in which the expression status of the gene is determined by its chromosomal position and the discovery of elements that define the boundaries of active chromatin domains in mammalian chromosomes (Chung et al. 1993; Chap. 11, this vol.), are examples of how chromatin structure can affect gene expression.

The positioning of nucleosomes provides the basis for local gene regulation, allowing or preventing *trans*-acting factors to gain access to their recognition elements (Wolffe 1994). This positioning is a dynamic process affected by modification or processing of the core histones, or modification of the DNA. Once the nucleosomal structure has been established, it is locked by the linker histone H1 or H5 (Pennings et al. 1994).

Only activators can destabilize the nucleosomal structure perhaps by removal of histone H1 (Bresnick et al. 1992). A major problem, which remains unsolved, concerns the heritability of the correct positioning of nucleosomes postreplication, especially in light of the observation that nucleosomes are disrupted during replication (Sogo 1995).

3
The Chromatin-DNA Methylation Connection

As discussed above, both DNA methylation and chromatin play a role in regulation of gene activity. It is clear that the two must work in concert when an inactive state of a gene has to be established and maintained. One striking example for interaction between chromatin and DNA methylation is the inactive X chromosome in eutherian female cells. The inactive X-chromosome is heavily methylated and displays a condensed chromatin structure. Experimental data indicate that a gene in its methylated state is wrapped by unusual nucleosome particles that migrate as large nucleoproteins on agarose gels (Keshet et al. 1986). This nuclease-resistant chromatin structure is assembled prior to inactivation of the gene (Buschhausen et al. 1987). This observation was consistent with the early finding that nucleosomal DNA has a threefold higher quantity of 5-methylcytosine than internucleosomal DNA (Razin and Cedar 1977) and that high levels of methylated CpGs correlate with nuclease resistance and transcriptional inactivity of endogenous chromosomes (Antequera et al. 1989).

Since methylation of CpGs takes place at the replication fork within a minute or two (Gruenbaum et al. 1983), it seems likely that DNA methylation plays a role in directing the assembly of chromatin structure. This conclusion is supported by an experiment in which methylated and unmethylated templates were injected into *Xenopus* oocytes. Both templates showed equivalent activities, until chromatin was assembled. At that time, the methylated template lost its DNase I hypersensitivity, RNA polymerase was dissociated, and the activity of the promoter repressed (Kass et al. 1997).

A more stable structure of chromatin can be established by histone H1 which has a role in sealing the nucleosome and forming a higher-order chromatin structure (Thomas et al. 1979). How is 5-methylcytosine related to histone H1? Firstly, 5-methylcytosine is located preferentially in mononucleosomes that contain histone H1 (Ball et al. 1983). Secondly, filter binding assays show preferential binding of histone H1 to methylated DNA (Levine et al. 1993) and similar observations were obtained by sedimentation assays and gel retardation assays (McArthur and Thomas 1996). Thirdly, in vitro experiments clearly show that DNA methylation facilitates inhibition of transcription by histone H1 (Levine et al. 1993; Johnson et al. 1995). This interaction between histone H1 and methylated CpGs may stabilize a higher-order chromatin structure which is inaccessible to the transcriptional machinery. This conclusion is consistent with the unmethylated state of CpG islands and their being devoid of histone H1 (Tazi and Bird 1990) and may explain the open chromatin structure and high accessibility to transcription factors of promoters positioned within CpG islands. Although histone H1 is crucial to lock the nucleosome and achieve complete repression, a strong activator such as GAL4-VP16 can counteract this repression (Laybourn and Kadonaga 1991). However, GAL4-VP16 was unable to counteract the repression of histone H1 locked chromatin when the DNA was methylated (Kass et al. 1997). This observation clearly testifies that a stable repressed state of a gene requires a high order of chromatin structure, working in concert with methylation. This stable and inherited state of inactive chromatin can probably be reversed by demethylation (Weiss et al. 1996).

In this respect, it is interesting how an inactive state of a gene can be established and maintained in an organism such as *Drosophila melanogaster* which is devoid of CpG methylation (Urieli-Shoval et al. 1982). One strategy which is used in *Drosophila* to establish an inactive state of a gene, is by transposing the gene to a heterochromatic surrounding (Position effect variegation, see Chap. 11, this vol.). Another way, which might be more common, would be to establish a heterochromatic structure at the gene locus by specific protein mediators. However, it is clear that in mammals, DNA methylation and chromatin collaborate in establishing and maintaining the inactive status of genes.

4
Epigenetic Regulation of Imprinted Genes

What epigenetic mechanism accounts for the active and inactive states of two alleles of the same gene in a given cell? First, it is clear that such a mechanism has to provide a mark to distinguish between the two alleles. Second, it should allow silencing of one allele while the other allele remains active. Third, the epigenetic mark has to be clonally inherited for many cell cycles and reprogrammed in the germline to establish a proper imprint corresponding to the gender of the offspring. As discussed in the first part of this chapter, DNA methylation and chromatin structure could fulfill the requirements for such an epigenetic control mechanism.

4.1
Involvement of Chromatin Structure in Imprinting

Chromatin structure and its plasticity could be considered as a key element in determining the regulation of monoallelic expression of imprinted genes. A convincing testimony to this argument is obtained by examining the X chromosome inactivation phenomenon. Promoters on the inactive X chromosome appear to be included in well-positioned nucleosomes, whereas the same promoters in the inactive X chromosome are devoid of nucleosomes, and transcription factors are free to bind to them (Riggs and Pfeifer 1992). Thus, specific chromatin structures clearly appear to play a role in regulating differential gene activity.

Another feature of chromatin structure which may be essential for the imprinting mechanism is segregation of parental nucleosomes to the daughter chromatin structures during chromosomal duplication (Perry et al. 1993). Thus, the transcriptional potential of the gene activity is stably transmitted during cell division (Patterton and Wolffe 1996).

Also the fact that paternal and maternal pronuclei in the zygote are very different environments for transcription may be relevant to imprinting. Immediately following fertilization, early development in the mouse is directed by maternally inherited mRNA and protein. The differences between the pronuclei are revealed by microinjection of template DNA. Promoters are repressed when injected into either the paternal or maternal pronucleus. However, this transcriptional repression is not relieved by the inhibition of histone deacetylases if injection of the DNA was into the paternal pronucleus, but is alleviated when injection was into the maternal pronucleus (Wiekowski et al. 1993). Changes in the pronuclei chromatin structure are known to occur immediately following the two-cell stage (Patterton and Wolffe 1996). It is therefore possible that the chromatin structure of imprinted genes which is inherited from the gametes escapes these changes and is maintained in a parent-specific manner, similarly to DNA methylation, where discrete loci in imprinted genes maintain their methylation pattern which was inherited from the gametes (vide infra).

It should also be noted that the structure of the nucleus is complex and not homogenous with evidence for subnuclear domains (Cremer et al. 1993; see also Chap. 14, this vol.). It has already been proposed that separate nuclear territories may play a role in epigenetic phenomena such as position effect variegation in *Drosophila* (Schlossher et al. 1994).

4.2
Involvement of DNA Methylation in Establishing and Maintaining Monoallelic Expression of Imprinted Genes

DNA methylation, being an epigenetic mark on the DNA, can serve as a local signal to mark the alleles. DNA methylation is suitable to play this role since it is known to affect protein-DNA interactions and to be clonally inherited as discussed in Section 1.1. DNA methylation, being a part player in establishing and maintaining an inactive state of a gene, can also participate in allele-specific expression of the imprinted genes. In accordance, allele-specific methylation patterns are, in fact, a common feature of all imprinted genes for which methylation analysis has been studied (Razin and Cedar 1994; Shemer and Razin 1996). Many of the differentially methylated regions (DMRs) in imprinted genes are located within CpG islands, contrary to the general rule that CpG islands are unmethylated.

The process of imprinting should involve the establishment of the imprint during gametogenesis, its maintenance during embryonic development and in somatic cells of the adult, and erasure of the mark during gametogenesis before a new imprint can be established (Fig. 4). As mentioned above DNA methylation can indeed play a major role in the establishment and maintenance of imprints.

Primordial germ cells (PGCs) emerge from the epiblast of the early embryo and are first seen in the base of the allantois at 7.5 days postcoitum (dpc) (McCarrey 1993). By 11.5 dpc, the PGCs colonize the developing genital ridges, where the imprinted genes are completely unmethylated (Brandeis et al. 1993). Since imprinted genes in cells of the epiblast show differential methylation, it must be assumed that methylation is erased by 11.5 dpc when the PGCs enter the genital ridge. The capacity to erase this methylation has been found to exist in PGCs at least until 12.5 dpc (Tada et al. 1997). The absence of methylation in PGCs is in accord with the biallelic pattern of expression of imprinted genes which had been observed in postmigrating PGCs (Szabo and Mann 1995a).

The establishment of the new parent-specific methylation patterns must begin during gametogenesis or at the latest in the zygote before syngamy when the maternal and paternal genomes are still separate entities in the fertilized egg (Brandeis et al. 1993; Kono et al. 1996; Shemer et al. 1997). Whether the primary imprint signal is indeed methylation or rather chromatin structure that guides methylation is still an open question.

The process of establishing parent-specific methylation which starts during gametogenesis is often not completed until the stage of implantation. This may explain the biallelic expression of genes such as *Igf2* and *Igf2r* seen in the preimplantation embryo (Latham et al. 1994; Szabo and Mann 1995b). In contrast, the mouse *Snrpn* gene which inherits its complete methylation pattern from the gametes (Shemer et al. 1997) is monoallelically expressed throughout preimplantation and postimplantation embryo development (Szabo and Mann 1995b).

A number of sites which inherit their methylation status from the gametes do not escape the global demethylation event that takes place in the preimplantation embryo. However, their unmethylated state is very brief and they immediately undergo allele-specific de novo methylation (Shemer et al. 1996). Similarly, all differentially methylated sites survive the wave of de novo methylation which occurs in the postimplantation embryo, implying that also the postimplantation embryo has the capacity to discriminate the two alleles of the imprinted sequences.

Several lines of evidence indicate that differential methylation must start in the gametes and cannot be initiated in embryonic cells. DNA methyltransferase (Dnmtase)-deficient ES cells show aberrant expression of imprinted genes, and die upon induction of differentiation (Tucker et al. 1996; Shemer et al. 1997). These cells can be rescued by the introduction of the Dnmtase gene, which restores the overall methylation level to normal, but does not remethylate the imprinted genes. Normal monoallelic methylation and expression of the imprinted genes is only restored when passed through the germ line (Tucker et al. 1996). Accordingly, transfection of the *Snrpn* gene into ES cells or injection into pronuclei of fertilized eggs reveal that embryonic cells lack the capacity to establish the differential methylation pattern of *Snrpn* (Shemer et al. 1997).

These observations indicate that imprinted genes can only be recognized and marked in the germline. This mark is then used by the early embryo for spreading the correct methylation pattern, which is used to maintain the monoallelic expression throughout development and adult life. This model is corroborated by gene expression experiments in Dnmtase-deficient mice. In these mice *H19*, *Snrpn*, *Znf127*, and *Xist* were expressed biallelically while *Igf2r* and *Igf2* were completely shut off (Li et al. 1993; Shemer et al. 1997; unpubl. data). The mode of expression of *H19* and *Igf2* in Dnmtase-deficient mice are in accordance with the proposed enhancer competition model (Leighton et al. 1995; see also Chap. 5, this vol.) and the repression of *Igf2r* in these mice can be explained by the recent observation of *Igf2r* antisense RNA (Wutz et al. 1997; see also Chap. ••, this vol.).

5
Concluding Remarks

Epigenetic regulation of gene expression involves alteration of chromatin structure without change in DNA sequence. Cytosine methylation is another epigenetic tool used in the control of gene expression. DNA methylation in most cases seems to stabilize and maintain a chromatin structure that determines an inactive status of a gene. It has clearly been shown that DNA methylation is essential for postimplantation embryo development, as well as in genomic imprinting. In the process of imprinting, DNA methylation plays a dual role: specific methylated CpG sites mark the parental chromosome in a process that takes place during gametogenesis and the final differentially methylated pattern, presumably in concert with chromatin structure, allows gene function to be adjusted according to parental origin.

Chromatin structure is the epigenetic model favored to explain somatic inheritance of yeast mating type, and position-effect variegation in *Drosophila*, in spite of the fact that the DNA of both these organisms has no detectable 5-methylcytosine. Yet, the mechanism involved in heritability of chromatin structure remains an open question. The inheritance of mammalian chromatin structure may be guided by specific protein factors, DNA methylation, or by a combination of both. *Trans*-acting factors can recognize the methylated CpG sites which, by themselves, are clonally inherited by a maintenance methyltransferase that methylates the nascent DNA strand promptly after replication.

References

Antequera F, Macleod D, Bird AP (1989) Specific protection of methylated CpGs in mammalian nuclei. Cell 58:509–517

Archer TK, Lefebre P, Wolford RG, Hager GL (1992) Transcription factor locking on the MMTV promoter: a bimodal mechanism for promoter activation. Science 255:1573–1576

Ball DJ, Gross DS, Garrard WT (1983) 5-methylcytosine is localized in nucleosomes that contain histone H1. Proc Natl Acad Sci USA 80:5490–5494

Benvenisty N, Mencher D, Meyuhas O, Razin A, Reshef L (1985) Methylation of cytosolic PEPCK gene: pattern associated with tissue specificity and development. Proc Natl Acad Sci USA 82:267–271

Bird AP (1986) CpG islands and the function of DNA methylation. Nature 321:209–213

Boyes J, Bird A (1991) DNA methylation inhibits transcription indirectly via a methyl-CpG binding protein. Cell 64:1123–1134

Boyes J, Bird A (1992) Repression of genes by DNA methylation depends on CpG density and promoter strength: evidence for involvement of a methyl-CpG binding protein. EMBO J 11:327–333

Brandeis M, Kafri T, Ariel M, Chaillet JR, McCarrey J, Razin A, Cedar H (1993) The ontogeny of allele-specific methylation associated with imprinted genes in the mouse. EMBO J 12:3669–3677

Brandeis M, Frank D, Keshet I, Siegfried Z, Mendelsohn M, Nemes A, Temper V, Razin A, Cedar H (1994) Sp1 elements protect a CpG island from de novo methylation. Nature 371:435–438

Bresnick EH, Bustin M, Marssand V, Richard-Foy H, Hager GL (1992) The transcriptionally-active MMTV promoter is depleted of histone H1. Nucleic Acids Res 20:273–278

Buschhausen G, Wittig B, Graessmann M, Graessmann A (1987) Chromatin structure is required to block transcription of the methylated herpes simplex virus thymidine kinase gene. Proc Natl Acad Sci USA 84:1177–1181

Busslinger M, Hurst J, Flavell RA (1983) DNA methylation and the regulation of globin genes expession. Cell 34:197–206

Chung JH, Whiteley M, Felsenfeld G (1993) A 5' element of the chicken β globin domain serves as an insulator in human erythroid cells and protects against position effect in *Drosophila*. Cell 74:505–514

Conklin KF, Groudine M (1984) Chromatin structure and gene expression. In: Razin A, Cedar H, Riggs AD (eds) DNA methylation biochemistry and biological significance. Springer, Berlin Heidelberg New York, pp 293–351

Cremer T, Kurz A, Zirbel R, Dietzel S, Rinke B, Schrock E, Speicher MR, Mathieu U, Jauch A, Emmerich P, Scherthan H, Ried T, Cremer C, Lichter P (1993) Role of chromosome territories in the functional compartmentalization of the cell nucleus. Cold Spring Harbor Symp Quant Biol 58:777–792

Gruenbaum Y, Cedar H, Razin A (1982) Substrate and sequence specificity of a eukaryotic DNA methylase. Nature 295:620–622

Gruenbaum Y, Szyf M, Cedar H, Razin A (1983) Methylation of replicating and postreplicated mouse L cell DNA. Proc Natl Acad Sci USA 80:4919–4921

Huang L-H, Wang R, Gama-Sosa MA, Shenoy S, Ehrlich M (1984) A protein from human placental nuclei binds preferentially to 5-methylcytosine-rich DNA. J Biol Chem 262:293–295

Johnson CA, Goddard JP, Adams RLP (1995) The effect of histone H1 and DNA methylation on transcription. J Biochem 305:791–798

Jones PA, Taylor SM, Mohandas T, Shapiro LJ (1982) Cell cycle specific reactivation of an inactive X-chromosome locus by 5azadeoxycytidine. Proc Natl Acad Sci USA 79:1215–1219

Jost J-P, Hofsteenge J (1992) The repressor MDBP-2 is a member of histone H1 family that binds preferentially to methylated DNA in vitro and in vivo. Proc Natl Acad Sci USA 89:9499–9503

Kafri T, Ariel M, Brandeis M, Shemer R, Urven K, McCarrey J, Cedar H, Razin A (1992) Developmental pattern of gene-specific DNA methylation in the mouse embryo and germline. Genes Dev 6:705–714

Kafri T, Gao X, Razin A (1993) Mechanistic aspects of genome-wide demethylation in the preimplantation mouse embryo. Proc Natl Acad Sci USA 90:10558–10562

Kass SU, Landsberger N, Wolffe AP (1997) DNA methylation directs a time-dependent repression of transcription initiation. Curr Biol 7:157–165

Keshet I, Yisraeli J, Cedar H (1985) Effect of regional DNA methylation on gene expression. Proc Natl Acad Sci USA 82:2560–2564

Keshet I, Lieman-Hurwitz I, Cedar H (1986) DNA methylation affects the formation of active chromatin. Cell 44:535–543

Kono T, Obata Y, Yoshimzu T, Nakahara T, Carroll J (1996) Epigenetic modifications during oocyte growth correlates with extended parthenogenetic development in the mouse. Nature Genet 13:91–94

Kruczek I, Doerfler W (1983) Expression of the chloramphenicol acetyltransferase gene in mammalian cells under the control of retinovirus type 12 promoter: effect of promoter methylation on gene expression. Proc Natl Acad Sci USA 80:7586–7590

Latham KE, Doherty AS, Scott CD, Schultz RM (1994) Igf2r and Igf2 gene expression in androgenetic, gynogenetic, and parthenogenetic preimplantation mouse embryos: absence of regulation by genomic imprinting. Genes Dev 8:290–299

Laybourn PJ, Kadonaga JT (1991) Role of nucleosomal cores and histone H1 in regulation of transcription by RNA polymerase II. Science 254:238–245

Leighton PA, Saam JR, Ingram RS, Stewart CL, Tilghman SM (1995) An enhancer deletion affects both H19 and Igf2 expression. Genes Dev 9:2079–2089

Levine A, Cantoni GL, Razin A (1991) Inhibition of promoter activity by methylation: possible involvement of protein mediators. Proc Natl Acad Sci USA 88:6515–6518

Levine A, Cantoni GL, Razin A (1992) Methylation in the preinitiation domain suppresses gene transcription by an indirect mechanism. Proc Natl Acad Sci USA 88:6515–6518

Levine A, Yeivin A, Ben-Asher E, Aloni Y, Razin A (1993) Histone H1-mediated inhibition of transcription initiation of methylated templates in vitro. J Biol Chem 268:21754–21759

Lewis JD, Meehan RR, Henzel WI, Manver-Fogy I, Jeppesen P, Klein F, Bird AP (1992) Purification sequence and cellular localization of a novel chromosomal protein that binds to methylated DNA. Cell 69:905–914

Li E, Beard C, Jaenisch R (1993) Role for DNA methylation in genomic imprinting. Nature 366:362–365

Lichtenstein M, Keini G, Cedar H, Bergman Y (1994) B cell-specific demethylation. A novel role for the intronic k chain enhancer sequences. Cell 76:913–923

Macleod D, Charlton J, Mullins I, Bird AP (1994) Sp1 sites in the mouse aprt gene promoter are required to prevent methylation of the CpG island. Genes Dev 8:2282–2292

McArthur M, Thomas JD (1996) A preference of histone H1 for methylated DNA. EMBO J 15:1705–1714

McCarrey JR (1993) Development of the germ cell. In: Desjardinsand C, Ewing LL (eds) The testis: cell and molecular biology of the testis, vol. 5, Chap. 3. Oxford University Press, New York

Meehan RR, Lewis JD, McKay S, Kleiner EL, Bird AP (1989) Identification of a mammalian protein that binds specifically to DNA containing methylated CpGs. Cell 58:499–507

Meehan RR, Lewis JD, Bird AP (1992) Characterization of MeCP2, a vertebrate DNA binding protein with affinity for methylated DNA. Nucleic Acids Res 20:5085–5092

Monk M, Boubelik M, Lehnert S (1987) Temporal and regional changes in DNA methylation in the embryonic, extraembryonic and germ cell lineages during mouse embryo development. Development 99:371–382

Nan X, Tate P, Li E, Bird AP (1996) DNA methylation specifies chromosomal localization of MeCP2. Mol Cell Biol 16:414–421

Nan X, Campoy J, Bird AP (1997) MeCP2 is a transcriptional repressor with abundant binding sites in genomic chromatin. Cell 88:471–481

Paroush Z, Keshet I, Yisraeli I, Cedar H (1990) Dynamics of demethylation and activation of the α actin gene in myoblasts. Cell 63:1229–1237

Patterton D, Wolffe AP (1996) Developmental roles for chromatin and chromsomal structure. Dev Biol 173:2–13

Pennings S, Meersseman G, Bradbury ME (1994) Linker histones H1 and H5 prevent the mobility of positioned nucleosomes. Proc Natl Acad Sci USA 91: 10275–10279

Perry CA, Allis CD, Annunziato AT (1993) Parental nucleosomes segregated to newly replicated chromatin are underacetylated relative to those assembled de novo. Biochemistry 32:13615–13623

Pina B, Brüggemeier U, Beato M (1990) Nucleosome positioning modulates accessibility of regulatory proteins to the mouse mammary tumor virus promoters. Cell 60:719–731

Pollack Y, Stein R, Razin A, Cedar H (1980) Methylation of foreign DNA sequences in eukaryotic cell. Proc Natl Acad Sci USA 77:6463–6467

Razin A (1998) CpG methylation, chromatins structure and gene silencing a three-way connection. EMBO J 17:4905–4908

Razin A, Cedar H (1977) Distribution of 5methylcytosine in chromatin. Proc Natl Acad Sci USA 74:2725–2728

Razin A, Cedar H (1994) DNA methylation and genomic imprinting. Cell 77:473–476

Razin A, Kafri T (1994) DNA methylation from embryo to adult. Prog Nucleic Acids Res Mol Biol 48:53–81

Razin A, Szyf M, Kafri T, Roll M, Giloh H, Scarpa S, Carotti D, Cantoni GL (1986) Replacement of 5 methylcytosine by cytosine: a possible mechanism for transient demethylation during differentiation. Proc Natl Acad Sci USA 83:2827–2831

Riggs AD, Pfeifer GP (1992) X-chromosome inactivation and cell memory. Trends Genet 8:169–174

Schaffer CD, Wallrath LL, Elgin SCR (1993) Regulating genes by packaging domains: bits of heterochromatin in euchromatin. Trends Genet 9:35–38

Schlossher J, Egert H, Paro R, Cremer S (1994) Gene inactivation in *Drosophila* mediated by the polycomb gene product or by position-effect variegation does not involve major changes in the chromatin fibre. Mol Gen Genet 243:453–462

Shemer R, Razin A (1996) Establishment of imprinted methylation patterns during development. In: Russo VEA, Martienssen RA, Riggs AD (eds) Epigenetic mechanisms of gene regulation. CSHL Press, New York, USA pp 215–229

Shemer R, Eisenberg S, Breslow JL, Razin A (1991) Methylation patterns of the human ApoAI-CIII-AIV gene cluster in adult and embryonic tissue suggest dynamic changes in methylation during development. J Biol Chem 266:23676–23681

Shemer R, Birger Y, Dean WL, Reik W, Riggs AD, Razin A (1996) Dynamic methylation adjustment and counting as part of imprinting mechanisms. Proc Natl Acad Sci USA 93:6371–6376

Shemer R, Birger Y, Riggs AD, Razin A (1997) A Structure of the imprinted mouse *Snrpn* gene and establishment of its parental-specific methylation pattern. Proc Natl Acad Sci USA 94:10267–10272

Sinsheimer RL (1955) The action of pancreatic deoxyribonuclease II. Isometric dinucleotides. J Biol Chem 215:579–583

Smith HO, Kelly SV (1994) Methylases of the type II restriction modification systems in DNA methylation. In: Razin A, Cedar H, Riggs AD (eds) Biochemistry and biological significance. Springer, Berlin Heidelberg New York, pp 66–71

Sogo LR (1995) Replication of transcriptionally active chromatin. Nature 374:276–280

Stein R, Razin A, Cedar H (1982) In vitro methylation of the hamster adenine phosphoribosyltransferase gene inhibits its expression in mouse L-cells. Proc Natl Acad Sci USA 79:3418–3422

Stoger R, Kubicka P, Liu C-G, Kafri T, Razin A, Cedar H, Barlow DP (1993) Maternal-specific methylation of the imprinted mouse *Igf2/M6pr* locus identifies the expressed locus as carrying the imprinting signal. Cell 73:61–71

Szabo PE, Mann RJ (1995a) Biallelic expression of imprinted genes in the mouse germ line: implications for erasure, establishment, and mechanisms of genomic imprinting. Genes Dev 9:1857–1868

Szabo PE, Mann RJ (1995b) Allele-specific expression and total expression levels of imprinted genes during early mouse development: implications for imprinting mechanisms. Genes Dev 9:1–12

Tada M, Tada T, Lefebvre L, Barton SC, Surani MA (1997) Embryonic germ cells induce epigenetic reprogramming of somatic nucleus in hybrid cells. EMBO J 16:6510–6520

Tate P, Bird AP (1993) Effects of DNA methylation on DNA binding proteins and gene expression. Curr Opin Genet Dev 3:226–231

Tate P, Skarnes W, Bird AP (1996) The methyl-CpG binding protein, MeCP2 is essential for embryonic development in the mouse. Nat Genet 12:205–208

Tazi J, Bird A (1990) Alternative chromatin structure at CpG islands. Cell 60:909–920

Thomas F, Koller T, Klug A (1979) Involvement of histone H1 in the organization of the nucleosome and of the salt-dependent superstructure of chromatin. J Cell Biol 83:403–427

Tucker KL, Beard C, Dansman J, Jackson-Grusby L, Laird PW, Lei H, Li E, Jaenisch R (1996) Germline passage is required for establishment of methylation and expression patterns of imprinted but not nonimprinted genes. Genes Dev 10:1008–1020

Urieli-Shoval S, Gruenbaum Y, Sedat J, Razin A (1982) The absence of detectable methylated bases in *Drosophila melanogaster* DNA. FEBS Lett 146:148–152

Vardimon L, Kressman A, Cedar H, Maechler M, Doerfler W (1982) Expression of a cloned adenovirus gene is inhibited by in vitro methylation. Proc Natl Acad Sci USA 79:1073–1077

Wang R-H, Zhang X-Y, Khan R, Zhou Y, Huang L-H, Ehrlich M (1986) Methylated DNA binding protein from human placenta recognizes specific methylated sites on several prokaryote DNAs. Nucleic Acids Res 14:9843–9860

Watt F, Molloy PL (1988) Cytosine methylation prevents binding to DNA of a Hela cell transcription factor required for optimal expression of the adenovirus major late promoter. Genes Dev 2:1136–1143

Weiss A, Keshet I, Razin A, Cedar H (1996) DNA demethylation in vitro: involvement of RNA. Cell 86:709–718

Wiekowski M, Miranda M, DePamphilis ML (1993) Requirements for promoter activity in mouse oocytes and embyros distinguish paternal pronuclei from maternal and zygotic nuclei. Dev Biol 159:366–378

Wigler M, Levy D, Perucho M (1981) The somatic replication of DNA methylation. Cell 24:33–40

Wolffe AP (1994) Transcription: in tune with the histones. Cell 77:13–16

Wutz A, Smrzka OW, Schweifer N, Schellander K, Wagner EF, Barlow DP (1997) Imprinted expression of the *Igf2r* gene depends on an intronic CpG island. Nature 389:745–749

Yeivin A, Razin A (1993) Gene methylation patterns and expression in DNA methylation. In: Jost JP, Salug HP (eds) Molecular biology and biological significance, Birkhäuser, Basel, pp 523–568

Yisraeli J, Frank D, Razin A, Cedar H (1988) Effect of in vitro DNA methylation on β globin gene expression. Proc Natl Acad Sci USA 85:4638–4642

Polycomb Silencing and the Maintenance of Stable Chromatin States

Vincenzo Pirrotta

1
Introduction

In the course of development, the expression pattern of many genes is regulated, refined, and restricted to generate differential distributions of gene expression essential for the patterning, morphogenesis and allocation of cellular identities in different parts of the embryo. These patterns are produced by the interaction of multiple transcription factors, repressor and activators present in different parts of the embryo or induced by diffusible factors or cell-cell communication. The signals are often transient, since they depend on the situations existing at the time developmental decisions are made, yet the ensuing pattern of gene expression, characteristic of different determined states, must be maintained in later development by the tissues and the cells that constitute them.

Chromatin mechanisms are now revealing how complex regulatory information can not only control gene expression but preserve a memory of earlier states and transmit it to cellular progeny or even in some cases to the next generation. Because these states are metastable and heritable but do not involve permanent changes in the DNA sequence, they are epigenetic mechanisms to convey genetic information. Unlike the simple activation of a gene by the direct binding of a *trans*-acting factor to a specific DNA sequence, chromatin mechanisms have less well-defined targets and control the state, activity, or properties of larger chromatin domains. Examples of such mechanisms are telomeric and mating type silencing in yeast, and the silencing induced by centric heterochromatin, in subtelomeric regions or at certain developmentally regulated loci. This chapter deals primarily with the last of these, in which decisions whether to express or repress a gene or set of genes determined at an early developmental state are maintained in the clonal descendants in a way that resembles closely the behavior of genes that are inserted in the vicinity of centric heterochromatin or of telomeres. In all these cases, a metastable state is established in the chromatin of the genes affected

Department of Zoology, University of Geneva, 30 quai Ernest Ansermet, CH1211 Geneva, Switzerland

which results in silencing to different degrees in different cells. The relatively stable, heritable silenced state is trasmitted to the cellular progeny for many cell divisions but can occasionally be released in some individual cells giving rise to clones of derepressed progeny. This can cause a characteristically variegated pattern of expression in which the degree of variegation is strongly dependent on the available supply of the chromatin proteins responsible.

Though we usually think of these as silencing mechanisms, it is unlikely that this is the primary purpose at telomeres or in heterochromatin, where they more likely have multiple roles related to chromosome mechanics, nuclear architecture, and centromere function. The similarities of the molecular mechanisms involved in these larger-scale phenomena and those involved in developmental gene silencing suggest that these too affect and are dependent in turn on the larger-scale organization of the genetic material in the nucleus. Thus, determined cell states, tissue differentiation, and stable patterns of gene expression might be associated with changes in nuclear architecture. Recent experiments support this surmise and reveal surprising consequences. The best-known case of developmental gene silencing is that of the homeotic genes in Drosophila and it is with them that this chapter must begin.

2
Homeotic Genes and Silencing in Drosophila

In the early embryo, homeotic genes are activated primarily in response to segmentation gene products which establish the embryonic segmental domains. The products of these genes are transcriptional regulators such as Fushi tarazu (FTZ) or Even-skipped (EVE) which activate homeotic genes by binding to sets of enhancer modules (Simon et al. 1990; Müller and Bienz 1991; Qian et al. 1991; Pirrotta et al. 1995). An essential feature of homeotic genes is that their expression is confined to specific segmental domains to which they will confer their segmental identity. Thus the expression of *Ultrabithorax* (*Ubx*) is restricted to the region posterior to parasegment 5 by the Hunchback (HB) segmentation protein, which is present in the anterior half of the early embryo. HB represses *Ubx* expression in this region by competing with the activators for binding to overlapping binding sites in the enhancer modules (Qian et al. 1991, 1993; Zhang et al. 1991). These interactions define the segmental pattern of *Ubx* expression for the first 4–5 h of embryonic development. At this time, the HB repressor in the anterior half stops being produced, yet the *Ubx* gene must continue to be restricted to its initial segmental domain for the rest of development when other enhancer elements activate its expression in other tissues. Thus, each cell must preserve a memory of whether its *Ubx* gene had been activated or repressed in the early embryo and the memory must be transmitted to the cellular progeny. The maintenance of the repressed state and of this cellular memory is due to the Polycomb Group (PcG) of proteins. These proteins are present in all cells but they repress *Ubx* only in those cells in which it was initially silent while they allow its expression in cells in which it had been activated in the early embryo.

The mechanism by which PcG proteins maintain repression bears similarities to heterochromatic silencing mechanisms in yeast and in Drosophila. Like heterochromatic silencing, PcG repression is a metastable state that is inherited epigenetically by progeny cells. It involves multiple components that are thought to interact to form silencing complexes. In all three cases, the establishment or maintenance of the silent state has an all-or-none quality that can cause variegated expression of the genes affected. While the silencing of homeotic genes is normally quite stable, it is sensitive to a reduction in the dosage of the PcG genes. Mutations of PcG genes therefore frequently have a dominant effect, causing derepression in particularly sensitive tissues or in clonal patches of cells, veritable homeotic tumors, which develop as homologous structures characteristic of more posterior segments. Antennal cells develop as legs; head or eye cells as thorax or wing. That is, the relatively stable, heritable repressive state may be established in some cells of a population but not others or, once established, may occasionally be released in some individual cells giving rise to clones of derepressed progeny.

This clonal derepression caused by lower levels of PcG proteins is strongly suggestive of an all-or-none mechanism and is the equivalent of the variegated expression observed in heterochromatic position-effect variegation (PEV). When a translocation places a chromosome segment in the vicinity of centric heterochromatin, the cytological heterochromatic appearance spreads to invade the euchromatic segment with a concomitant silencing of the genes it contains (Reuter and Spierer 1992). A silenced gene is inactivated in a mosaic fashion so that it may still be active in clonal patches of cells while it is silenced in other patches. PEV is relieved by mutations in a group of genes, the *Suppressor of variegation* or *Suvar* genes, a number of which have now been shown to encode proteins associated with heterochromatin. The Drosophila PcG and heterochromatic silencing systems appear to be independent but their effects and probably their mechanisms are very similar and both involve many proteins whose level determines the extent of the silenced chromatin region and the degree or stability of silencing.

In yeast, genes introduced in the subtelomeric region are subject to a heritable, metastable silencing that results in variegated colonies and is equivalent to the Drosophila PEV. Telomeric silencing is caused by a multiprotein complex assembled from the telomeres and involving the SIR2, 3, and 4 proteins (Sherman and Pillus 1997). The same silencing proteins are used to repress the silent mating type genes, which could be considered the yeast equivalent of the fly homeotic genes. The comparison of the Drosophila and yeast silencing systems has led to the concept that PcG protein complexes associate with silenced homeotic genes, spread cooperatively to involve large regions, taking advantage of the fact that homeotic genes are clustered and arranged in an order such that derepression always progresses from one end of the cluster.

For example, in the bithorax complex, all three genes, *Ubx*, *abd-A*, and *Abd-B*, are silent in the anterior half of the embryo and become sequentially derepressed as we move towards the posterior end of the embryo. Since heterochromatin is highly condensed, late replicating and, in Drosophila polytenic tissues, strongly underreplicated, the comparison has encouraged the view that PcG silencing involves the packaging of chromatin in a condensed state that makes its DNA inaccessible to the transcriptional machinery.

3
PcG Genes and Their Interactions

At least 13 PcG genes have been identified in Drosophila by the effects of their mutations in derepressing homeotic genes. Many more genes affect the stability of homeotic gene silencing and the precise definition of the group remains somewhat vague (Jürgens 1985). Most of the 13 better characterized genes have been cloned and both the genetic and biochemical properties of the proteins suggest that they act together, most likely by forming multiprotein complexes. The function of this group of genes is very sensitive to their dosage. In most cases a mutation inactivating one copy of a PcG gene results in partial derepression of the homeotic genes. Mutations in different PcG genes have synergistic effects and, in some cases, additional copies of one gene can compensate for the decrease in dosage of another (Cheng et al. 1994). Increasing evidence indicates that the PcG silencing system is substantially conserved in vertebrates. Several of the Drosophila genes have mammalian counterparts sharing considerable homology in addition to significant structural motifs (Schumacher and Magnuson 1997). However, no similarity can be detected between PcG genes and the yeast silencing proteins.

The predicted amino acid sequences of the PcG proteins shows that they constitute a heterogeneous group, not a family of genes. Collectively, they contain a set of structural motifs that are found in other chromatin proteins and are implicated in protein-protein interactions. The Polycomb protein (PC) contains a chromodomain (Paro and Hogness 1991), first identified by comparison of the PC sequence with that of the heterochromatin protein HP1, the product of the *Suvar (2)5* gene involved in PEV in *Drosophila*. Well-conserved *Pc* homologues have been identified in *Xenopus*, mouse, and man and the mouse homologue, *M33*, can rescue in part the effect of *Pc* mutations in the fly (Müller et al. 1995; Reijnen et al. 1995). The PSC and SU(Z)2 proteins contain a region of homology to one another consisting of a RING finger motif and a helical region, also found in a number of mammalian nuclear proteins (Brunk et al. 1991). Two of these in particular, the mouse Bmi-1 and Mel-18 oncoproteins, are implicated in the regulation of *Hox* genes since their loss of function results in posterior homeotic transformations of the axial skeleton while overexpression causes anterior transformations (van der Lugt et al. 1994; Akasaka et al. 1996). The SCM and PH proteins are related and share a zinc finger motif and another C-terminal domain, SPM, conserved in the mouse Rae28 protein (Bornemann et al. 1996). The E(Z) protein has two human homologues and contains a SET domain, found also in a heterochromatin protein, SUVAR(3)9 and in the TRX protein (Jones and Gelbart, 1993; Laible et al. 1997). The ESC protein consists mostly of a set of WD40 repeats, a motif found in many other proteins, not all of them nuclear, and thought to mediate protein-protein interactions (Gutjahr et al. 1995; Sathe and Harte, 1995; Simon et al. 1995). A mouse homologue of *esc* is *eed*, essential in early development even before gastrulation but causing also posterior axial transformations in hypomorphic mutants (Schumacher et al. 1996). In *Drosophila*, *esc* function is required only in the first few hours of development, when PcG silencing is thought to be established, and may therefore be involved in the initial assembly of the complex.

Direct interactions between PcG proteins were confirmed by the fact that immunoprecipitation of one PcG protein brings down a large complex containing other members of the group (Franke et al. 1992). More recently, direct interactions of PSC with PC and PH proteins have been demonstrated using the yeast two-hybrid system and *in vitro* by co-immunoprecipitation of the proteins expressed in bacteria (Kyba and Brock, 1998). Similar *in vitro* interactions have been demonstrated between the *Xenopus* PC and the mouse Bmi-1 protein (Reijnen et al. 1995). While the anatomy of some PcG complexes is beginning to be elucidated, the situation is likely to be complicated. Antibody staining of polytene chromosomes allows the visualization of some 100 cytological loci that are binding sites for PcG proteins; many of these are common to several proteins but some sites appear to bind some but not other members of the group (Franke et al. 1992; Rastelli et al. 1993; Martin and Adler, 1993). The binding of PC, PSC, SU(Z)2, and PH proteins at most, but not all, of these cytological sites is abolished by mutations in the *E(z)* gene. Genetic evidence also shows that some of the proteins are implicated in processes in which other PcG proteins are not involved (Pelegri and Lehmann, 1994). While these results suggest that the formation of stable PcG complexes at most sites requires the participation of a whole set of PcG proteins, they also imply that this is not always the case and that the composition of PcG complexes can vary from one target site to another. In particular, PSC and SU(Z)2, but not PC or PH, are associated with telomeric regions. In recent experiments, the insertion of a reporter gene in a subtelomeric region resulted in its variegated expression which depended on the *Psc* and *Su(z)2* genes but not on *Pc* or on classical *Su(var)* mutations (L. Wallrath and S. Elgin, pers. comm.).

A second set of genes, the trithorax Group (trxG) is often thought to be the positive counterpart of the PcG genes. Mutations in trxG genes counteract the effects of PcG mutations by reducing the level of expression of homeotic genes and this effect was in fact used in genetic screens to identify many of the members of this group. Many of these genes are likely therefore to represent components that stimulate gene expression in a variety of ways not necessarily related to the action of PcG proteins. The trxG includes *brahma*, whose product is homologous to the yeast SWI2/SNF2 protein and to mammalian BRG1 and like them is part of an ATP-dependent chromatin remodeling complex that facilitates access of transcriptional activators. Other members of the group include *trx, ash1, zeste,* and *Trl*. TRX and ASH1 are large proteins containing a SET domain (Tripoulas et al. 1996), also found in the PcG protein E(Z) and in SUVAR(3)9. Specific DNA binding or enzymatic activities have not been demonstrated. A TRX mammalian homologue, MLL, causes anterior homeotic transformations in knockout mice, formally equivalent to trx mutations in the fly, and in man, MLL is implicated in infantile leukemias (Gu et al. 1992; Yu et al. 1995). The *Trl* product is the GAGA factor (Farkas et al. 1994), first identified as a transcriptional activator binding to various promoter regions, including the *Ubx* and heat shock promoters (Biggin and Tjian, 1988; Lee et al. 1992). GAGA factor contains a zinc finger and a POZ domain, reponsible for protein-protein interactions, and has specific DNA-binding activity in vitro. Purified GAGA factor bound to a chromatin template has been shown to direct the nucleosome remodeling activity of a multisubunit complex called NURF (Tsukiyama et al. 1994). The *zeste* gene encodes a specific DNA-binding protein associated with pairing -dependent phenomena generally grouped under the name transvection. A particular mutation, z^1 causes a pairing-dependent silencing of the *white* gene which is suppressed by mutations in some PcG genes such as *Psc, Su(z)2,* or *E(z)* (Pirrotta and Rastelli, 1994). Each of these trxG proteins is found at 100 or more sites on polytene chromosomes, often but not always corresponding to PcG sites (Rastelli et al. 1993; Chinwalla et al. 1995). These connections with PcG silencing suggest the possibility that at least some trxG proteins act in concert, in association with or antagonistically to the PcG proteins, but a direct interaction has not yet been demonstrated (however, see below for further evidence).

4
Polycomb Response Elements

The silencing of the *Ubx* gene encompasses a region of more than 100 kb (Fig. 1), containing multiple enhancers scattered over nearly the same length, some upstream and some downstream of the promoter (Pirrotta et al. 1995). Two other homeotic genes are part of bithorax cluster, *abd-A* and *Abd-B*, which are active in more posterior regions of the embryo. The *abd-A* and *Abd-B* genes are also subject to PcG silencing, so that in cells of the anterior half of the embryo, the entire cluster is repressed, most likely increasing the stability of the silenced state.

When the isolated early enhancers of the *Ubx* gene are tested in a reporter construct consisting of the *Ubx* promoter driving the expression of the *lacZ* gene, they initiate their pattern of expression correctly, and are correctly silenced by the HB repressor in the anterior part of the embryo. Shortly after

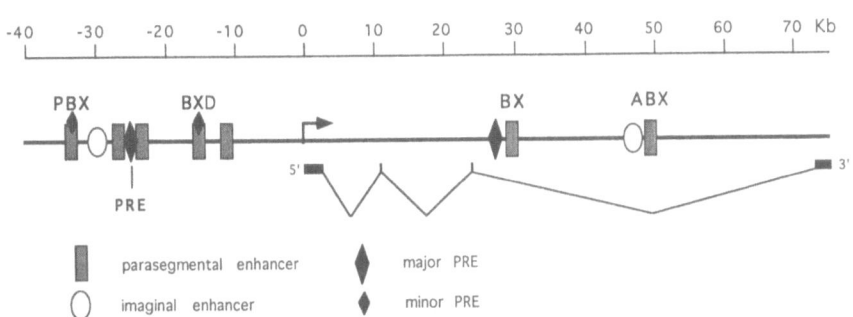

Fig. 1. Regulatory elements of the Ultrabithorax gene. The Ubx gene contains multiple enhancer modules located at great distances from the promoter. Some of these are activated early during embryonic development, others much later to specify adult structures (imaginal enhancers). In addition to two major PREs, two minor sites with weak PRE activity have been identified, though more probably exist

gastrulation, however, when the HB repressor fades away, they become depressed and expression extends to all segments, indicating that they are unable to recruit or respond to the PcG silencing system. Sequences conferring the ability to respond to PcG silencing, the Polycomb response element or PRE, were identified in two major sites of the *Ubx* gene, one about 24 kb upstream of the promoter and one 30 kb downstream (Chan et al. 1994; Chiang et al. 1995). Detailed analysis of the *Ubx* regulatory region has shown that other, weaker PRE-like sequences exist within the gene and, though unable to initiate silencing by themselves, they can cooperate with one another and probably contribute to the stability of silencing of the entire gene (Müller and Bienz, 1991; Poux et al. 1996). PREs have been identified in other homeotic genes, in the *engrailed* gene, in the *escargot* gene and even in the promoter region of *ph*, itself one of the PcG genes, suggesting that PcG genes may regulate their own expression (Zink and Paro 1989; Fauvarque and Dura 1993; Kassis 1994; Gindhart and Kaufman 1995).

When a PRE is included in a *Ubx-lacZ* reporter construct under the control of an early *Ubx* enhancer, the correct segmental expression pattern is maintained throughout embryonic development even at later stages when the HB repressor is no longer present (Chan et al. 1994). Experiments with *Ubx* enhancers activated later in embryonic development suggest that effective PRE silencing sets in just after gastrulation though many of the PcG products are present earlier and have in fact a maternal component. If the construct is driven by a late-acting *Ubx* enhancer, the PRE prevents its activation but, if both an early and a late enhancer are present, the late enhancer is not completely silenced but functions only in those segments of the embryo in which the early enhancer had previously activated the gene (Poux et al. 1996). It appears then that early transcriptional activity protects the gene from silencing, preserving the open or potentially active state.

The presence of a PRE in a transposon can affect multiple genes contained in the construct, indicating that the silencing effect operates over a distance of many kilobases. In some cases the PRE in the transposon has been shown to affect also genes flanking its insertion site. The silencing caused by the PRE is most easily visualized from its effect on the expression of the *white* gene, which is usually included in the transposon constructs as a transformation marker. The *white* gene is required for the pigmentation of the eye, which serves therefore as a sensitive indicator of many PRE properties. When silencing occurs, the *white* gene can be completely repressed but more often repression is partial, resulting in a characteristic streaky, sectored or salt-and-pepper eye pigmentation (Fauvarque and Dura 1993; Chan et al. 1994; Kassis 1994). This variegated expression is very similar to heterochromatic PEV, suggesting that the repressed state is established early in development when the primordia of the eye consist of a few cells, and is then maintained in the clonal descendants of the repressed cells. The salt-and-pepper variegation suggests that in these cases the repression is unstable and derepression occurs in random cells at various times in later development.

PREs are physical targets of the PcG proteins as shown by the fact that the silencing they induce is dependent on PcG genes and that PcG protein binding can be detected on polytene chromosomes at the insertion site of the transposon that contains them. DNA fragments containing PRE activity are generally over 1 kb in length. When they are dissected, residual PRE activity is found in several subfragments, indicating that multiple sequences contribute to the recruitment and stability of a PcG complex. Moreover, the *Ubx* PRE also contains a target site for binding of TRX, as shown both by staining of polytene chromosomes and by a tissue culture transfection assay (Chang et al. 1995; Chinwalla et al. 1995). The action of PcG proteins and of TRX is then targeted to the same response element, suggesting that their two functions are in some way coordinated.

5
Position Effects and PRE Interactions

Both the degree of silencing and the binding of PcG proteins at a PRE-containing construct are highly variable and strongly dependent on the genomic site of insertion, suggesting that surrounding sequences or nuclear environment participate in some way in the formation of the complex or its stabilization. At one chromosomal site, the same construct may be strongly repressed and create a new chromosomal binding site for PcG proteins while at another site it may be only slightly silenced or not at all. This could be due to contributing weak PRE sites in the flanking sequences that cooperate with the PRE or to the state of activity of the chromosomal region that prevents the establishment of the silent state. Another way in which the nuclear environment can affect the function of a PRE is shown by the pairing effect. When flies are made homozygous for the insertion of a PRE-containing transposon, the pairing of the two copies of the PRE can result in the dramatic enhancement of silencing (Fauvarque and Dura 1993; Kassis 1994; Chan et al. 1994). A similar interaction is probably involved in the homing phenomenon: the tendency of a PRE-containing transposon to integrate preferentially near genomic sites that contain endogenous PREs. The implication is that the complexes formed at PREs are "sticky" and tend to interact with other PRE complexes, achieving greater stability. In some cases, such *trans*-interactions can occur between PREs inserted at different sites and even on different chromosomes (Sigrist and Pirrotta 1997). Thus, a PRE inserted at one chromosomal site may cause partial silencing of its associated reporter gene but, in the presence of another similar construct inserted at a different genomic site, the silencing effect is greatly increased leading to a lower expression of both copies of the reporter gene. These effects imply that the repressing ability of a PRE depends on the sum of the interactions with other PREs that are accessible from its genomic position and may explain why a given PRE construct is strongly silenced at some insertion sites but not others.

Similar proximity effects have been observed for heterochromatic silencing of the *brown-Dominant* (bw^D) allele in which a large block of heterochromatin has transposed near the *brown* gene. This tends to silence not only the nearby *bw* allele but, in heterozygotes also the wild type copy of the gene that is homologously paired with it. In situ hybridization shows that the bw^D locus tends to associate with centric heterochromatin and that silencing increases in proportion to this association (Csink and Henikoff 1996; Dernburg et al. 1996). One interpretation proposed for this effect is that the interactions between the block of heterochromatin at bw^D and centric heterochromatin drags the *bw* gene into a heterochromatic nuclear compartment where no transcriptional activity is possible. Another interpretation suggested by the *trans*-interactions between PREs is that the proximity to related silencing complexes strengthens the stability of the complex and consequently the silencing effect. Yeast silencing complexes also show such proximity effects, since the silencing activity of a mating type silencer is greatly decreased if it is placed at internal chromosomal positions, distant from the telomeres which represent the major binding sites of the yeast silencing proteins (Maillet et al. 1996).

In Drosophila, heterochromatin forms a more or less contiguous block in the pericentric region of each chromosome, hence the degree of heterochromatic silencing of a gene increases with its distance from the centromere. PREs, on the other hand, are scattered throughout the genome, yet they can in many cases find and interact with one another even if they are on different chromosomes. This indicates a degree of mobility of parts of the genome with respect to others that allows a search for interacting sites. Such searching would require time as the chromosomes relax from their alignment during mitosis and would be made possible by longer interphases. Even the pairing of homologues occurs gradually during the course of development, apparently because rapid cell divisions limit the time available for search and alignment (Hiraoka et al. 1993). As cell division slows down, both chromosome pairing and *trans*-interactions between distant PREs would become possible, resulting in stabler PcG complexes and stronger silencing. In Drosophila, most differentiated larval tissues are polytenized: DNA replication occurs without cell division or separation of the chromatids, which remain in a perfectly aligned bundle. An interesting speculation would be that in polytenic tissues PRE silencing would be much more effective than in the early embryo or in nonpolytenic cells like those of the nervous system or the pigment-containing cells of the eye in which the white gene is expressed.

The ability of PREs to search for and interact with other PREs might be responsible for another surprising phenomenon recently reported. While studying the expression of chimeric transgenes containing the *Adh* coding region under the control of the *white* regulatory region, Pal-Bhadra et al. (1997) found that increasing the number of transgene copies inserted at different genomic sites did not increase the overall level of expression but, on the contrary, resulted in a progressive silencing of the transgenes as well as of the endogenous *Adh* gene. This cosuppression effect was dependent on PcG genes and was accompanied by the emergence of PcG protein binding at the transgene sites, visible on polytene chromosomes. While cosuppression in plants is thought to be a posttranscriptional event (Metzlaff et al. 1997), these results strongly suggest that the *white* or *Adh* genes contain some weak proto-PRE sequences which cannot by themselves induce silencing. When the number of copies in the nucleus increases, they would begin to interact with one another, favored by their sequence homology, creating larger and more stable silencing complexes. Interactions would be much easier between transgene copies inserted at the same site on homologous chromosomes and in fact paired copies are more effective in silencing one another than dispersed copies.

6
PRE Complex Formation

Although PcG complex formation recognizes and initiates at specific sequences, the PREs, no sequence motif has been detected that is common to the many PREs that have now been identified. Furthermore, none of the isolated PcG proteins has so far been shown to have DNA binding activity or to possess an identifiable DNA-binding structural motif. It is possible that a DNA binding activity is generated when they assemble into a multiprotein complex. Or, weak interactions of different PcG proteins with specific DNA sequences may be stabilized by cooperative interactions and add up to a strong PcG complex when the sequences are clustered together at a PRE. However, on the analogy with yeast silencing, the most likely scenario is that PcG proteins are recruited by interactions with specific DNA binding proteins. To understand how this may happen, it is worth examining in more detail how the recruitment of the silencing complex occurs at the yeast mating type silencers. The silencer region contains binding sites for RAP1, ABF1 and the ORC replication origin recognition complex (Boscheron et al. 1996; Marcand et al. 1996; Triolo and Sternglanz 1996). These proteins have other functions at many other genomic sites where they can even act as transcriptional activators but at the mating type silencers they act cooperatively and additively to recruit the SIR proteins. RAP1, for example, binds SIR4, which is involved in a complex that includes SIR3. The ORC complex, which is not involved in telomeric silencing, binds SIR1, which contributes to the recruitment by interacting with SIR4. These recruiting elements can replace one another to a large extent and weak sites that have no silencing activity by themselves can cooperate to result in a functional silencer.

The efficient binding of one PcG protein to a target site is, in fact, sufficient to recruit an entire silencing complex. In the fly, a PcG protein fused to the GAL4 or lexA DNA binding domain can silence a reporter gene provided with the corresponding binding sites (Müller 1995; S. Poux and V. Pirrotta, unpubl.). This silencing is not intrinsic to the PcG protein but is dependent on the function of the other endogenous PcG genes, indicating that the other PcG proteins are recruited by the tethered protein leading to a functional silencing complex. One of the interactions that permits this recruitment is mediated by the chromodomain of PC. This was shown by a chromodomain swap experiment in which the chromodomain to the HP1 protein was replaced by that of PC (Platero et al. 1995). The resulting protein still binds to heterochromatin, like the parent HP1 but, in addition it becomes targeted to PcG sites in the euchromatin. At the same time, the hybrid protein recruits the endogenous PcG proteins to heterochromatin, where they are not normally found.

Recent experiments suggest the identity of at least one of the proteins involved in the recruitment of PcG complexes to a PRE. GAGA factor, usually thought of as an activator both in vitro and in vivo, has a consensus DNA binding sequence GAGAG that is frequently, though not always found clustered at PRE regions. Clusters of GAGAG sites are also found at many promoters, including the *Ubx* promoter itself, where GAGA binding has been shown to important for normal promoter activity (Biggin and Tjian 1988; Laney and Biggin 1992). Nevertheless, GAGA mutations appear to interfere with the silencing action of some PREs (Hagstrom et al. 1997). Chromatin crosslinking experiments confirm that GAGA factor and PC protein binding sites coexist in vivo at some PREs (Strutt et al. 1997). In vitro binding experiments show that the ability of PcG complexes in nuclear extracts to bind to some, but not all, PRE fragments is mediated by GAGA factor and is abolished by mutations in the GAGA consensus sequences (B. Horard and V. Pirrotta unpubl.). The fact that GAGA factor at promoters stimulates transcription but at PREs recruits silencing complexes indicates that the PRE context converts the activator into a repressor, probably by the adjacent binding of other factors. Other DNA binding proteins that function as activators can be converted to repressors at specific DNA targets in certain contexts by recruiting cofactors that do not themselves bind to DNA. The *Drosophila* protein DORSAL, homologous to the mammalian NF-$\varkappa$B, is a transcriptional activator but can be converted to a repressor by the proximity of other DNA binding factors and by the interaction with GROUCHO, a WD repeat proteins which does not itself bind to DNA (Dubnicoff et al. 1997). The mammalian transcriptional activator E2F1 recruits the retinoblastoma protein Rb, which in turn binds to RbAp48, another WD repeat protein, part of a complex that includes the histone deacetylase HDAC1, leading to transcriptional repression (Brehm et al. 1998; Magnaghi-Jaulin et al. 1998). The yeast α2 DNA binding protein recruits a global repressor complex containing TUP1, another WD repeat protein, leading to a reorganization of adjacent nucleosomes and to transcriptional repression (Cooper et al. 1994; Herschbach et al. 1994). The recurrence of WD repeat proteins as part of a silencing complex is reminiscent of the PcG protein ESC, also a WD repeat protein, required for the establishment of PcG silencing at homeotic genes. PcG silencing then might be a version of a general repressing mechanism widely used in both lower and higher eukaryotes, adapted to involve more proteins in a larger cooperative complex complex that persists through mitosis.

If different parts of the complex can be recruited by different DNA-binding proteins, no single sequence motif need be shared by all PREs and the exact composition and properties of the PcG complex formed might also vary from one site to another. Chromatin crosslinking experiments show that the distribution of different PcG proteins does not always coincide even at the resolution of a few hundred base pairs, the maximum permitted by the technique (Strutt and Paro 1997). These results are consistent with the genetic data and with the distribution of binding sites on polytene chromosomes and indicate that the recruitment of one PcG protein does not always automatically lead to a recruitment of all the others. Most likely, the participation of different PcG proteins is dictated by the number and kind of recruiting proteins that bind to different PREs.

7
Mechanisms of Repression

Once a core complex is recruited at a PRE, how does it interfere or compete with the transcriptional activity in the surrounding region and in particular, how does it silence enhancers or promoters many tens of kb away? The simplest and oldest view, derived from models of heterochromatic silencing, is that the protein complexes spread cooperatively to coat the chromatin, induce a more compact state, and block the access of activators and transcriptional machinery to the DNA. PcG-silenced loci have, in fact, a more condensed appearance on polytene chromosomes though it is not clear whether this is a cause or a consequence of the silencing. In yeast, silencing is accompanied by a decreased sensitivity to restriction enzymes in vitro or to specific DNA metylases in vivo and and is accompanied by the establishment of a more ordered nucleosome array (Gottschling 1992; Loo and Rine 1994). In *Drosophila*, when a reporter transposon is inserted in a heterochromatic position, its silencing is associated with reduced accessibility at certain restriction enzyme sites and an increased ordering of the nucleosome array (Wallrath and Elgin 1995). However, PcG silencing or PEV had little effect on restriction enzyme cleavage in another study (Schlossherr et al. 1994). Probably, then, the interference with restriction enzyme access involves specific positions rather than a global inaccessibility of the entire silenced region. Another approach was used by McCall and Bender (1996) to ask if silencing of the *Ubx* gene prevented the access of RNA polymerases to a part of the downstream regulatory region. A GAL4-dependent promoter inserted within the *Ubx* gene can be activated at early stages but not when the PcG silencing sets in. On the other hand, T7 RNA polymerase can still initiate trancription from its promoter situated in the same region. This difference might be due to the great difference in size between T7 polymerase and the eukaryotic transcriptional machinery or it may reflect the different way in which GAL4 recruits the PolII promoter complex.

The *Ubx* gene contains two major PREs, one more than 20 kb upstream of the promoter and one some 30 kb downstream while enhancer elements are scattered over 100 kb. Once initiated at the PREs, does the silencing complex spread to flanking sequences to block the promoter and all the enhancers? Chromatin crosslinking experiments show that in yeast telomeric silencing, the SIR complexes recruited at the telomeric repeats can spread for a distance of more than 15 kb when the SIR3 protein is overexpressed (Hecht et al. 1996: Strahl-Bolsinger et al. 1997). That PcG complexes spread over large distances has been a widely held assumption that went along with the concept of a progressive derepression along the DNA of the entire bithorax complex and with the concept of heterochromatic spreading in chromosomal rearrangements exhibiting PEV. Chromatin crosslinking experiments show, however, that PcG proteins are linked primarily to the vicinity of known PREs and decrease nearly to background levels within one or two kb (Strutt et al. 1997). This does not exclude a possible physical presence of PcG proteins over a larger chromatin region but if such spreading occurs, it involves interactions with DNA that are more subtle or indirect than those involved at the PRE itself and are not detected by crosslinking experiments.

A useful approach to understand the effect of silencing at a distance is to consider to what extent they could be viewed as negative enhancers. Enhancer elements also act at a distance which in some genes can be over 100 kb from the promoter. They are generally thought to interact with the promoter region by looping to contact promoter complexes, facilitating their recruitment. While short-range looping over a few kilobases is a reasonable mechanism, long-range looping over tens of kilobases strains credibility and poses difficulties in explaining the action of chromatin insulators and the ability of enhancers to identify their own promoter. One possible solution is to imagine that looping proceeds by a series of short range interactions between an enhancer module which assembles a set of factors to a tightly clustered set of binding sites and a number of weaker binding sites for individual factors that might occur frequently on the way between distant enhancers and promoters (Fig. 2). According to this model, weak but frequently occurring binding sites would not permit stable binding of transacting factors by themselves but might be stabilized by cooperative interactions with a enhancer module looping over a distance of a few kilobases. A similar model can be applied to silencing complexes assembled at a PRE which could loop over a few kilobases to interact with weak proto-PRE sequences. Thus, an increasingly larger and more stable complex could implicate larger regions, leading eventually to the vicinity of the promoter. We do not know so far of any interactions between PcG proteins and promoter complexes, but it is interesting to note that many, though not all, promoters that are repressed by PREs contain binding sites for GAGA factor. In a silencing context, these binding sites might mediate a close contact between PRE complexes and the promoter region.

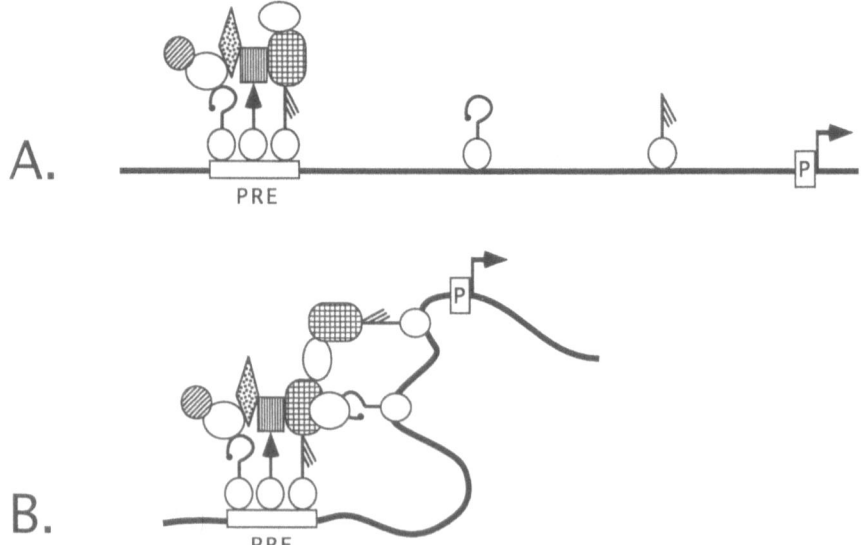

Fig. 2A–B. Short-range looping model. **A** A PcG complex is assembled by a set of recruiting proteins that bind at the PRE site. **B** The PRE complex is then envisioned to interact and stabilize transient complexes formed at isolated binding sites for individual recruiting proteins, bringing the silencing complex within reach of the promoter (P)

While yesterday's favorite mechanism for the activation of promoters was the contact between enhancer factors and components of the promoter complex, today's favorite is nucleosome remodelling. Two categories of remodeling machines are known: those that alter the state of acetylation of the histones and those that use ATP hydrolysis to alter nucleosome structure. Both histone acetylation and nucleosome remodeling increase the accessiblity of DNA to transcriptional activators or promoter complexes while also facilitating the progress of the transcribing RNA polymerase. In yeast, Drosophila and mammals, several types of ATP-dependent nucleosome remodelling machines have been identified (SWI/SNF, NURF, CHRAC), all containing homologues of the trxG protein BRAHMA or related proteins (Coté et al. 1994; Tsukiyama and Wu, 1995; Tsukiyama et al. 1995; Varga-Weisz et al. 1997). The effect of these machines is to alter nucleosome structure and enhance the accessibility to the DNA while mutations in their components decrease the expression of many genes. Histone acetylases have been found to be components of coactivator complexes recruited by a variety of trans-acting factors such as CREB, STAT-1, or nuclear hormone receptors, and are required to activate transcription by these factors (Bannister and Kouzarides, 1996; Kamei et al. 1996). These two types of chromatin opening devices appear to be major components of transcriptional activation, supporting the notion that the principal role of enhancers is to relieve the constitutive repression due to chromatin, that is, the assembly of the DNA into nucleosome arrays (Walters et al. 1996). If so, then silencing would only require deacetylating the histones and reverting the nucleosome remodelling. We do not know what mechanism reverts the nucleosome remodeling effected by NURF, SWI/SNF and related complexes, but several transcriptional repressors are now known to recruit histone deacetylases. Thus the retinoblastoma protein Rb binds to the activator E2F and converts it to a repressor by recruiting a complex containing HDAC1, a human homologue of the yeast and Drosophila Rpd3 histone deacetylase (Brehm et al. 1998; Magnaghi-Jaulin et al. 1998). The MAD/MAX repressor also recruits an HDAC1-containing complex, as do the unliganded nuclear receptors (Laherty et al. 1997; Heinzel et al. 1997). Interestingly, transcriptional activators can become repressors depending on their context or conformational state, which determines whether an acetylating or deacetylating complex is recruited.

The involvement of histones is clear in yeast telomeric silencing where the silencing SIR proteins interact with the N-terminal tails of histones H3 and H4, which protrude from the nucleosome core (Hecht et al. 1995). A role of histones and their acetylation is very likely in *Drosophila* heterochromatic silencing, which is decreased by reducing the dosage of histone genes and modulated by mutations or treatments that affect histone acetylation. Although so far we know very little about the role of histones or acetylation in PcG silencing, an interaction between the PcG complex and nucleosomes remains the most likely mechanism to explain its silencing effect. Acetylation of histones H3 and H4 occurs at lysines in the N-terminal tails. This would loosen their association with DNA and make them more accessible to DNA-binding proteins such as activators or transcriptional complexes. However this might also make them more available for interaction with other chromatin proteins, including silencing proteins, with surprising consequences. In contrast to the general trend of increased acetylation leading to higher expression levels, mutations in the Rpd3 histone deacetylase cause increased heterochromatic silencing in Drosophila and increased telomeric silencing and acetylation at H4 lys5 and 12 in yeast (De Rubertis et al. 1996; Rundlett et al. 1996). Moreover, mutations of H4 at position 12 show that this acetylation is required for efficient silencing (Braunstein et al. 1996). This has led to the hypothesis that the SIR complex recruits a lys12 acetylase that acts on the neighboring nucleosome leading in turn to better interaction with the Sir proteins thus propagating the complex. Another possibility is that different patterns of acetylation are differentially recognized by activators and by silencing proteins: acetylation at H4 lys12 promoting silencing while acetylation at lys 5, 8, and 16 would promote activator binding.

8
Long-Term Stability and Cellular Memory

The most characteristic feature of PcG silencing is that the decision to repress or not a target gene is made at early times and is maintained in subsequent cellular generations. This cellular memory implies two things: (1) the repressed state cannot be established de novo at later developmental stages and (2) the complex or some part of it either persist at the site of repression through DNA replication and mitosis or must mark the PRE site for preferential reassembly of the complex in the daugher cells. Another possible way to interpret the cellular memory is that activation of a gene in the early embryo marks it in such a way that is incompatible with assembly of a repressive complex which otherwise would reform in every cell at all PRE sites.

Recent experiments by Cavalli and Paro (1998) shed some light on the properties of the cellular memory. To understand their experiments, we must start with the observation that transcriptional activation is, in fact, physically incompatible with the binding of PcG proteins. Zink and Paro (1995) placed the *Fab7* PRE from the abdominal region of the bithorax complex, in a construct in which a reporter gene is under the control of a GAL4 binding site. The construct also contains the *white* gene as a transformation marker, whose expression is repressed by the PRE in the familiar variegated fashion. Transcription of the reporter gene is also repressed but massive production of GAL4 activator (from a heat shock promoter) can induce expression. When this happens, the binding of PcG proteins at the chromosomal insertion site of the construct is abolished. In some way then, just as the PcG complex normally prevents the binding of GAL4, successful binding of GAL4 displaces a preexist-ing PcG complex from the site. An explanation for this antagonism might be that the histone acetylation induced by binding of the activator is incompatible with the assembly of the silencing complex. When the pulse of GAL4 is produced in larvae, the derepression of the construct is transient and repression returns when the GAL4 protein begins to decay. Cavalli and Paro (1998) discovered, however, that if the GAL4 activation takes place during embryonic development, repression failed to be reestablished. As the activator fades, the reporter gene returns to a low but detectable basal level but PcG silencing does not return, as indicated by the strong expression of the *white*, resulting in flies with red eyes. It appears then that some critical event takes place in the early embryo, when the silencing complex is first established, and cannot be reenacted at later stages. Not only the derepressed state but also hyper-repressed states can be preserved by the cellular memory. Such states can be produced by raising the embryos at higher temperatures, which favors PcG silencing. When they are returned to normal temperature after embryonic development, the hyperrepressed state persists and results in increased silencing of the *white* gene in the adult eye. Interestingly, a very similar effect is obtained with heterochromatic PEV where however the temperature dependence is reversed: embryo development at higher temperature leads to decreased heterochromatic silencing in the adult but the memory effect is similar and once again indicates that some durable state is imprinted on the chromatin of the affected genes during embryonic development (Spofford 1976).

The most remarkable result of Cavalli and Paro's experiments, however, is that up to 25% of the progeny of flies that developed red eyes as a consequence of embryonic derepression is still derepressed. Similar results have been obtained using a different PRE and a heat shock promoter instead of a GAL4 activated promoter (G. I. Dellino and V. Pirrotta, unpubl.). The fact that the derepressed state can be inherited implies not only that it survives in the germ line through gametogenesis, meiosis, and fertilization but that repression cannot be reinstated during the ensuing embryonic development at the time PcG complexes normally form. The PRE in the germ line cells has been stripped of PcG proteins by the massive pulse of GAL4 activator and this "unsilenced" state prevents formation of new PcG complexes in the next generation. To account for this, we must suppose that the open state is somehow marked. One possibility is that some protein associated with the open state, perhaps a trxG protein, remains bound to the "unsilenced" PRE through mitosis and meiosis and prevents reassembly of the silencing complex. Conversely, some PcG protein might normally remain bound to a silenced PRE to mark it for rapid reinstatement of silencing.

Another possible marker is the state of acetylation of the nucleosomes. During DNA replication, the old nucleosomes are partitioned randomly between the two daughter DNA molecules. If, for example, selective acetylation at H4 lys12 is both a consequence of PcG silencing and the cause of preferential reassembly of a silencing complex at the PRE, the silenced state could be perpetuated from one cell generation to another. Nucleosomes marked by acetylation at other histone lysines would, on the other hand, prevent the reassembly of the silencing complex. To make such an active configuration persist through many cell divisions, the region containing acetylated nucleosomes would need to continue to recruit acetylating complexes to mark the newly deposited nucleosomes (Fig. 3). This could be the role of proteins such as TRX which have been shown to bind to the PRE (Chang et al. 1995; Chinwalla et al. 1995). In this model, then, the semiconservative partitioning of acetylated nucleosomes would provide the continuity through the cell cycle that is necessary for preserving the cellular memory. Needless to say, no evidence yet proves or disproves this model; a variety of alternatives are still very much in the running and more surprises may be in store.

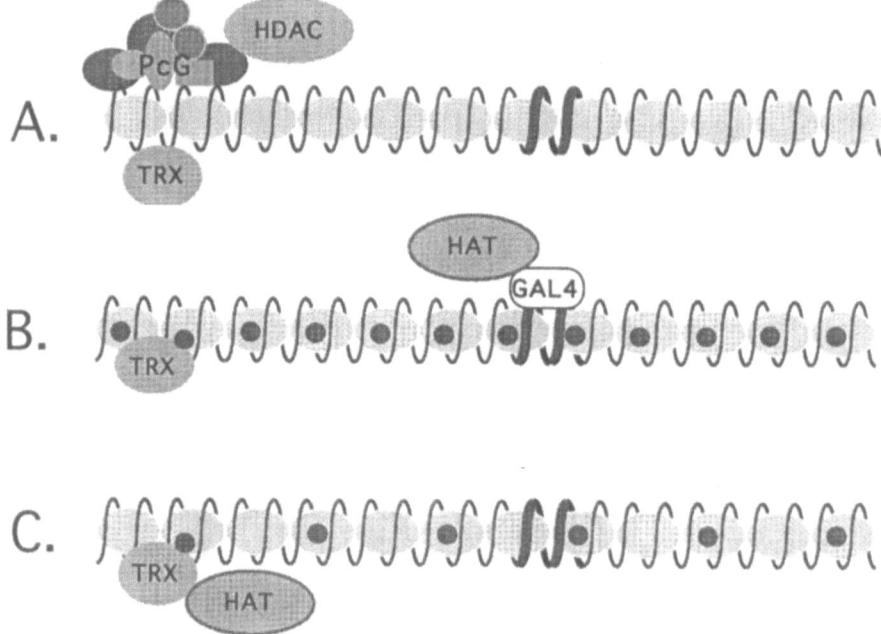

Fig. 3A–C. Scenario for the maintenance of an active chromatin state. **A** The PcG complex formed at the PRE is conjectured to recruit a histone deacetylase complex (*HDAC*), thus inhibiting the binding of activators. **B** GAL4 activator, expressed at high levels, succeeds in binding to its target sequences and, in turn, recruits a histone acetylase (*HAT*), acetylating the nucleosomes (*black dot*) permitting the activation of transcription and the activation of TRX, constitutively bound to the PRE. **C** The semiconservative partitioning of nucleosomes during DNA replication produces partly acetylated daughter chromatin strands. The activated TRX protein now recruits a maintenance acetylase (*HAT*), whose job is to acetylate the newly deposed nucleosomes and maintain a fully acetylated and "open" chromatin state

9
Is There Any Relationship Between PEV and Genomic Imprinting?

Chromatin silencing mechanisms closely related to the Drosophila PcG mechanisms are clearly used in mammals to affect not only homeotic gene expression but also, judging from the phenotypes of mice with various PcG mutations, other processes such as cell proliferation, pregastrulation events, terminal differentiated hematopoietic functions. As the Drosophila evidence implies, certain targets may involve subsets of the PcG, probably depending on the recruiting proteins involved in each particular case. Of the many mammalian phenomena that are likely to involve PcG-related mechanisms, one in particular is highly reminiscent of the inherited chromatin state that can be established by PcG/trxG proteins: could chromatin silencing mechanisms be implicated in mammalian genetic imprinting? A number of loci in mouse or man show allele-specific silencing depending on the parent of origin. In some cases, the silenced gene is the paternally derived allele, in other cases, the maternal allele. Promoter competition may be implicated in the detailed working out of the mechanism in each case (Wutz et al. 1997; Webber et al. 1998) but the fact remains that some promoters are active if they are inherited from the mother but silent if inherited from the father, or viceversa, depending on the locus. Clearly, some chromatin mark or imprint must be set down during gametogenesis and preserved through early embryonic development. It was recognized very early that CpG methylation has the right properties to be the imprinting signal, since it is handed down to the daughter cells every cell division by the semiconservative replication of DNA and is then fully regenerated by the maintenance methylase which recognizes hemimethylated DNA (Razin and Cedar 1994). Although imprinting is often accompanied by changes in the methylation pattern, they do not always correspond to silencing in their pattern or timing. An interesting possibility is that methylation patterns are secondary to a primary event that involves chromatin complexes of the PcG type. Such complexes would be recruited during gametogenesis (or earlier in the parental germline) and establish a chromatin state that is maintained through maturation of the gametes, fertilization and embryonic development. In an effort to determine whether such complexes exist and are evolutionarily conserved, the critical regions from two imprinted mammalian loci, the mouse H19 and the human SNRPN loci, both containing CpG islands, were placed in front of a reporter gene in a Drosophila transposon construct (Lyko et al. 1997, 1998). In both cases, when introduced in the fly, the reporter gene was expressed at very low levels and the *white* gene tranformation marker was also strongly repressed. Selection for higher levels of expression showed that loss of repression resulted from deletions of the mammalian sequences. It is too early to draw conclusions from these experiments: repression of the constructs in *Drosophila* is independent of parent of origin and therefore shows no evidence of imprinting or epigenetic maintenance; no effect of PcG or *Su(var)* mutations could be demonstrated. Nevertheless, one possible interpretation is that a conserved silencing mechanism is involved, one that, in mammals, becomes modified and overridden in one of the two parental germlines so that in the resulting gametes, the locus has an open, derepressed chromatin state that is inherited in the next generation much as the derepressed states described in the preceding section. Clearly the *Drosophila* system could permit a mutational analysis of the mechanism and provide valuable clues about the nature of the proteins involved.

Acknowledgments. I am grateful to Giacomo Cavalli for information and discussion of his results. The work in my laboratory was supported by grants from the Swiss National Science Foundation, from the Donation Georges and Antoine Claraz and from the Human Frontier Science Program.

References

Akasaka T, Kanno M, Balling R, Mieza MA, Taniguchi M, Koseki H (1996) A role for mel-18, a Polycomb group-related vertebrate gene, in the anteroposterior specification of the axial skeleton. Development 122:1513–1522

Bannister AJ, Kouzarides T (1996) The CBP co-activator is a histone acetyltransferase. Nature 384:641–643

Biggin MD, Tjian R (1988) Transcription factors that activate the *Ultrabithorax* promoter in developmentally staged extracts. Cell 53:699–711

Bornemann D, Miller E, Simon J (1996) The *Drosophila* Polycomb group gene *Sex comb on midleg* (*Scm*) encodes a zinc finger protein with similarity to polyhomeotic protein. Development 122:1621–1630

Boscheron C, Maillet L, Marcand S, Tsai-Pflugfelder M, Gasser SM, Gilson E (1996) Cooperation at a distance between silencers and proto-silencers at the yeast HML locus. EMBO J 15:2184–2195

Braunstein M, Sobel RE, Allis CD, Turner BM, Broach JR (1996) Efficient transcriptional silencing in *Saccharomyces cerevisiae* requires heterochromatin histone acetylation pattern. Mol Cell Biol 16:4349–4356

Brehm A, Miska EA, McCance DJ, Reid JL, Bannister AJ, Kouzarides T (1998) Retinoblastoma protein recruits histone deacetylase to repress transcription. Nature 391:597–601

Brunk BP, Martin EC, Adler PN (1991) *Drosophila* genes *Posterior Sex Combs* and *Suppressor two of zeste* encode proteins with homology to the murine BMI-1 oncogene. Nature 353:351–353

Cavalli G, Paro R (1998) The *Drosophila Fab-7* chromosomal element conveys epigenetic inheritance during mitosis and meiosis. Cell 93:505–518

Chan CS, Rastelli L, Pirrotta V (1994) A Polycomb response element in the *Ubx* gene that determines an epigenetically inherited state of repression. EMBO J 13:2553–2564

Chang Y, King BO, O'Connor M, Mazo A, Huang D (1995) Functional reconstruction of trans regulation of the *Ultrabithorax* promoter by the products of two antagonistic genes, *trithorax* and *Polycomb*. Mol Cell Biol 15:6601–6612

Cheng NN, Sinclair DAR, Campbell RB, Brock HW (1994) Interactions of *polyhomeotic* with Polycomb group genes of *Drosophila melanogaster*. Genetics 138:1151–1162

Chiang A, O'Connor MB, Paro R, Simon J, Bender W (1995) Discrete Polycomb binding sites in each parasegmental domain of the bithorax complex. Development 121:1681–1689

Chinwalla V, Jane EP, Harte PJ (1995) The *Drosophila* trithorax protein binds to specific chromosomal sites and is co-localized with Polycomb at many sites. EMBO J 14:2056–2065

Cooper JP, Roth SY, Simpson RT (1994) The global transcriptional regulators, SSN6 and TUP1, play distinct roles in the establishment of a repressive chromatin structure. Genes Dev 8:1400–1410

Coté J, Quinn J, Workman JL, Peterson CL (1994) Stimulation of GAL4 derivative binding to nucleosomal DNA by the yeast SWI/SNF complex. Science 265:53–60

Csink AK, Henikoff S (1996) Genetic modification of heterochromatic association and nuclear organization in *Drosophila*. Nature 381:529–531

Dernburg AF, Broman KW, Fung JC, Marshall WF, Philips J, Agard DA, Sedat JW (1996) Perturbation of nuclear architecture by long-distance chromosome interactions. Cell 85:745–759

De Rubertis F, Kadoshi D, Henchoz S, Pauli D, Reuter G, Struhl K, Spierer P (1996) The histone deacetylase RPD3 counteracts genomic silencing in *Drosophila* and yeast. Nature 384:589–591

Dubnicoff T, Valentine SA, Chen G, Shi T, Lengyel JA, Paroush Z, Courey AJ (1997) Conversion of Dorsal from an activator to a repressor by the global corepressor Groucho. Genes Dev 11:2952–2957

Farkas G, Gausz J, Galloni M, Reuter G, Gyurkovics H, Karch F (1994) The *Trithorax-like* gene encodes the *Drosophila* GAGA factor. Nature 371:806–808

Fauvarque M-O, Dura J-M (1993) *polyhomeotic* Regulatory sequences induce developmental regulator-dependent variegation and targeted P-element insertions in *Drosophila*. Genes Dev 7:1508–1520

Franke A, DeCamillis M, Zink D, Cheng N, Brock HW, Paro R (1992) Polycomb and polyhomeotic are constituents of a multimeric protein complex in chromatin of *Drosophila melanogaster*. EMBO J 11:2941–2950

Gindhart JG Jr, Kaufman TC (1995) Identification of Polycomb and trithorax group responsive elements in the regulatory region of the *Drosophila* homeotic gene *Sex combs reduced*. Genetics 139:797–814

Gottschling DE (1992) Telomere-proximal DNA in *Saccharomyces cerevisiae* is refractory to methyltransferase activity in vivo. Proc Natl Acad Sci USA 89:4062–4065

Gu Y, Nakamura T, Alder H, Prasad R, Canaani O, Cimino G, Croce CM, Canaani E (1992) The t(4;11) chromosome translocation of human acute leukemias fuses the ALL-1 gene, related to *Drosophila trithorax*, to the AF-4 gene. Cell 71:701–708

Gutjahr T, Frei E, Spicer C, Baumgartner S, White RAH, Noll M (1995) The Polycomb-group gene, *extra sex combs*, encodes a nuclear member of the WD-40 repeat family. EMBO J 14:4296–4306

Hagstrom K, Müller M, Schedl P (1997) A Polycomb- and GAGA-dependent silencer adjoins the *Fab7* boundary in the *Drosophila* bithorax complex. Genetics 146:1365–1380

Hecht A, Laroche T, Strahl-Bolsinger S, Gasser SM, Grunstein M (1995) Histone H3 and H4 N-termini interact with SIR3 and SIR4 proteins: a molecular model for the formation of heterochromatin in yeast. Cell 80:583–592

Hecht A, Strahl-Bolsinger S, Grunstein M (1996) Spreading of transcriptional repressor SIR3 from telomeric heterochromatin. Nature 383:92–96

Heinzel T, Lavinsky RM, Mullen T, Söderstrom M, Laherty CD, Torchia J, Yang W, Brard G, Ngo SD, Davie JR, Seto E, Eisenman RN, Rose DW, Glass CK, Rosenfeld MG (1997) A complex containing N-CoR, mSin3 and histone deacetylase mediates transcriptional repression. Nature 387:43–48

Herschbach BM, Arnaud M, Johnson AD (1994) Transcriptional repression directed by the yeast α2 protein in vitro. Nature 370:309–311

Hiraoka Y, Dernburg AF, Parmelee SJ, Rykowski MC, Agard DA, Sedat JW (1993) The onset of homologous chromosome pairing during *Drosophila melanogaster* embryogenesis. J Cell Biol 120:591–600

Jones RS, Gelbart WM (1993) The *Drosophila* Polycomb-group gene *Enhancer of zeste* contains a region with sequence similarity to *trithorax*. Mol Cell Biol 13:6357–6366

Jürgens G (1985) A group of genes controlling the spatial expression of the Bithorax complex in *Drosophila*. Nature 316:153–155

Kamei Y, Xu L, Heinzel T, Torchia J, Kurokawa K, Gloss B, Lin S, Heyman RA, Rose DW, Glass CK, Rosenfeld MG (1996) A CBP integrator complex mediates transcriptional activation and AP-1 inhibition by nuclear receptors. Cell 85:403–414

Kassis JA (1994) Unusual properties of regulatory DNA from the *Drosophila engrailed* gene: three "pairing-sensitive" sites within a 1.6 kb region. Genetics 136:1025–1038

Kyba M, Brock HW (1998) The *Drosophila* Polycomb Group protein *Psc* contacts *ph* and *Pc* through specific conserved domains. Mol Cell Biol 18:2712–2720

Laherty CD, Yang W, Sun J, Davie JR, Seto E, Eisenman RN (1997) Histone deacetylases associated with the mSin3 corepressor mediate Mad transcriptional repression. Cell 89:349–356

Laible G, Wolf A, Dorn R, Reuter G, Nislow C, Lebersorger A, Popkin D, Pillus L, Jenuwein T (1997) Mammalian homologues of the Polycomb-group gene *Enhancer of zeste* mediate gene silencing in *Drosophila* heterochromatin and at *S. cerevisiae* telomeres. EMBO J 16:3219–3232

Laney JD, Biggin MD (1992) *zeste*, a nonessential gene, potently activates *Ultrabithorax* transcription in the *Drosophila* embryo. Genes Dev 6:1531–1541

Lee H-s, Kraus KW, Wolfner MF,Lis JT (1992) DNA sequence requirements for generating paused polymerase at the start of *hsp70*. Genes Dev 6:284–295

Loo S, Rine J (1994) Silencers and domains of generalized repression. Science 264:1768–1771

Lyko F, Brenton JD, Surani MA, Paro R (1997) An imprinting element from the mouse H19 locus functions as a silencer in *Drosophila*. Nat Genet 16:171–173

Lyko F, Buiting K, Horsthemke B, Paro R (1998) Identification of a silencing element in the human 15q11-q3 imprinting center by using transgenic *Drosophila*. Proc Natl Acad Sci USA 95:1698–1702

Magnaghi-Jaulin L, Groisman R, Naguibneva I, Robin P, Lorain S, LeVillain JP, Troalen F, Trouche D, Harel-Bellan A (1998) Retinoblastoma protein represses transcription by recruiting a histone deacetylase. Nature 391:601–605

Maillet L, Boscheron C, Gotta M, Marcand S, Gilson E, Gasser SM (1996) Evidence for silencing compartments within the yeast nucleus: a role for telomere proximity and Sir protein concentration in silencer-mediated repression. Genes Dev 10:1796–1811

Marcand S, Buck SW, Moretti P, Gilson E, Shore D (1996) Silencing of genes at nontelomeric sites in yeast is controlled by sequestration of silencing factors at telomeres by Rap1 protein. Genes Dev 10:1297–1309

Martin EC, Adler PN (1993) The Polycomb group gene *Posterior Sex Combs* encodes a chromosomal protein. Development 117:641–655

McCall K, BenderW (1996) Probes for chromatin accessibility in the *Drosophila* bithorax complex respond differently to Polycomb-mediated repression. EMBO J 15:569–580

Metzlaff M, O'Dell M, Cluster PD, Flavell RB (1997) RNA-mediated RNA degradation and chalcone synthase A silencing in *Petunia*. Cell 88:845–854

Müller J (1995) Transcriptional silencing by the Polycomb protein in *Drosophila* embryos. EMBO J 14:1209–1220

Müller J, Bienz M (1991) Long range repression conferring boundaries of *Ultrabithorax* expression in the *Drosophila* embryo. EMBO J 10:3147–3155

Müller J, Gaunt S, Lawrence PA (1995) Function of the Polycomb protein is conserved in mice and flies. Development 121:2847–2852

Pal-Bhadra M, Bhadra U, Birchler JA (1997) Cosuppression in *Drosophila*:gene silencing of Alcohol dehydrogenase by *white-Adh* transgenes is Polycomb dependent. Cell 90:479–490

Paro R, Hogness DS (1991) The Polycomb protein shares a homologous domain with a heterochromatin-associated protein of *Drosophila* Proc Natl Acad Sci USA 88:263–267

Pelegri F, Lehmann R (1994) A role of polycomb group genes in the regulation of gap gene expression in *Drosophila*. Genetics 136:1341–1353

Pirrotta V, Rastelli L (1994) *white* Gene expression, repressive chromatin domains and homeotic gene regulation. BioEssays 16:549–556

Pirrotta V, Chan CS, McCabe D, Qian S (1995) Distinct parasegmental and imaginal enhancers and the establishment of the expression pattern of the *Ubx* gene. Genetics 141:1439–1450

Platero JS, Hartnett T, Eissenberg JC (1995) Functional analysis of the chromodomain of HP1. EMBO J 14:3977–3986

Poux S, Kostic C, Pirrotta V (1996) Hunchback-independent silencing of late *Ubx* enhancers by a Polycomb Group Response Element. EMBO J 15:4713–4722

Qian S, Capovilla M, Pirrotta V (1991) The bx region enhancer, a distant cis-control element of the *Drosophila Ubx* gene and its regulation by *hunchback* and other segmentation genes. EMBO J 10:1415–1425

Qian S, Capovilla M, Pirrotta V (1993) Molecular mechanisms of pattern formation by the BRE enhancer of the *Ubx* gene. EMBO J 12:3865–3877

Rastelli L, Chan CS, Pirrotta V (1993) Related chromosome binding sites for *zeste*, suppressors of *zeste* and Polycomb group proteins in *Drosophila* and their dependence on *Enhancer of zeste* function. EMBO J 12:1513–1522

Razin A, Cedar H (1994) DNA methylation and genomic imprinting. Cell 77:473–476

Reijnen MJ, Hamer KM, den Blaauwen JL, Lambrechts C, Schoneveld I, van Driel R, Otte AP (1995) Polycomb and bmi-1 homologs are expressed in overlapping patterns in *Xenopus* embryos and are able to interact with each other. Mech Dev 53:35–46

Reuter G, Spierer P (1992) Position effect variegation and chromatin proteins. BioEssays 14:605–612

Rundlett SE, Carmen AA, Kobayashi R, Bavykin S, Turner BM, Grunstein M (1996) HDA1 and RPD3 are members of distinct yeast histone deacetylase complexes that regulate silencing and transcription. Proc Natl Acad Sci USA 93:14503–14508

Sathe SS, Harte PJ (1995) The *extra sex combs* protein contains WD motifs essential for its function as a repressor of the homeotic genes. Mech Dev 52:77–87

Schlossherr J, Eggert H, Paro R, Cremer S, Hack RS (1994) Gene inactivation in *Drosophila* mediated by the Polycomb gene product or by position-effect variegation does not involve major changes in the accessibility of the chromatin fibre. Mol Gen Genet 243:453–462

Schumacher A, Magnuson T (1997) Murine Polycomb- and trithorax-group genes regulate homeotic pathways and beyond. Trends Genet 13:167–170

Schumacher A, Faust C, Magnuson T (1996) Positional cloning of a global regulator of anterior-posterior patterning in mice. Nature 383:250–253

Sherman JM, Pillus L (1997) An uncertain silence. Trends Genet 13:308–313

Sigrist CJA, Pirrotta V (1997) Chromatin insulator elements block the silencing of a target gene by the *Drosophila* Polycomb Response Element (PRE) but allow *trans* interactions between PREs on different chromosomes. Genetics 147:209–221

Simon J, Peifer M, Bender W, O'Connor M (1990) Regulatory elements of the bithorax complex that control expression along the anterior-posterior axis. EMBO J 9:3945–3956

Simon J, Bornemann D, Lunde K, Schwartz C (1995) The *extra sex combs* product contains WD40 repeats and its time of action implies a role distinct from other Polycomb group products. Mech Dev 53:197–208

Spofford JB (1976) Position-effect variegation in *Drosophila*. In: Ashburner M, Novitski E (eds) The genetics and biology of *Drosophila*, vol 1c. Academic Press, London, pp 955–1018

Strahl-Bolsinger S, Hecht, A, Luo K, Grunstein M (1997) Sir2 and Sir4 interactions differ in core and extended telomeric heterochromatin in yeast. Genes Dev 11:83–93

Strutt H, Paro R (1997) The Polycomb Group protein complex of *Drosophila melanogaster* has different composition at different target genes. Mol. Cell Biol 17:6773–6783

Strutt H, Cavalli G, Paro R (1997) Co-localization of Polycomb protein and GAGA factor on regulatory elements responsible for the maintenance of homeotic gene expression. EMBO J 16:3621–3632

Triolo T, Sternglanz R (1996) Role of interactions between the origin recognition complex and SIR1 in transcriptional silencing. Nature 381:251–253

Tripoulas N, LaJeunesse D, Gildea J, Shearn A (1996) The *Drosophila* ash1 gene product, which is localized at specific sites on polytene chromosomes, contains a SET domain and a PHD finger. Genetics 143:913–928

Tsukiyama T, Wu C (1995) Purification and properties of an ATP-dependent nucleosome remodeling factor. Cell 83:1011–1020

Tsukiyama T, Becker PB, Wu C (1994) ATP-dependent nucleosome disruption at a heat-shock promoter mediated by binding of GAGA transcription factor. Nature 367:525–532

Tsukiyama T, Daniel C, Tamkun J, Wu C (1995) ISWI, a member of the SWI2/SNF2 ATPase family, encodes the 140-kDa subunit of the nucleosome remodeling factor. Cell 83:1021–1026

van der Lugt NMT, Domen J, Linders K, van Roon M, Robanus-Maandag E, te Riele H, van der Valk M, Deschamps J, Sofroniew M, van Lohuizen M, Berns A (1994) Posterior transformation, neurological abnormalities, and severe hematopoietic defects in mice with a targeted deletion of the bmi-1 proto-oncogene. Genes Dev 8:757–769

Varga-Weisz PD, Wilm M, Bonte E, Dumas K, Mann M, Becker PB (1997) Chromatin-remodelling factor CHRAC contains the ATPases ISWI and topoisomerase II. Nature 388:598–602

Wallrath LL, Elgin SCR (1995) Position effect variegation in *Drosophila* is associated with an altered chromatin structure. Genes Dev 9:1263–1277

Walters MC, Magis W, Fiering S, Eidemiller J, Scalzo D, Groudine M, Martin DIK (1996) Transcriptional enhancers act in cis to suppress position-effect variegation. Genes Dev 10:185–195

Webber AL, Ingram RS, Levorse JM, Tilghman SM (1998) Location of enhancers is essential for the imprinting of H19 and Igf2 genes. Nature 391:711–715

Wutz A, Smrzka OW, Schweifer N, Schellander K, Wagner EF, Barlow DP (1997) Imprinted expression of the Igf2r gene depends on an intronic CpG island. Nature 389:745–749

Yu BD, Hess JL, Horning SE, Brown GAJ, Korsmeyer SJ (1995) Altered Hox expression and segmental identity in Mll-mutant mice. Nature 378:505–508

Zhang C-C, Müller J, Hoch M, Jäckle H, Bienz M (1991) Target sequences for *hunchback* in a control region conferring *Ultrabithorax* expression boundaries. Development 113:1171–1179

Zink B, Paro R (1989) *In vivo* binding pattern of a trans-regulator of homoeotic genes in *Drosophila melanogaster*. Nature 337:468–471

Zink D, Paro R (1995) *Drosophila* Polycomb-group regulated chromatin inhibits the accessibility of a trans-activator to its target DNA. EMBO J 14:5660–5671

Domains and Boundaries in Chromosomes

Tatiana I. Gerasimova and Victor G. Corces

1
Introduction

Current models to explain regulatory mechanisms underlying eukaryotic gene expression rely heavily on ideas derived from prokaryotic systems. These models assume that transcriptional enhancers account for the temporal and spatial specificity of eukaryotic gene expression by serving as binding sites for transcription factors that can then activate transcription by looping out intervening sequences and contacting components of the promoter-bound transcription complex (see, for example, Thompson and McKnight 1992; Franklin, Chap. 8, this vol.); but many phenomena, observed in a variety of systems, suggest that this picture is too simplistic to realistically explain the complexity of the eukaryotic genome and the variety of patterns of gene expression that are required for proper development of higher eukaryotes. DNA in the eukaryotic nucleus wraps around histone complexes to form nucleosomes, and the resulting primary chromatin fiber is organized into higher-order domains that arise from further compaction and attachment to specific subnuclear structures. This complex organization of the DNA within the nucleus would appear to preclude the type of interactions necessary to activate transcription, which require contacts among proteins bound to enhancers and promoters. Some of the regulatory input that controls eukaryotic gene expression must thus be directed to the establishment and maintenance of this higher-order domain organization, and transcriptional activation might involve the alteration of this organization in a manner that permits enhancer-promoter interactions.

Results from both cytological and molecular studies suggest the existence of a structural organization of the DNA within the nucleus. For example, the reproducible banding pattern of insect polytene chromosomes is suggestive of an underlying structural organization perhaps imposed by the DNA sequence on the higher-order organization of chromatin. This specific structural layout was thought to have a functional significance based on the correlation between transcriptional activation and decondensation of particular polytene bands (Tissières et al. 1974; Lewis et al. 1975). Similarly, the finding of

Department of Biology, The Johns Hopkins University, 3400 N. Charles St., Baltimore, Maryland 21218, USA

active genes in the loops of lampbrush chromosomes was taken as an early indication of a direct relationship between activation of gene expression and location within a specific structural chromosomal domain (Callan 1986). More recently, biochemical studies have identified DNA sequences involved in this structural organization. When histones and other chromosomal proteins were extracted from nuclei of interphase cells, loops of DNA containing negative unrestrained supercoils were observed. The bases of these loops are attached to a matrix or scaffold through sequences termed MARs (matrix attachment regions) or SARs (scaffold attachment regions) (reviewed by Ludérus and van Driel 1997). Although some of these sequences have been found to play a role in the expression of particular genes, the question of whether they are merely structural components or whether they play a functional role is still unanswered. Functional assays have also been used to identify DNA sequences that affect transcription by establishing domains of organization of the DNA within the nucleus. These sequences have been named insulator or boundary elements. The structural attributes and cellular properties of proteins that bind to these sequences are beginning to give insights into the role of nuclear architecture in the regulation of gene expression, and might explain aspects of eukaryotic transcription difficult to comprehend simply based on prokaryotic models.

2
The Nature of Insulator DNA

Insulator elements have been identified based on two different properties: their ability to interfere with enhancer-promoter interactions and their capacity to buffer a transgene from chromosomal position effects by conferring position-independent expression to the transformed sequences. Figure 1 summarizes the first property of insulators. When an insulator is placed between an enhancer and a promoter, activation of transcription by this enhancer is repressed, although expression from a second enhancer located downstream of the promoter can still take place (Fig. 1A). If the insulator is placed between the downstream enhancer and the promoter, transcription from this enhancer is now repressed and activation by the upstream enhancer is now possible (Fig. 1B). When the insulator is placed downstream from the second enhancer, transcription is normal, in spite of the fact that the insulator is located in an intron of the gene (Fig. 1C). The observation that an insulator represses transcription only when located between an enhancer and a promoter distinguishes these sequences from classical repressors. An important characteristic of insulator action is that, although an insulator represses activation of transcription by an enhancer on a promoter when placed in between, the enhancer is still functional and able to activate transcription from a second promoter located on the other side (Fig. 1D). As a consequence of their effects on enhancer-promoter interactions, insulators are able to buffer a transgene

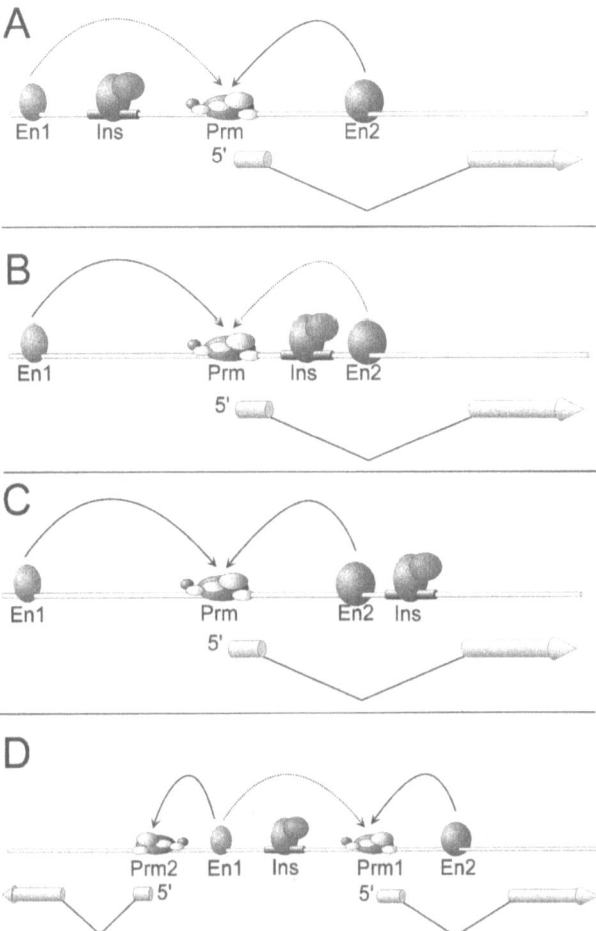

Fig. 1A–D. Polar effect of an insulator on enhancer-promoter interactions. *En1* and *En2* represent two different enhancers and their associated transcription factors bound to nucleosomal DNA. *Prm* is the promoter of the gene where the different components of the transcription complex are present. *Ins* is an insulator element with its associated proteins. *Solid arrows* indicate a positive activation of transcription by the enhancer element; *dotted arrows* indicate a repression of this effect. **A.** An insulator located in the 5' region of the gene inhibits its transcriptional activation by an upstream enhancer (*En1*) without affecting the function of a second enhancer (*En2*) located in the intron of the gene. **B.** When the insulator is located in the intron, expression from the downstream enhancer (*En2*) is blocked, whereas the upstream enhancer (*En1*) is active. **C.** When the insulator is located in the intron but distal to the *En2* enhancer, both enhancers are active and transcription of the gene is normal. This property distinguishes an insulator from a typical silencer. **D.** If a second gene is located upstream of the *En1* enhancer, although this enhancer cannot act on the *Prm1* promoter, it is still functional and able to activate transcription from the upstream *Prm2* promoter

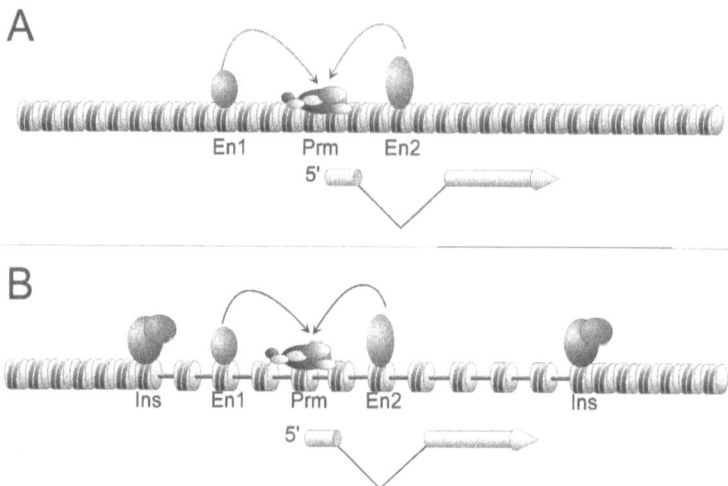

Fig. 2A–B. Insulator elements buffer gene expression from repressive effects of adjacent chromatin. Symbols are as in Fig. 1. **A.** A transgene integrated in the chromosome in a region of condensed chromatin is not properly expressed; the repressive chromatin structure of the surrounding region presumably spreads into transgene sequences, inhibiting enhancer-promoter interactions. **B.** If the transgene is flanked by insulator elements, these sequences inhibit the spreading of the repressive chromatin conformation, allowing an open chromatin conformation and normal transcription of the gene

inserted into the genome from the repressive effect of adjacent sequences. The repression of transcription observed in many transgenes is thought to be a consequence of their insertion into regions of the genome in which the DNA is assembled into a compact chromatin structure (Fig. 2A); but if the transgene is flanked by insulator sequences, transcription is normal; this observation can be interpreted on the basis of an effect of the insulator on the spreading of the compact chromatin conformation into sequences of the transgene (Fig. 2B). Using as an assay system one or both of these properties of insulators, these novel regulatory sequences have been identified in a variety of systems; three cases have been analyzed in detail: the insulator present in the chicken β-globin gene, the boundary elements found in one of the *Drosophila hsp70* loci and the insulator of the gypsy retrovirus of *Drosophila*.

Perhaps the best-studied example of an insulator element in a vertebrate system is the case of the 5′ constitutive hypersensitive site present in the chicken β-globin locus adjacent to the locus control region (LCR) (Chung et al. 1993). This LCR acts as an activator of expression of all globin genes present in the domain by decondensing the higher-order chromatin structure of this region in erythroid tissues (Elder et al. 1990). This decondensation extends for several hundred kilobases in the 3′ direction, but is limited in the 5′ direction by the location of a strong constitutive hypersensitive site that is present in all tissues examined, regardless of the transcriptional state of the globin gene. This property suggests that this site, named 5'HS4, might be the natural boundary of the β-globin domain. This hypothesis was tested by analyzing the ability of a 1.2-kb DNA fragment containing the 5'HS4 site to insulate a gene from the effects of the LCR. When a reporter gene containing the γ-globin promoter and the coding region of the bacterial neomycin gene is stably transformed into human erythroleukemia cells, the presence of the LCR results in an increase in reporter gene expression, as well as a 30–100-fold increase in the number of resistant colonies obtained (Chung et al. 1993). Insertion of the 5'HS4 fragment between the LCR and the reporter gene causes a tenfold reduction in the number of resistant colonies, and this reduction is even more dramatic when additional copies of the 5' element are used, leading to an almost complete insulation of LCR activation. This effect is observed only when the 5' element is placed between the LCR and the reporter gene but not when located outside, supporting the premise that this sequence behaves as a typical insulator. The same results were obtained with the 5' hypersensitive site of the human β-globin gene (Li and Stamatoyannopoulos 1994). The precise location of insulator sequences within the 1.2-kb fragment of the chicken β-globin gene tested originally has been determined by deletion analysis (Chung et al. 1997). Deletion of a 250-bp region containing the 5'HS4 site results in almost complete loss of insulator activity; this core sequence has some activity itself, and its strength increases with copy number. The core insulator sequence has a 70% GC content and contains multiple small internal repeats. Nuclear extracts from the K562 human erythroleukemic cell line give rise to a complex footprint pattern, suggesting that multiple proteins factors might interact with the globin insulator. In particular, the core sequence contains consensus binding sites for Sp1 and the yeast α2 repressor, although the role of these proteins in insulator activity has not been tested (Chung et al. 1997).

A second insulator that has been studied in detail is that found at the boundaries of one the *Drosophila* heat shock protein 70 gene loci. The *Drosophila* heat shock locus at 87A7 contains two divergently transcribed *hsp70* genes flanked by two unusual chromatin structures that display properties expected of domain boundaries. These specialized chromatin structures, termed scs and scs', are located flanking the puff that arises upon induction of gene expression by heat shock, suggesting that they might demarcate the extent of chromatin that decondenses consequent to induction of transcription by temperature elevation (Udvardy et al. 1985). This hypothesis is supported by the chromatin organization of the region containing these sequences. Each of these elements contains two strong nuclease hypersensitive sites surrounding a nuclease-resistant core. This central structure is flanked by additional weaker nuclease cleavage sites present at intervals corresponding to the length of a nucleosome (Udvardy et al. 1985). The finding of strong hypersensitive sites at the location of the scs and scs' elements correlates with the finding of a similar chromatin organization at the site of the chicken β-globin insulator and suggests that these sequences might have similar functional properties. The insulator properties of the scs and scs' sequences have been confirmed by studying their ability to insulate the *Drosophila hsp70* promoter from the effect of the yolk protein *yp1* enhancer in a hybrid *hsp70-lacZ* reporter gene construct. Activation of transcription of the *lacZ* gene by the the *yp1* enhancer in the female fat body is abolished when scs sequences are interposed between the enhancer and the promoter (Kellum and Schedl 1991, 1992). The insulator role of specific sequences, and particular chromatin structures contained within the originally defined scs element, have been determined by testing the ability of specific subfragments to repress the activation of a downstream promoter by the eye and testis enhancers of the *white* gene (Vazquez and Schedl 1994). Results from this analysis have led to the conclusion that sequences associated with DNase I hypersensitive sites are essential for complete blocking activity of enhancer function, whereas the central nuclease-resistant A/T rich region is dispensable for this effect. Deletion of sequences associated with some hypersensitive sites leads to a reduction in enhancer blocking, whereas multimerization of subfragments with partial activity restores full boundary function (Vazquez and Schedl 1994). Sequences responsible for the boundary function of the scs' element have been characterized in more detail thanks to the identification of insulator-binding proteins (see below). A series of CGATA repeats that interact with the BEAF-32 proteins are responsible for the insulator activity of the scs' sequences (Zhao et al. 1995; Hart et al. 1997). Mutations in this sequence that interfere with binding of the BEAF-32 protein also abolish insulator activity, whereas multimers containing several copies of the sequence display boundary function. The latter results are similar to those obtained with the *Drosophila* scs sequences and the chicken β-globin insulator, and suggest that the effect of boundary elements on transcription might require the binding of a critical number of proteins that somehow cause chromatin alterations as a consequence of their interaction with DNA.

A second insulator element of *Drosophila* is found in the *Fab-7* region of the bithorax complex. *Fab-7* separates *cis*-acting regulatory sequences, named *iab-6* and *iab-7*, that control expression of the *Abdominal-B* (*Abd-B*) gene in parasegments PS11 and PS12. The *Fab-7* region contains a Polycomb-response element (PRE) and a sequence that has the properties of a chromatin boundary or insulator (Hagstrom et al. 1996; Zhou et al. 1996; Mihaly et al. 1997). The latter sequences behave as a typical insulator in an enhancer-blocking assay: when the *Fab-7* element is placed between the eye and testes-specific enhancer and the promoter of the *Drosophila white* gene, transcription is repressed in these two tissues. In addition, deletion of the insulator in the chromosomal DNA results in cross-talk between the *iab-6* and *iab-7* regulatory regions, causing homeotic phenotypes in the adult fly. These results indicate that the *Fab-7* region contains an insulator element that is involved in the normal regulation of the *Abd-B* gene. The location of the insulator has been narrowed down to a 1.2-kb DNA fragment using the enhancer blocking assay described above. This fragment contains one weak and two strong DNaseI hypersensitive sites (Hagstrom et al. 1996; Mihaly et al. 1997).

A third insulator element found in *Drosophila* is associated with the gypsy retrotransposon. This element is 350 bp in length and is located in the 5' transcribed untranslated region of gypsy, upstream from the start of the *gag* open reading frame (reviewed in Gdula et al. 1996). The gypsy insulator was originally identified by the nature of the mutations induced by insertion of this retroelement. In most instances, gypsy inserts into noncoding regions of genes and causes a tissue-specific mutant phenotype due to the inactivation of specific enhancers. The inactivated enhancers are always located distal to the insertion site with respect to the promoter. This polar effect on transcription is one of the characteristic properties of insulator elements, and the mutagenic effect of gypsy on adjacent genes can be explained exclusively by the presence of the internal insulator element (Holdridge and Dorsett 1991; Jack et al. 1991; Geyer and Corces 1992). The gypsy insulator can also buffer the expression of a transgene from position effects from adjacent sequences in the genome (Roseman et al. 1993), and it protects the replication origin of the *Drosophila* chorion genes from similar position effects (Lu and Tower 1997). This insulator contains 12 copies of a 26-bp sequence composed of a core element similar to the octamer motif found in various vertebrate enhancers and promoters; this core element is flanked by an AT-rich region that induces DNA bending and is required for insulator function. Mutations that interfere with bending of the DNA also impair the ability of the insulator to interfere with enhancer-promoter interactions. The strength of the insulator depends on the number of copies of the 26-bp basic motif: one copy causes a very small effect on enhancer activation of transcription, and additional copies result in a stronger effect, with an apparent linear relationship between number of copies and transcriptional repression (Spana and Corces 1990). As in other boundary elements, the gypsy insulator also contains a series of five strong DNaseI hypersensitive sites indicative of a special chromatin organization (D. Gdula and V. Corces, unpubl. observ.).

Several general conclusions can be drawn from these results on the structure of insulator DNA. Insulators are composed of relatively large DNA sequences containing binding sites for multiple proteins or tandem arrays of a binding site for a single specific protein. These sequences determine the establishment of a specific chromatin structure that is manifested in the formation of multiple DNaseI hypersensitive sites. Whether this chromatin conformation is required for, or is simply a consequence of, insulator function is not known at this time.

3
Protein Components of Insulators

With the exception of the scs' and gypsy insulators, the characterization of proteins that interact with insulator DNA has been limited. The chicken β-globin insulator contains Sp1 consensus sequences and it is likely that this protein participates in insulator function, but this has not been demonstrated. The *Fab-7* element has been shown to bind Polycomb (Pc) and GAGA, but the role of these proteins in the function of insulator activity has not been verified. A more extensive analysis has been carried out in the case of the scs' sequences (Zhao et al. 1995; Hart et al. 1997). Two related 32-kDa proteins termed BEAF-32A and BEAF-32B (for boundary element associated factor of 32 kDa) have been purified from nuclei of a *Drosophila* cell line. These proteins bind with high affinity to a site containing three copies of the CGATA motif that flanks the two hypersensitive regions in the scs' sequence. The DNA binding activity resides in the amino-terminal region, which is different in the two proteins; the carboxy terminus is shared and it is involved in heterocomplex formation between the two proteins. The sequence containing BEAF-32 binding sites acts as a typical boundary element in an enhancer-blocking assay involving stable transformation into cultured cells of a reporter gene containing CAT under the control of ecdysone response and heat shock regulatory elements. The BEAF-32 binding site blocks the activity of both heat shock and ecdysone responsive enhancers in stably transfected cells (Zhao et al. 1995). Immunolocalization of BEAF-32 using antibodies shows the presence of this protein in specific subnuclear regions and its exclusion from the nucleolus. BEAF-32 is present in the interband region that separates the highly reproducible and characteristic polytene bands of *Drosophila* third instar larval chromosomes. Interbands contain lower amounts of DNA than bands, and are presumed to be regions of partial unfolding of the 30-nm chromatin fiber. As expected, BEAF-32 is present at the scs'-containing border of the 87A7 chromomere, and is also found at one of the edges of many developmental puffs typically seen in polytene chromosomes at this stage of larval development (Zhao et al. 1995). This observation suggests that BEAF-32 might have general structural and functional roles in defining many boundary elements throughout the *Drosophila* genome.

The gypsy insulator is perhaps the best-studied system with respect to the characterization of protein components that interact with insulator DNA. One of these components, the suppressor of Hairy-wing [su(Hw)] protein, was originally identified on the basis of the observation that mutations in the *su(Hw)* gene reverse the phenotypic effect of gypsy-induced mutations. This observation can now be interpreted in light of the fact that the mutant effect of gypsy is caused by the presence of the insulator and its effect on enhancer-promoter interactions; in the absence of su(Hw) protein, the insulator is not able to repress enhancer function, suggesting that su(Hw) is an essential component of the insulator. The su(Hw) protein contains 12 zinc fingers involved in DNA binding; interaction of the su(Hw) protein with its target sequence is a prerequisite for proper insulator function, as mutations that disrupt the zinc fingers hamper the ability of the insulator to repress enhancer-promoter interactions (Harrison et al. 1993). In addition, su(Hw) contains two acidic domains, located in the amino- and carboxy-terminal ends of the protein. These two domains are dispensable for the effect of the insulator on enhancer function and its ability to buffer against chromosomal position effects. An α-helical region homologous to the second helix-coiled coil region of basic HLH-bzip proteins is absolutely required for insulator function (Harrison et al. 1993). Since leucine zipper regions usually mediate protein-protein interactions, and su(Hw) does not interact with itself, this observation suggests the involvement of other proteins, in addition to su(Hw), in the formation of the boundary element.

A second component of the su(Hw) insulator has been identified by searching for mutations that alter gypsy-induced phenotypes (Georgiev and Gerasimova 1989). Mutations in the *modifier of mdg4* [*mod(mdg4)*] gene reverse the polar effect of the insulator on enhancer function by causing a partial bi-directional repression of all enhancers, independent on whether they are located distal or proximal to the promoter with respect to the location of the boundary element. The *mod(mdg4)* gene encodes at least three different proteins, arising from alternatively spliced RNAs, that contain a BTB domain (Dorn et al. 1993; Gerasimova et al. 1995). The BTB domain is present in various transcription factors (Zollman et al. 1994) including GAGA, a transcriptional activator encoded by the *Trithorax-like* (*Trl*) gene (Farkas et al. 1994). Genetic and molecular analyses suggest that mod(mdg4) proteins interact directly with su(Hw) and therefore constitute a second component of the gypsy insulator (Gerasimova et al. 1995).

4
Effects of Insulators on Chromatin Structure

Insulators affect the ability of enhancers to interact with the promoter and the mechanism of this effect should be intimately linked to that by which enhancers activate transcription. Current views of enhancer function suggest that these sequences mediate gene expression by serving as binding sites for transcription factors that can then interact with protein components of the transcription complex and activate gene expression (see Franklin Chap. 8, this vol.). Based on this, the simplest explanation for the effect of insulators on enhancer-promoter interactions is that insulators act as sinks for enhancer-bound transcription factors that are then unable to interact with the transcription complex. This type of model agrees well with the observation that an enhancer that cannot activate a downstream promoter due to the presence of an insulator can nonetheless activate a second promoter located upstream

(Fig. 1D). This observation has now been confirmed in two different systems (Cai and Levine 1995; Scott and Geyer 1995). When the gypsy insulator or the scs element are interposed between the two enhancers that control expression of the *even-skipped* gene in stripes 2 and 3 during *Drosophila* embryogenesis, transcription from a promoter located on the other side of the insulator is repressed whereas expression from a different promoter located on the same side of the enhancer is still active (Cai and Levine 1995). The same effect has been observed for the shared enhancers present in the intergenic region of the *Drosophila* yolk protein (yp) genes that control transcription of the divergently expressed *yp1* and *yp2* genes in the fat body of the fly. When the gypsy insulator is interposed between the *yp1* promoter and the enhancer, this gene is not expressed but the *yp2* gene is transcribed normally (Scott and Geyer 1995). This conclusion suggests that the presence of an insulator does not disrupt the interaction of transcription factors with enhancer sequences and that its effect must be at the level of direct interference with enhancer-promoter communication; but a wealth of information supports the view that repression of enhancer function by insulators might not be direct but might take place through changes in chromatin conformation, rather than the simple attraction and binding of transcription factors present on the enhancer.

The first direct evidence indicating an effect of insulators on the structure of the primary chromatin fiber came from studies on the chicken β-globin gene insulator. The effect of the LCR on the expression of the genes in the β-globin domain is mediated by changes in chromatin that create an altered nucleosome structure or nucleosome-free region over the promoters of these genes. In particular, an ApaI restriction site present in the γ-globin promoter is inaccessible to the enzyme in HeLa cells, where the gene is inactive presumably due to the presence of a nucleosome over these sequences. In cells in which the γ-globin gene is active, the restriction site becomes accessible due to the displacement of this nucleosome. A similar analysis carried out on the promoter of a transformed reporter gene under the control of the LCR shows a decrease in ApaI accessibility when the reporter gene is separated from the LCR by insulator sequences. These results indicate that the interference of the chicken globin insulator with LCR-promoter interactions involves alterations of chromatin structure that block nucleosome displacement over the promoter region (Chung et al. 1993).

Additional evidence for an involvement of chromatin in insulator function is mostly indirect, but the large number of independent observations builds a supportive body of evidence to bolster this claim. Insulators can buffer a transgene from chromosomal position effects (Roseman et al. 1993); since these effects are thought to be caused by the spreading into the transgene of a condensed chromatin structure from adjacent sequences, this observation suggests that the insulator can interfere with the transmission of this transcriptionally repressive chromatin conformation. Additional evidence suggesting a relationship between the function of the gypsy insulator and chromatin structure comes from the observation that insulator elements can affect dosage compensation in *Drosophila*. In this organism, equal expression of X-linked genes in both sexes is due to a two-fold increase in the transcription of genes on the male X chromosome. This is accomplished by the assembly of multimeric male-specific lethal (MSL) protein complexes on the X chromosome of the male that remodel nucleosome structure through the specific acetylation of Lys 16 on histone H4 (Bone et al. 1994). Transgenes containing X chromosome loci inserted in autosomal regions fail to dosage compensate completely, suggesting an inhibitory effect of autosomal chromatin on the compensation process. Nevertheless, when the copy of the transgene is flanked on both sides by the gypsy insulator, almost 90% of autosomal insertions show proper dosage compensation (Roseman et al. 1995). A third example of the involvement of chromatin structure in insulator function is illustrated by the similarities between dominant position effects in the *brown-Dominant* (*bw^D*) mutation of *Drosophila* caused by adjacent heterochromatic sequences (Dreesen et al. 1991) and trans-effects caused by *mod(mdg4)* mutations. The effect of the gypsy insulator in the absence of mod(mdg4) can be transmitted to the paired homologous chromosome causing repression of specific enhancers (Georgiev and Corces 1995). This tissue specificity suggests an effect on enhancer function rather than a general effect on the promoter. In an otherwise wild type fly, a female heterozygous for the gypsy-induced *y^2* mutation shows normal coloration, since expression from one of the copies of the *yellow* gene is normal. Nevertheless, in the background of a mutation in *mod(mdg4)*, the flies have abnormal bristle pigmentation, suggesting that the bristle enhancer is not functional. This effect is not observed when one of the copies of the *yellow* gene is moved by a chromosomal rearrangement that disrupts pairing between the two homologues. The simplest explanation for this observation is that the silencing effect of su(Hw) in the absence of the mod(mdg4) protein can be transmitted to the homologous paired chromosome and interferes with the function of the bristle enhancer (Georgiev and Corces 1995). This trans-effect is similar to the dominant position-effect variegation observed at the *Drosophila brown* (*bw*) locus in the *bw^D* allele (Dreesen et al. 1991). In this case, heterochromatic sequences adjacent to the *bw* gene in one chromosome cause *trans*-inactivation of the normal gene located in the paired homologue, suggesting that the inhibitory effect of heterochromatin can be transmitted between chromosomes.

Much of the evidence supporting an effect of the gypsy insulator on chromatin structure comes from the analysis of the mod(mdg4) protein itself. The *mod(mdg4)* gene has two important properties: it acts as an enhancer of position effect variegation and it is a member of the *trithorax-Group* (*trxG*) gene family. Mutations in *mod(mdg4)* enhance the phenotype of the variegated *white-apricot* (*w^a*) allele caused by a rearrangement of the *white* gene close to

centromeric heterochromatin. The phenotype of this mutation is characterized by patches of eye tissue pigmented orange/brown in a background of cells
pigmented white/yellow. This phenotype is caused by spreading of a repressive
chromatin conformation from the centric heterochromatin that results in partial or complete inactivation of *white* gene expression in a clonal manner.
Mutations in mod(mdg4) behave as typical enhancers of position-effect variegation, suggesting that the lack of mod(mdg4) protein allows the spreading of
heterochromatin and, therefore, this protein is normally involved in repressing the formation of heterochromatin (Dorn et al. 1993; Gerasimova et al.
1995). This role for mod(mdg4) is also evident in the effect of *mod(mdg4)*
mutations on the function of the gypsy insulator, which loses its directional
effect on enhancer function and acquires the properties of a silencer in the
absence of this protein. For example, in the background of mutations in the
mod(mdg4) gene, the effect of the gypsy insulator on the upstream body
cuticle enhancer is partially suppressed in some cells and partially enhanced in
others, giving rise to a variegated phenotype similar to that observed in
mutations caused by a rearrangement of the *yellow* gene near centromeric heterochromatin. This effect can be interpreted as due to the
heterochromatization of sequences adjacent to the insulator in the absence of
the mod(mdg4) protein (Gerasimova et al. 1995). Since changes in gene expression caused by position-effect variegation are concomitant with alterations in chromatin structure (Wallrath and Elgin 1995), the enhancer silencing
observed in *mod(mdg4)* mutants might also be caused by changes in the
packaging of nucleosomes.

A second property of the *mod(mdg4)* gene suggestive of an involvement in
chromatin phenomena is the finding that this gene has the hallmarks of members of the trxG family: mutations in mod(mdg4) enhance the phenotype of
other *trxG* mutants, reverse the dominant phenotype of *Polycomb* (*Pc*) and
result in lack of expression of homeotic genes (Gerasimova and Corces 1998).
trxG proteins are thought to act by antagonizing the effect of PcG proteins,
which repress the expression of homeotic and presumably other genes in a
manner that can be transmitted through cell division. PcG proteins appear to
form multi-protein complexes that permanently repress transcription of genes
in the absence of other factors originally required to turn the genes off
(Kennison 1995; Paro 1995; Simon 1995). The mechanism for this stable repression is not well understood, but it is thought to involve alterations of
chromatin structure (Peterson and Tamkun 1995; Paro and Harte 1996). The
observation of a shared pathway in the function of a chromatin insulator and
trxG and *PcG* genes is suggestive of a possible commonality in their mechanism of action. Other *trxG* members, such as the *Trl* gene that encodes the
GAGA factor, have also been involved in regulating gene expression at the level
of chromatin. Like *mod(mdg4)*, mutations in *Trl* act as enhancers of position-
effect variegation. Mutations in *Trl* result in lower levels of expression of
homeotic genes by a mechanism involving chromatin packaging (Farkas et al.

1994), and the GAGA protein itself has been shown to remodel chromatin in vitro (Pazin et al. 1994; Tsukiyama et al. 1994). The similarities among components of the gypsy insulator, modifiers of position-effect variegation and *trxG* genes suggest that the mechanisms by which these proteins affect transcription might have a common molecular and cellular basis.

5
Mechanisms of Insulator Function: Establishment of Higher-Order Chromatin Domains

Some of the characteristics of insulators and the biological properties of their protein components suggest that some of their effects on chromatin structure might take place not only at the level of the organization of the primary chromatin fiber, but also by regulating the establishment of higher order domains of chromatin organization. Figure 3 summarizes a model explaining insulator function at this level. Insulator-binding proteins, through interactions among themselves and perhaps with specific subnuclear structures, establish functional domains of gene activity. Enhancer elements present in one domain can activate transcription from promoters located in the same domain (Fig. 3A). When a second insulator is inserted between the two enhancers, the new insulator separates the original domain into two topologically isolated ones (Fig. 3B). Under these conditions, the enhancer present in one domain is unable to activate transcription from a promoter located in the second one. Evidence supporting this model comes from the analysis of the properties and distribution of the mod(mdg4) protein.

The mod(mdg4) protein is present in approximately 400 sites on polytene chromosomes of *Drosophila* larvae. Half of these sites overlap with sites of su(Hw) localization, such that mod(mdg4) is present in all su(Hw) sites but the two proteins do not overlap in the rest. These sites do not correspond to locations of the gypsy retroelement, as most fly strains only contain a few copies of this transposon. It is thought that these sites correspond to endogenous insulators that play a role in the normal organization of the genome. In a *su(Hw)* mutant fly, neither su(Hw) nor mod(mdg4) are present at common sites, suggesting that the two proteins interact and that su(Hw) is necessary for the binding of mod(mdg4) to the chromosomes. In addition, mod(mdg4) overlaps with other trxG and PcG proteins at some but not all chromosomal locations (Gerasimova and Corces 1998).

An important clue to explain the mechanism of insulator action came from the observation that mutations in *trxG* genes enhance the effect of the gypsy insulator whereas mutations in *Pc* have the opposite effect and reverse the effect of the insulator on enhancer-promoter interactions. Interestingly, trxG and PcG proteins such as trx, GAGA, ash1, and Pc are not present at gypsy insulator sites, suggesting that their effect on the insulator is indirect. The effect of these proteins on insulator function appears to be at the level of the

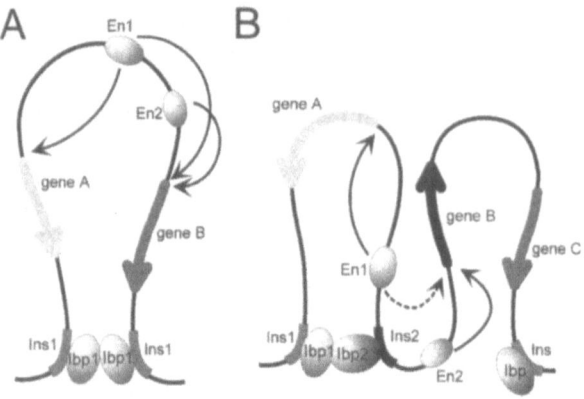

Fig. 3A–B. Mechanism of insulator effect on enhancer function at the level of higher order domains of chromatin organization. **A.** Diagram of two genes, *A* and *B*, located within a chromosomal domain defined by two insulator sequences (*Ins1*) and their associated proteins (*Ibp1*). Enhancers located between the two genes (*En1* and *En2*) can activate transcription from the promoter of either gene. **B.** If a boundary element such as the gypsy insulator (*Ins2*) is inserted between the two enhancers, a new chromosomal domain forms, leaving gene *A* in one domain and gene *B* outside. One of the insulators forming the original domain is now free to form other domains with alternative boundary elements (in this case containing genes *C* and *D*). Enhancer 1 (*En1*) is now unable to act on the promoter of gene B due to the new location of the second insulator (*Ins2*). Nevertheless, this enhancer is still functional and competent to activate transcription from the promoter of gene *A*, located within the same chromosomal domain

subnuclear distribution of insulator components. The su(Hw) and mod(mdg4) proteins are not distributed uniformly in the nuclei of interphase diploid cells. Rather, they appear to be localized in a punctated pattern, with dots dispersed throughout the nucleus but concentrated mainly in the nuclear lamina (Gerasimova and Corces 1998). The number of dots per nucleus varies between 10 and 20, and the patterns of su(Hw) and mod(mdg4) localization are identical. This suggests that 20–40 individual DNA binding sites for these two proteins come together in specific subnuclear regions. Interestingly, trxG and PcG proteins are also distributed in a punctated pattern that is relegated mostly to the central region of the nucleus and does not overlap with that of su(Hw)/mod(mdg4). The distribution of trxG and PcG proteins is not altered by mutations in *su(Hw)* or *mod(mdg4)*, but the punctated pattern of su(Hw)/mod(mdg4) localization changes dramatically in the presence of mutations in *trxG* or *PcG* genes (Gerasimova and Corces 1998). In *trxG* mutants, the su(Hw) and mod(mdg4) proteins loose their nuclear localization and appear distributed in a diffuse pattern mostly in the cytoplasm. In the background of *Pc* mutations, the localization of these two proteins is nuclear, but the punctated patter disappears to give rise to a uniform nuclear localization. These results

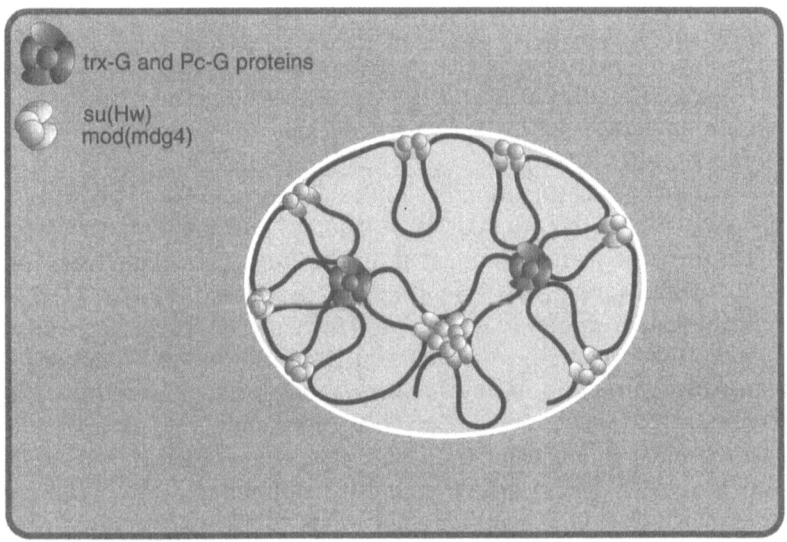

Fig. 4. Schematic model explaining the role of trxG and PcG proteins in the function of the gypsy insulator. The diagram represents a cell (*dark gray square*) with a nucleus (*light gray oval*). The chromatin fiber is represented as a *black line* and proteins are represented as *ovals colored in light gray* [mod(mdg4) and su(Hw)] *and dark gray* (various trxG and/or PcG proteins). The nuclear lamina is represented as *a white line surrounding the nucleus*

suggest a model in which both insulator components and members of the trxG and PcG proteins function to organize the chromatin fiber within the nucleus by attaching the DNA to specific nuclear compartments (Fig. 4). The su(Hw) and mod(mdg4) proteins are mostly involved in organizing the DNA around the nuclear periphery, perhaps by attaching the chromatin fiber to the nuclear lamina. trxG and PcG proteins appear to play a similar role in the central region of the nucleus. The relative effects of mutations in the genes encoding these proteins suggest that trxG and PcG proteins play a primary role in this organization in that the structure set up by these proteins is required for the establishment of insulators. The role of su(Hw) and mod(mdg4) is secondary and it is not required for the primary level of organization established by trxG and PcG proteins (Gerasimova and Corces 1998).

6
Insulators Might Correspond to MARs/SARs

The localization of insulator components to the nuclear lamina and their possible role in the organization of the chromatin fiber within the nucleus are properties characteristic of scaffold or matrix attachment regions (SARs or MARs) (Mirkovitch et al. 1984; Cockerill and Garrad 1986). The similarity between properties of the gypsy insulator described above and those of MARs/SARs suggest that these two types of sequences might play similar or identical roles in the biology of the eukaryotic nucleus. MARs/SARs are AT-rich DNA sequences, often containing topoisomerase II cleavage sites, that mediate the anchoring of the chromatin fiber to the chromosome scaffold or nuclear matrix and delimit the boundaries of discrete and topologically independent higher order domains (Laemli et al. 1992). Proteins bound to these sequences tether the chromatin fiber into structural loops held together by the components of the chromosome scaffold. It is easy to visualize how these structural domains that organize the chromosome could also serve a functional purpose to organize genes into specific hierarchies of regulated transcription by acting as boundary or insulator elements. Although some SARs have failed to confer position-independent expression on reporter genes (Poljack et al. 1992), MARs from genes such as the chicken lysozyme and human β-interferon can ensure elevated expression of reporter genes when stably integrated in the chromosome independent of their location (Bonifer et al. 1997). MARs flanking the human apolipoprotein B gene have also been shown to have similar properties (Kalos and Fournier 1995). The nuclear attachment sites in the apoB gene are located 5 kb upstream and 43 kb downstream from the transcription start site; these sequences contain DNase I hypersensitive sites at the junctions between regions of closed versus open chromatin. These properties suggest that the 5′ and 3′ MARs are the physical boundaries of the apoB chromatin domain, and recent results indicate that they also have a functional role in the control of apoB expression unrelated to classical enhancer activity. The apoB MARs can direct proper expression of a reporter gene in stably transformed but not transiently transfected human hepatoma cells. Transformed cells containing a single copy of the reporter gene support low and variable transcription. In contrast, expression of the reporter flanked by 5′ and 3′ MARs was 200-fold higher and independent of its integration site (Kalos and Fournier 1995). The same type of results has been obtained using transgenic mice (Wang et al. 1996). Whether the ability of MARs/SARs to act as insulators is a general phenomenon is not clear at this time, but the localization of protein components of the gypsy insulator to the nuclear matrix strongly suggests similar biological roles and mode of action for these two types of regulatory elements.

7
Insulators and Imprinting

Based on the properties of insulators and their mechanism of action, the establishment of the imprinted state could be explained by the presence of insulator sequences at imprinted loci. The *H19* and *Igf2* genes of the mouse

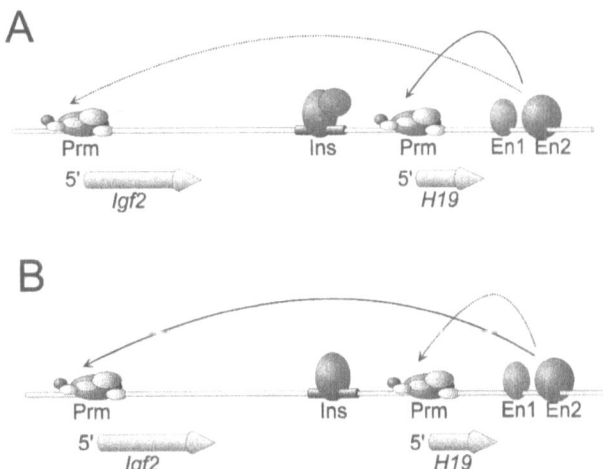

Fig. 5. Effect of insulators on the imprinting state at the *H19* locus. Symbols are as in Fig. 1. **A.** In the maternally derived chromosome, a functional insulator (*Ins*) located upstream of *H19* inhibits the interaction of enhancers (*En1* and *En2*) with the promoter (*Prm*) of the *Igf2* gene, whereas transcription of *H19* is normal. **B.** In the paternally derived chromosome, one of the insulator components is lacking, causing the formation of a promoter-specific silencer. In this case, transcription of *H19* is repressed and expression of *Igf2* is normal Department of Biology, The Johns Hopkins University, 3400 N. Charles St., Baltimore, Maryland 21218, USA

perhaps best illustrate this model. The expression of these two genes is controlled by tissue-specific enhancers located 3' to *H19*. Expression of *Igf2* is activated in the paternally derived chromosome, whereas transcription of *H19* takes place from the maternal chromosome. This imprinting effect depends on sequences located between *Igf2* and *H19*; methylation of these sequences correlates with repression of *H19* in the paternal chromosome, but in the absence of methylation in the maternal chromosome *H19* is expressed and *Igf2* is repressed (see Bartolomei and Tilghman 1997 for a recent review). Methylation of imprinted sequences is required to maintain the imprinted state of these genes but it is not clear whether methylation constitutes the initial causative agent required to allow distinction between the parental alleles. One possible model to explain these results is that sequences required for imprinting, located downstream of the *Igf2* gene and upstream of *H19*, contain an insulator (Fig. 5). The activity of this insulator must be dependent on the parental origin of the chromosome. In the maternally derived chromosome, the insulator is functional and the enhancers located downstream from *H19* can activate this gene but are unable to turn on *Igf2* transcription. In the paternally derived chromosome, the insulator is nonfunctional and the *Igf2* gene is expressed. This simple model, based on the basic properties of insulators, does not explain why *H19* is not transcribed in the paternally derived chromosome. In addition, the model requires the same insulator to be functional in one chromosome but not in the other. Both objections could be explained by assuming decreased levels of one of the insulator proteins in the paternally-derived chromosome to form an altered insulator. This explanation is based on the properties observed for the gypsy insulator formed in flies carrying a hypomorphic mutation in the *mod(mdg4)* gene. In this mutant, the amount of mod(mdg4) protein present in the insulator is decreased with respect to wild type, and the insulator shows altered properties in its interactions with enhancers and promoters. In a *mod(mdg4)* mutant, the gypsy insulator behaves as a promoter-specific silencer that can repress expression of some genes but not others, depending on the structure of their promoters (Cai and Levine 1997). Thus, if we assume that a mod(mdg4)-like component of the *H19* insulator is present in the male primordial germ cells at lower levels than in the female gamete progenitors, the properties of the *H19* insulator that forms in these two cell types will be different. The function of the insulator will be normal in the female gametes, whereas in the male gametes the insulator will behave as a promoter-specific silencer. This model explains the transcriptional behavior of the *Igf2* and *H19* genes in each of the two chromosomes without a requirement for ovary- and sperm-specific de novo methylases capable of recognizing sequence differences between the female and male-derived genes. The altered properties of the gypsy insulator assembled in the presence of low levels of mod(mdg4) protein in a *mod(mdg4)* mutant fly are maintained through many cell division, from embryonic to adult stages, in the wild type progeny of this fly (Gerasimova and Corces, 1998). Therefore, the assembly of an altered insulator at the *H19* locus in the male germ line only requires limiting amounts of the mod(mdg4)-like protein in these cells, and its properties will be maintained in somatic cells even in the presence of normal amounts of the mod(mdg4) homolog. Interestingly, and in support of this model, sequences required for *H19* imprinting that show hypermethylation in the paternally-derived chromosome have been shown to function as a parent of origin-specific silencer in *Drosophila* (Lyko et al. 1997). Since *Drosophila* does not methylate its DNA, these sequences conserve their silencing properties in the absence of methylation, suggesting that this modification might not be the primary cause of the mechanism underlying imprinting. DNA methylation of imprinted sequences might simply be a secondary mark to the silencing phenomenon, perhaps required in higher eukaryotes by the complexity of their genome.

Acknowledgments. Work carried out in the authors' laboratory is supported by United States Public Health Service award GM35463 from the National Institutes of Health.

References

Bartolomei MS, and Tilghman, SM (1997) Genomic imprinting in mammals. Annu Rev Genet 31:493–525

Bone JR, Lavender J, Richman R, Palmer MJ, Turner BM, and Kuroda MI (1994) Acetylated histone H4 on the male X chromosome is associated with dosage compensation in *Drosophila*. Genes Dev 8:96–104

Bonifer C, Faust N, Huber MC, Saueressig H, and Sippel AE (1997) The chicken lysozyme domain. In: van Driel R, Otte AP (eds) Nuclear organization, chromatin structure, and gene expression. Oxford University Press, Oxford, pp 116–128

Cai H, and Levine M (1995) Modulation of enhancer-promoter interactions by insulators in the *Drosophila* embryo. Nature 376:533–536

Cai HN, and Levine M (1997) The gypsy insulator can function as a promoter-specific silencer in the *Drosophila* embryo. EMBO J 16:1732–1741

Callan HG (1986) Lampbrush chromosomes. Springer, Berlin Heidelberg New York

Chung JH, Whiteley M, and Felsenfeld G (1993) A 5′ element of the chicken β-globin domain serves as an insulator in human erythroid cells and protects against position effects in *Drosophila*. Cell 74:505–514

Chung JH, Bell AC, and Felsenfeld G (1997) Characterization of the chicken β-globin insulator. Proc Natl Acad Sci USA 94:575–580

Cockerill PN, and Garrad WT (1986) Chromosomal loop anchorage of the kappa immunoglobulin gene occurs next to the enhancer in a region containing topoisomerase II sites. Cell 44:273–282

Dorn R, Krauss V, Reuter G, and Saumweber H (1993) The enhancer of position-effect variegation of *Drosophila E(var)3-93D* codes for a chromatin protein containing a conserved domain common to several transcriptional regulators. Proc Natl Acad Sci USA 90:11376–11380

Dreesen TD, Henikoff S, and Loughney K (1991) A pairing-sensitive element that mediates transinactivation is associated with the *Drosophila brown* gene. Genes Dev 5:331–340

Elder JT, Forrester WC, Thompson C, Mager D, Henthorn P, Peretz M, Papayannopoulou T, and Groudine M (1990) Translocation of an erythroid-specific hypersensitive site in deletion-type hereditary persistence of fetal hemoglobin. Mol Cell Biol 10:1382–1389

Farkas G, Gausz J, Galloni M, Reuter G, Gyurkovics H,and Karch F (1994) The *Trithorax-like* gene encodes the *Drosophila* GAGA factor. Nature 371:806–808

Gdula DA, Gerasimova TI, and Corces VG (1996) Genetic and molecular analysis of the gypsy chromatin insulator of *Drosophila*. Proc Natl Acad Sci USA 93:9378–9383

Georgiev PG, and Corces VG (1995) The *su(Hw)* protein bound to gypsy sequences in one chromosome can repress enhancer-promoter interactions in the paired gene located in the other homolog. Proc Natl Acad Sci USA 92:5184–5188

Gerasimova TI, and Corces VG (1998) Polycomb and thritorax group proteins mediate the funciton of a chromatin insulator. Cell 92:511–521

Georgiev PG, and Gerasimova TI (1989) Novel genes influencing the expression of the *yellow* locus and mdg4 (gypsy) in *Drosophila melanogaster*. Mol Gen Genet 220:121–126

Gerasimova TI, Gdula DA, Gerasimov DV, Simonova O, and Corces VG (1995) A *Drosophila* protein that imparts directionality on a chromatin insulator is an enhancer of position-effect variegation. Cell 82:587–597

Geyer PK, and Corces VG (1992) DNA position-specific repression of transcription by a *Drosophila* Zn finger protein. Genes Dev 6:1865–1873

Hagstrom K, Muller M, and Schedl P (1996) Fab-7 functions as a chromatin domain boundary to ensure proper segment specification by the *Drosophila bithorax* complex. Genes Dev 10:3202–3215

Harrison DA, Gdula DA, Coyne RS, and Corces VG (1993) A leucine zipper domain of the *suppressor of Hairy-wing* protein mediates its repressive effect on enhancer function. Genes Dev 7:1966–1978

Hart C, Zhao K, and Laemmli UK (1997) The scs' boundary element: characterization of boundary element-associated factors. Mol Cell Biol 17:999–1009

Holdridge C, and Dorsett D (1991) Repression of hsp70 heat shock gene transcription by the *suppressor of Hairy-wing* protein of *Drosophila melanogaster*. Mol Cell Biol 11:1894–1900

Jack J, Dorsett D, Delotto Y, and Liu S (1991) Expression of the *cut* locus in the *Drosophila* wing margin is required for cell type specification and is regulated by a distal enhancer. Development 113:735–748

Kalos M, and Fournier REK (1995) Position-independent transgene expression mediated by boundary elements from the apolipoprotein B chromatin domain. Mol Cell Biol 15:198–207

Kellum R, and Schedl P (1991) A position-effect assay for boundaries of higher order chromatin domains. Cell 64:941–950

Kellum R, and Schedl P (1992) A group of scs elements function as domain boundaries in an enhancer-blocking assay. Mol Cell Biol 12:2424–2431

Kennison JA (1995) The Polycomb and trithorax group proteins of *Drosophila*: *trans*-regulators of homeotic gene function. Annu Rev Genet 29:289–303

Laemli UK, Käs E, Poljak L, and Adachi Y (1992) Scaffold-associated regions: *cis*-acting determinants of chromatin structural loops and functional domains. Curr Opin Genet Dev 2:275–285

Lewis M, Helmsing PJ, and Ashburner M (1975) Parallel changes in puffing activity and patterns of protein synthesis in salivary glands of *Drosophila*. Proc Natl Acad Sci USA 72:3604–3608

Li Q, and Stamatoyannopoulos G (1994) Hypersensitive site 5 of the human beta locus control region functions as a chromatin insulator. Blood 84:1399–1401

Lu L, and Tower J (1997) A transcriptional insulator element, the su(Hw) binding site, protects a chromosomal DNA replication origin from position effects. Mol Cell Biol 17:2202–2206

Ludérus MEE, and van Driel R (1997) Nuclear matrix-associated regions. In: van Driel R and Otte AP (eds) Nuclear organization, chromatin structure, and gene expression. Oxford University Press, Oxford, pp 99–115

Lyko F, Brenton JD, Surani MA, and Paro R (1997) An imprinting element from the mouse H19 locus functions as a silencer in *Drosophila*. Nat Genet 16:171–173

Mihaly J, Hogga I, Gausz J, Gyurkovics H, and Karch F (1997) In situ dissection of the Fab-7 region of the *bithorax* complex into a chromatin domain boundary and a Polycomb-response element. Development 124:1809–1820

Mirkovitch J, Mirault M-E, and Laemmli UK (1984) Organization of the higher-order chromatin loop: specific DNA attachment sites on nuclear scaffold. Cell 39:223–232

Paro R (1995) Propagating memory of transcriptional states. Trends Genet 11:295–297

Paro R, and Harte PJ (1996) The role of Polycomb group and trithorax group chromatin complexes in the maintenance of determined cell states. In: Russo VEA, Martienssen RA and Riggs AD (eds) Epigenetic mechanisms of gene regulation, Cold Spring Harbor Laboratory Press, Cold Spring Harbor, pp 507–528

Pazin MJ, Kamakaka RT, and Kadonaga JT (1994) ATP-dependent nucleosome reconfiguration and transcriptional activation from preassembled chromatin templates. Science 266:2007–2011

Peterson CL, and Tamkun JW (1995) The SWI-SNF complex: a chromatin remodeling machine? Trends Biochem Sci 20:143–146

Poljak L, Seum C, Mattioni T, and Laemmli UK (1994) SARs stimulate but do not confer position independent gene expression. Nucleic Acids Res 22:4386–4394

Roseman RR, Pirrotta V, and Geyer PK (1993) The *su(Hw)* protein insulates expression of the *Drosophila melanogaster white* gene from chromosomal position-effects. EMBO J 12:435–442

Roseman RR, Swan JM, and Geyer PK (1995) A *Drosophila* insulator protein facilitates dosage compensation of the X chromosome *mini-white* gene located at autosomal insertion sites. Development 121:3573–3582

Scott KS, and Geyer PM (1995) Effects of the *Drosophila* su(Hw) insulator protein on the expression of the divergently transcribed yolk protein genes. EMBO J 14:6258–6279

Simon J (1995) Locking in stable states of gene expression: transcriptional control during *Drosophila* development. Curr Opin Cell Biol 7:376–385

Spana C, and Corces VG (1990) DNA bending is a determinant of binding specificity for a *Drosophila* zinc finger protein. Genes Dev 4:1505–1515

Thompson CC, and McKnight SL (1992) Anatomy of an enhancer. Trends Genet 8:232–236

Tissières A, Mitchell HK, and Tracy UM (1974) Protein synthesis in salivary glands of *Drosophila melanogaster*: relation to chromosome puffs. J Mol Biol 84:389–398

Tsukiyama T, Becker PB, and Wu C (1994) ATP-dependent nucleosome disruption at a heat shock promoter mediated by binding of GAGA transcription factor. Nature 367:525–532

Udvardy A, Maine E, and Schedl P (1985) The 87A7 chromomere: identification of novel chromatin structures flanking the heat shock locus that may define the boundaries of higher order domains. J Mol Biol 185:341–358

Vazquez J, and Schedl P (1994) Sequences required for enhancer blocking activity of scs are located within two nuclease-hypersensitive regions. EMBO J 13:5984–5993

Wallrath LL, and Elgin SC (1995) Position effect variegation in *Drosophila* is associated with an altered chromatin structure. Genes Dev 9:1263–1277

Wang DM, Taylor S, and Levy-Wilson B (1996) Evaluation of the function of the human apolipoprotein B gene nuclear matrix associated regions in transgenic mice. J Lipid Res 37:2117–2124

Zhao K, Hart CM, and Laemmli UK (1995) Visualization of chromosomal domains with boundary element-associated factor BEAF-32. Cell 81:879–889

Zhou J, Barolo S, Szymanski P, and Levine M (1996) The Fab-7 element of the *bithorax* complex attenuates enhancer-promoter interactions in the *Drosophila* embryo. Genes Dev 10:3195–3201

Zollman S, Godt D, Prive GG, Couderc JL, and Laski FA (1994) The BTB domain, found primarily in zinc finger proteins, defines an evolutionarily conserved family that includes several developmentally regulated genes in *Drosophila*. Proc Natl Acad Sci USA 91:10717–10721

A Role for Modifier Genes in Genome Imprinting

C. Cristofre Martin[1] and Carmen Sapienza[1,2]

1
Introduction

It is, by now, an old observation that the phenotype elicited by a mutant allele at a particular locus is not the sole result of the DNA sequence of that allele. The truth of this statement may be easily recognized by comparing different individuals within the same family, each of whom has the same genetic disease. Siblings who have neurofibromatosis, for example, all share the same disease allele at the *NF1* locus but the manifestations of disease (number and size of cafe au lait spots, number and size of neurofibromas, presence of neurofibrosarcoma, etc.) among the siblings may vary dramatically (Easton et al. 1993). These variations in disease phenotype between individuals who carry the same mutant allele have been subsumed under the mechanistically vague concept that many diseases exhibit variable expressivity (Thompson et al. 1991). Such differences in disease phenotype may result from many factors, including stochastic processes and environmental influences, but at least some of these differences are thought to result from the collective action of additional genetic factors that differ between individuals. These additional genetic factors are said to modify the phenotype and are, therefore, called modifier genes.

For much of this century, the major focus in the study of modifier genes revolved around the population genetic and evolutionary implications of R.A. Fisher's controversial treatise, "The possible modification of the response of the wild type to recurrent mutations" (Fisher 1928a). The substance of this debate, which continues to the present day (Mayo and Bürger 1997), is not over whether modifier genes exist or whether they may engender significant phenotypic consequences, but over whether the dominant or recessive nature of mutations is a result of the action of natural selection operating on modifier genes. Although this debate is of great interest and importance in genetics and evolutionary biology, and may have particular importance in the study of

[1] Fels Institute for Cancer Research and Molecular Biology
[2] Department of Pathology and Laboratory Medicine, Temple University School of Medicine, 3307 N. Broad Street, Philadelphia, Pennsylvania 19140, USA

genome imprinting, the study of modifier genes at this level is one step removed from the interests of those who wish to know where in the genome modifier genes reside and just how the products of these genes act to modify the phenotype of an allele at an unlinked locus. There has been substantial progress made toward this end in both yeast and *Drosophila*, but in mammalian systems the study of modifier genes has, until recently, occupied approximately the same role in genetics as did the weather in Mark Twain's famous quip, "Everybody talks about the weather but nobody does anything about it."

This comparative neglect has not been due to lack of interest (in either case), but stems from the difficulty inherent in identifying and characterizing genes that modify any particular phenotype. This difficulty has appeared almost insurmountable in some outbred populations, such as the human, in which multiple modifiers of the same phenotype may be segregating independently. Given these problems, it is small wonder that once the primary locus involved in generating a particular disease phenotype is identified, a rash of studies on genotype/phenotype correlation soon follows. Demonstrating that at least some of the phenotypic variablity observed between individuals in different families with the same disease may be ascribed to different mutant alleles at the disease locus is a far more tractable problem than that posed by identifying the unlinked genetic factors involved in variable expressivity. Unfortunately, it is one of nature's cruelties that the operation of modifier genes is likely to have the most interesting consequences and be most important in generating phenotypic diversity in just such genetically heterogeneous populations.

Despite these complexities, it is likely that the application of sophisticated genetic epidemiological techniques will uncover the chromosomal location and identification of modifiers of many human disease phenotypes in the relatively near future (Risch and Merikangas 1996). Major modifiers of many mutant phenotypes in the mouse are also likely to be identified in the near future as more geneticists turn their attention to the analysis of quantitative traits. Extrapolation of studies on the mouse model system to the human is also likely to identify additional modifiers of complex human traits. In this way, the identification, isolation and biochemical characterization of a collection of modifier genes and their products may reveal unsuspected pathways of disease pathogenesis, as well as yield important clues as to the major mechanisms by which phenotypic diversity is generated within populations.

In the context of this chapter, modification is defined at the level of phenotype. It is worth stressing that phenotype encompasses a wide range of characters. It may be a trait observed at a gross level, such as whether a mouse is large or small, or it may be a more subtle, laboratory-derived representation of a biochemical character, such as an X-ray film exposure of a Southern blot, showing the relative proportion of *Hpa*II sites in which a cytosine in the recognition site is methylated. While this chapter is not intended to be a comprehensive review, we hope to accomplish two things in the following discussion: (1) to review some published examples on the importance and mode of action of modifier genes; and (2) to provide some historical perspective and suggest a general mechanism by which modifier genes might exert an effect on phenotypes resulting from imprinted genes.

2
Modification of Disease Phenotype

Susceptibility or resistance to many diseases is thought to be multigenic. It is therefore not surprising that genetic analysis has identified many modifier loci for any particular disease (Table 1). By inference, the action of modifier genes is likely to account for the high variability in penetrance and severity of a number of human diseases, including some cancers. Because rapid progress in the techniques of genetic screening now allows the determination of whether individuals harbor mutations at some primary disease loci, the role of modifier genes in both diagnosis and disease prognosis will become increasingly important. The example of one modifying gene identified recently serves to illustrate this point.

Mutations in the adenomatous polyposis coli gene (Apc/APC) lead to the formation of intestinal neoplasms in both the mouse and human. The tumorigenic effects of one mutation of this gene in the mouse (Apc^{Min}) was found to be influenced strongly by genetic background (Moser et al. 1992). A major modifier ($Mom1$) was later mapped to a region of mouse chromosome 4, while a major modifier of the human disease was mapped to a syntenic region of chromosome 1 (Dietrich et al. 1993; Tomlinson et al. 1996; Dobbie et al. 1997). The gene for secretory phospholipase A2 was suggested as a possible candidate for this modifier when it was discovered that $Mom1$-sensitive mouse strains carried a germline mutation in this gene (MacPhee et al. 1995). Secretory phospholipase A2 is involved in synthesizing hormone-like fatty acid prostaglandins, believed to promote tumor cell growth and proliferation (Prescott and White 1996). Support for this hypothesis was reported by Cormier et al. (1997) when they produced Apc^{Min} mice that were transgenic for the $Pla2g2a$ gene. Overexpression of this gene resulted in significant reduction in both the number and size of intestinal tumors. Subsequent analysis of the human disease also indicates that the $PLA2G2A$ gene is the major modifier of colon cancer due to APC mutations (Tomlinson et al. 1996).

3
Strain Background Effects on the Phenotype Induced by Specific Genetic Changes

Much of the evidence for the existence and operation of modifier genes in mice comes from observation of the phenotype elicited by the same genetic

Table 1. Evidence of Genetic modifier loci

Modified locus or trait	Variant phenotype	Genetic evidence	Modifier mapped	Reference
Androgenetic development	Timing of developmental arrest	DBA/2 oocyte: arrest prior to 16 cell stage C57BL/6 oocyte: arrest at blastocyst stage	Two independently segregating loci	Latham and Solter (1991) Latham and Sapienza (1998)
Numerous Drosophila mutations*	Position-effect variegation (mosaic)	genetic analysis, modifier duplication	Many modifiers Su(var) and E(var) enhancers and suppressors	Sass and Henikoff (1998) Locke et al. (1988) see Pirotta, chap. 10, this vol.
Troponin I transgene*	Methylation	DBA/2 background: ↓ methylation C57BL/6 background: ↑ methylation	n.a.	Sapienza et al. (1989)
Tg4-22 transgene*	Expression/methylation (mosaic)	C57BL/6 background: ↑ expression, ↓ methylation BALB/c background: ↓ expression, ↑ methylation	n.a.	McGowan et al. (1989)
Hsp 70.1 transgene	Expression/methylation	C3H background: ↑ expression, ↓ methylation BALB/c background: ↓ expression, ↑ methylation	n.a.	Chastant et al. (1996)
pHRD (1991) transgene*	Methylation	DBA/2 background: ↓ methylation C57BL/6 background: ↑ methylation	Chromosome 4 (Ssm-1)	Engler et al.
TKZ 751 transgene*	Expression/methylation	DBA/2 background: ↑ expression, ↓ methylation BALB/c background: ↓ expression, ↑ methylation	n.a.	Allen et al. (1990)

CMZ 12 transgene*	Expression	DBA/2 background: ↑ expression	n.a.	Surani et al. (1990)
RSVIg myc transgene*	Expression/methylation	BALB/c background: ↓ expression FVB/N background: ↑ expression, ↓ methylation C57BL/6 background: ↓ expression, ↑ methylation	n.a.	Weichman and Chaillet (1997)
pRSVCAT transgene (zebrafish)*	Expression/methylation (mosaic)	various wildtype and semi-inbred stains	n.a.	Martin and McGowan (1995a,b)
hpt transgenic Arabidopsis	Expression/methylation	various mutant strains	n.a.	Scheid et al. (1988)
Disorganization mutant	Skeletal and organ deformaties	129 background: 9% penetrance C3H background: 89% penetrance	n.a.	Hummel (1958) Hummel (1959)
curly tail mutant	Neural tube defects	DBA/2 background: 2% penetrance BALB/c background: 8% penetrance	Chromosome 3 and 5 Chromosome 17 (mct 1)	Neumann et al. (1994) Letts et al. (1995)
Blind mutant	Failure of eyelid closure	C57BL/6 background: 18% penetrance	n.a.	Teicher and Caspari (1978)
Fused mutant (Fu)* (now Axin)	Shortened, kinky tail, neural tube defects, inner ear defects	DBA/2 background: ↑ penetrance BALB/c background: ↓ penetrance C57BL/6 background: 8% penetrance other backgrounds: 90–100% penetrance	n.a.	Ruvinsky and Agulnik (1990)
Tme deletion*	Tail length	various inbred and wild-type mouse strains	Chromosome 7	Forejt et al. (1995)
Juvenile cystic kidneys (jck) mutation	Variability in the degree of polycystic kindey disease (size of kidney)	C57BL/6 background: mild disease severity DBA/2 background: increased disease severity	Chromosome 1 and 10	Iakoubova et al. (1995)

Table 1. *Continued*

Modified locus or trait	Variant phenotype	Genetic evidence	Modifier mapped	Reference
Min mutation (adenomatous polyposis coli gene) (*Apc*)	No. of intestinal adenomas	AKR background: average 4–8 tumors C57BL/6 background: average 29 tumors	Chromosome 4 *Pla2g2a* (secretory phospholipase)	Dietrich et al. (1993) Cormier et al. (1997)
Keratin 8 knockout	Embryonic lethality	C57B/6 background: 1.6% viable FVB/N background: 55% viable	n.a.	Baribault et al. (1994)
Epidermal growth factor receptor knockout	Time of embryonic lethality	CF1 background: preimplantation 129/SV background: mid-gestation CD1 background: 3 weeks	n.a.	Threadgill et al. (1995) Sibilia and Wagner (1995)
Cystic fibrosis transmembrane regulator knockout	Time of lethality	DBA/2 background: prior to 10 days from birth C57BL/6 background: weaning period BALB/c background: prolonged survival	Chromosome 7	Rozmahel et al. (1996)
TGFβ1 knockout	Time of lethality	C57BL/6 background: prenatal lethality NIH background: three weeks post-partum	Chromosome 5	Bonyadi et al. (1997)
Type 1 von Willebrand disease (VWD) (mouse model)	Expression of plasma von Willebrand factor(VWF)	VWF levels are 20-fold higher in RIIIS/J parental strain compared to CASA/Rk parental strain	Chromosome 11	Mohlke et al. (1996)

Disease/condition	Phenotype	Pedigree analysis / background	Chromosome	Reference
Familial adenomatous polyposis (Apc) (human)	No. of intestinal adenomas	Pedigree analysis	Chromosome 1 (syntenic to mouse chromosome 4)	Tomlinson et al. (1996); Dobbie et al. (1997)
Lung carcinogen resistance	Resistance (No. of induced tumors)	A/J background: very susceptible; BALB/c background: 14X more resistant; C3H and C57BL/6 background: 20X more resistance	Chromosome 18 (Par2); Chromosome 6, 9, 17, 19 (Fas loci)	Obata et al. (1996); Devereux et al. (1994); Festing et al. (1994)
Colon carcinogen resistance	Resistance	BALB/c and STS background: 13–53% incidence; Ccs-19 background: 83% incidence	Chromosome 1, 2, 17, 18 (Scc loci)	Moen et al. (1996); Wezel et al. (1996)
Chemical hepato-carcinogenesis	Resistance	C57BL/6 background: resistant; DBA/2 background: 20X more susceptible; C3H background: 50X more susceptible	Chromosome 4 and 10 (Hcr loci)	Lee et al. (1995); Lee and Drinkwater (1995)
K14-HPV16 transgene (squamous cell carcinoma)	Progression	C57BL/6 and SENCAR background: hyperplasia; BALB/c background: dysplasia; FVB/n background: malignant carcinoma	n.a.	Coussens et al. (1996)
Neurofibromatosis type I (NF1) (human)	Expression	Monozygotic twins: minor phenotypic variation; first-degree and distant relatives: high variation	n.a.	Easton et al. (1993)
Huntington disease (human)	Age of onset	Paternal origin: juvenile onset; maternal origin: later onset	n.a.	Ridley et al. (1992)
Tourette Syndrome (human)	Male: female sex ratio	3:1 male preponderance for disease	Modeled to X-chromosome	Comings and Comings (1986)

* Denotes evidence of parent-of-origin modification (imprinting).

alteration in different genetic backgrounds. The study of spontaneous mutations, chemically induced mutations and, more recently, transgenic animals
and targeted mutations have implicated the action of modifier genes in determining the phenotype produced by these genetic changes. For example, the
Blind mutation that causes failure of eyelid closure has a significantly higher
penetrance when placed on a DBA/2 strain background compared to that on a
BALB/c background (Teicher and Caspari 1978). The curly-tail (*ct*) mutant
that develops neural tube defects similar to those found in human cases of
spina bifida shows a 30-fold variance in penetrance between various backcross
strains. For this mutation, highest penetrance was observed on a C57BL/6
background while lowest penetrance was observed on a DBA/2 background.
The extreme variation in penetrance of neural tube defects in humans and
these mice has suggested a multifactoral etiology. Analysis using recombinant-
inbred strains derived from C57BL/6 and DBA/2 suggest that at least three
modifier loci influence the incidence of *ct*-associated neural tube defects
(Neumann et al. 1994; Letts et al. 1995).

The use of wild-derived *Mus musculus musculus* strains of mice in the
analysis of mutations has also provided evidence for genetic variability in the
modification of phenotype. The *Tme* deletion on mouse chromosome 17 is
normally lethal when inherited from the mother. Forejt and Gregorova (1992)
demonstrated that this effect could be overcome when male mice of the PWB,
PWD, and PWK strains were used to sire the offspring of *Tme* transmitting
females. Forejt and colleagues subsequently mapped one of the modifying loci
to mouse chromosome 7 (Forejt et al. 1995).

Strain-specific modifiers have also been observed to influence the DNA
methylation phenotype and expression of a number of transgene loci in the
mouse (McGowan et al. 1989; Sapienza et al. 1989; Allen et al. 1990; Surani et al.
1990; Engler et al. 1991). It is difficult to determine whether these modifications
occur as a result of factors present in the egg cytoplasm or whether they occur
at later stages in development. However, some examples suggest that at least
some modification occurs due to factors present in the oocyte while other
examples suggest the modifications are accomplished by factors synthesized
after fertilization. For example, the *CMZ 12* transgene is highly expressed when
DBA/2 eggs are fertilized by transgenic males, while enhancement in expression is not observed when eggs from transgenic females are fertilized by DBA/
2 sperm (Surani et al. 1990). The expression of another mouse transgene, *TKZ
751*, is suppressed when BALB/c eggs are fertilized by transgenic males, but
expression does occur when transgenic females are fertilized by BALB/c males
(Allen et al. 1990). Further, an inverse correlation was observed between the
level of transgene expression and DNA methylation. The HSP 70.1 gene is one
of the first genes expressed in the mouse zygote. Recent data obtained by
Chastant et al. (1996) indicate that HSP 70.1 expression at the two-cell stage is
twofold higher in embryos whose maternal cytoplasm is from the mouse C3H
strain compared to embryos whose maternal cytoplasm is from the BALB/c

strain. All of these examples argue for the inclusion of modifying factors in the egg cytoplasm. On the other hand, the segregation of transgene methylation phenotype into two distinct categories following fertilization of eggs from F_1 females by sperm from transgenic males (Sapienza et al. 1989) indicates that at least some modifying factors may act after formation of ova.

These experiments indicate that factors present in the egg cytoplasm or factors synthesized early in development can act to enhance or suppress the expression and methylation of a number of transgene loci. Several of these modifiers have been mapped to different chromosomes in the mouse: a strain-specific modifier (*Ssm1*) located on chromosome 4 modifies the methylation and expression of a pHRD transgene (Engler et al. 1991). Interestingly, some of these transgenes also display gamete-of-origin dependent modification (i.e., imprinting). This type of modification of transgene methylation phenotype is not unique to mammals. Such parent-of-origin effects have been observed in plants (Scheid et al. 1998) and in zebrafish (Martin and McGowan 1995a,b). These observations suggest that this type of modification is wide-spread among evolutionarily diverse phyla.

The condition in zebrafish is particularly interesting because the modification observed appears to be nearly identical to that observed for imprinted transgenes in the mouse. However, unlike the situation in mouse, viable gynogenetic and androgenetic zebrafish embryos have been produced (although androgenetic zebrafish larvae are smaller than biparental controls) (Corley-Smith et al. 1996) implying that any parental origin effect on gene expression is incomplete in the zebrafish. Perhaps of relevance in this regard is the fact that the modifications observed in zebrafish did not result in the complete inactivation of the hemizygous transgene loci (Martin and McGowan 1995a,b), suggesting that the genome imprinting-like phenomenon in these fish could be more plastic than that observed in mammals.

At least two transgenes, one from mouse and one from zebrafish, show both parent-of-origin and strain-specific effects on DNA methylation and expression, and each is expressed in a mosaic pattern (McGowan et al. 1989; Martin and McGowan 1995a,b). In both cases, the level of methylation between different tissues was similar, suggesting that the methylation mosaicism was established in the last common ancestor of these tissues, likely during gastrulation. The facts that individual cells of identical descent display variable patterns of methylation and variegated expession suggests that mosaicism may be a common mechanism by which quantitative changes in phenotype may be achieved (Hall 1988; Sapienza and Hall 1995; see below).

Knockout mice show similar variation in phenotype depending on the genetic background in which the mutation is examined. Homozygotes for a targeted mutation in the keratin 8 gene shows embryonic lethality such that only 1.6% of offspring survive on a C57BL/6 background while a remarkable 55% of these embryos are viable on an FVB/N genetic background (Baribault et al. 1994). Two other targeted mutations have been reported to show wide variation in the timing of lethality and the age of onset. Mice homozygous for a null mutation in the epidermal growth factor receptor (*Egfr*) are perimplantation lethal on a CF1 background, midgestation lethal on a 129/SV background, and can survive for over 3 weeks on an outbred (CD1) background (Sibilian and Wagner 1995; Threadgill et al. 1995). Similarly, a disruption of the cystic fibrosis transmembrane regulator (*Cftr*) gene results in death 10 days following birth when the mutation is maintained on a DBA/2 genetic background, but shows prolonged survival on C57BL/6 and BALB/c background (Rozmahel et al. 1996).

These few examples provide ample demonstration that a number of different alleles of modifier loci have been fixed in different inbred mouse strains. The fact that dramatic modification of many different phenotypes is observed by using a relatively small collection of inbred strains argues that the modifier genes affecting these phenotypes may achieve their effect through a limited number of common pathways. The process of strain establishment and selection for coat color, fecundity or other strain-specific traits appears to have resulted in the concomitant selection of modifier genes that are able to act on many loci. When using such mice as models for human disease, it will be increasingly important to analyze the effects of genetic background on phenotype (Erickson 1996).

4
Modifiers in Genomic Imprinting

The pronuclear transplantation experiments of McGrath and Solter (1984) demonstrated that both paternal and maternal genetic contributions are required for the completion of normal mouse development. This requirement for both parental genomes is the result of genome imprinting that produces a requirement for both parental contributions to the embryo.

Apart from the parent-of-origin effects of the donor genome in uniparental embryos, there are also significant differences in the development of these embryos depending on the genetic background of the donor oocyte. Oocytes from C57BL/6 mice support a higher frequency of androgenetic development to the blastocyst stage than do oocytes derived from DBA/2 mice, regardless of the origin of the male pronuclei. Even transient exposure of donor nuclei to DBA/2 egg cytoplasm negatively affects the ability of subsequently produced androgenotes to develop to the blastocyst stage (Latham and Solter 1991). These experiments suggested the presence of some factor that was able to modify the donor male pronucleus. Presumably, these modifications can occur at any number of loci in the sperm genome during normal breeding and may be at least in part responsible for other observations of strain-specific modifiers. An interesting conclusion that can be drawn from these observations is that there is no requirement for the marking of paternal and maternal genomes to occur during gametogenesis (as opposed to in the zygote). A similar conclusion has recently been made regarding the parental marking of the genome by DNA methylation during gametogenesis (Jaenisch 1997). Latham (1994), by assaying the ability of eggs to support androgenetic development derived from (B6D2)F1 X DBA/2 backcrosses, concluded that these strain-specific effects of egg cytoplasm are the result of two independently segregating modifier loci. Mapping experiments have suggested that these loci are on chromosomes 1 and 2 (Latham and Sapienza 1998).

5
Fisher's Theory of Dominance and Modifiers of Imprinting

R. A. Fisher proposed his theory on the origin of dominance in 1928 (Fisher 1928a,b). This theory was an attempt to reconcile the observation that the wild-type form of a gene was, in most but not all cases, dominant to its mutant counterparts. The existence of multiple mutant alleles at some loci and the fact that some mutant alleles were only partially dominant convinced Fisher that dominance and recessivity could not be explained simply by the presence or absence of a gene product. Fisher put forward the idea that the dominance of wild-type alleles was not the sole function of the particular structure of those alleles, but was a result of the action of one or many unlinked "modifier" genes to decrease (or in the case of advantageous mutations, increase) the penetrance and expressivity of mutant alleles. During evolution these modifiers would act almost exclusively on organisms that were heterozygous for particular mutations and of intermediate phenotype. Natural selection would, over time, result in a collection of modifier genes in which the mutant allele (or the wild-type allele, in the case of an advantageous mutation) would have no phenotype in the heterozygous state. Fisher proposed that this modification by enhancement or suppression of the expressivity of the mutant allele would ultimately lead not only a deleterious allele to become recessive but an advantageous allele to become dominant.

Fisher based his conjecture on observations of the behavior of mutant alleles in both plants and animals: *Drosophila* mutants cultured for long periods of time were found to display their mutant characteristics to a considerably lower degree than those observed at their initial appearance (Morgan et al. 1925). This effect was reversible by breeding the "modified" mutants to unrelated fly stocks, demonstrating that the same mutant allele could result in a dominant trait in one case and a recessive trait in another. The behavior of the *crinkled dwarf* mutation in Sea Island cottons provided an additional argument in favor of Fisher's theory. This mutation normally behaves as a complete recessive. However, when crossed with a New World cotton (in which the mutation was unknown) and the F_1 progeny selfed, no clear segregation of phenotype into a $3:1$ ratio (an expectation of Mendelian segregation of a recessive trait among the progeny of an F_1 intercross) was observed. Instead, the progeny of these crosses displayed a wide range of phenotypes with no clear dominance relationship between wild-type and mutant alleles (J. B. Hutchinson and S. C. Harland in Fisher 1928b). Finally, Fisher himself conducted experiments using domestic fowl that initially appeared to carry mutant alleles that acted in a dominant fashion. After several generations of backcrossing these domestic fowl to wild jungle cocks, the mutant characters began to behave recessively. In this case, the process of domestication had lead to the selection of modifiers that have pushed the mutant allele to a position of dominance.

Experimental evidence for the operation of Fisherian selection does exist. For example, Carvalho et al. (1998) experimented with a population of *Drosophila mediopunctata* that showed a female-biased sex ratio. Fisher predicts that the 1:1 sex ratio found in most sexually reproducing species is the result of a process of natural selection acting on modifiers of sex ratio. In this experimental population of *Drosophila*, sex ratio evolved over a period of 6 years and 49 generations from the original 16% males to 32% males. Mathematical treatment of these results suggest that it would take 330 generations and 29 years for the proportion of males to reach 49%. While the number of generations required to select modifiers and thereby alter the sex ratio of offspring appears enormous when considering the generation time of *Drosophila* compared to human (for example), in either case this time represents a mere blink of the eye in evolutionary terms.

Fisher's proposal on the nature of dominance led to one of the most protracted debates in the history of genetics. In the intervening 70 years, there have been many arguments made against Fisher's theory (Wright 1929; Haldane 1930; Charlesworth 1979; Orr 1991), and at least one study that claims to have disproved it (Orr 1991). Indeed, to current era students in molecular biology, Fisher's explanation for why the wild-type is dominant and most mutations are recessive seems almost completely unnecessary; a generation weaned on DNA sequences accepts, more or less, the notion that most mutations are recessive because they result in varying degrees of loss of function (see Sapienza 1995 for discussion), despite the variable penetrance shown by the same mutant allele on different genetic backgrounds.

Although it might be argued that Fisher's theory is an idea whose time has passed, there are some cases in which it may still be a useful tool (Mayo and Bürger 1997). The study of genome imprinting is likely to be one such instance. The reason is that Fisher's theory simultaneously addresses two complex issues; the fundamental nature of dominance, and the evolutionary constraints under which unlinked genes might be selected to alter the dominance relationship of one allele to another at the same locus. The first of these issues has been dealt a serious blow by the analysis of Orr (1991). Fisher's theory predicts that recessiveness is a result of the action of unlinked modifier genes that have been selected to lessen the phenotypic consequences of mutant alleles and that this selection occurs while they are in the heterozygous state. Orr analysed mutations in *Chlamydomonas*, an organism that is normally haploid but can be induced to enter a vegetative diploid phase, and demonstrated that mutations in *Chlamydomonas* were generally recessive, even though they had never been subject to selection as heterozygotes. Thus, the dominance or recessiveness of a mutation does not appear to arise as a result of selection of modifier genes in the heterozygote, as a general rule. However, the process of imprinting provides an opportunity for Fisherian selection to play an important role in determining whether any particular locus may remain amongst the cadre of imprinted genes.

In order for selection of the type proposed by Fisher to occur, a mutation must have a phenotypic effect in the heterozygous state, i.e., it must be dominant or semidominant. Even if most mutations are recessive by nature (Orr 1991), mutant alleles at imprinted loci will be dominant when inherited from the parent that normally transmits the expressed allele. Even if the mutation results in loss of function, the allele contributed by the other parent is epigenetically inactivated, resulting in a complete loss of gene product even though the individual is a heterozygote.

If there is a selective advantage to be gained by modifying the mutant phenotype towards wild-type, then any changes that suppress the inactivation of the wild-type allele will be favored. This type of modification results, effectively, in a failure to imprint the wild-type allele. This may occur as a result of changes that occur in the imprinted allele itself or the selection of variant alleles of unlinked modifier genes, such that there is a failure to inactivate the normally imprinted allele. Both of these possibilities are expected to result, at minimum, in population level polymorphism in the inactivation of alleles at any imprinted locus, i.e., when outbred populations are examined, one fraction of the population is expected to exhibit parental-origin dependent expression of the imprinted gene, while the remaining individuals will show no, or partial (see below) parental-origin dependence in the expression of that gene. Such appears to be the case for both *IGF2R* (Xu et al. 1993) and *WT1* (Jinno et al. 1994) in the human population. The imprinting of the *IGF2R* gene shows the additional interesting characteristic that maternal allele-specific expression appears to be the rule in the mouse but the exception in the human. Such differences within and between species provide the opportunity to examine whether the differences are the result of changes to the imprinted allele, itself, or to the action of unlinked modifier genes.

An interesting possible consequence of imprinting polymorphism, or recurrent mutations that affect the imprinting of an allele, is that individuals who express both alleles at a locus that is normally imprinted may be more or less susceptible to additional genetic or epigenetic changes that result in disease. For example, loss of imprinting (LOI) of *IGF2* and/or *H19* is one of the most common epigenetic changes observed in Wilms tumors (Rainier et al. 1993; Ogawa et al. 1993; see also Tycko, Chap. 7. this vol.). Such LOI may occur as the result of a random somatic event (Feinberg 1995) but it may also be the result of an inherited inability to inactivate, or incompletely inactivate, the normally silenced allele as a result of variant *cis* – or *trans*-acting genetic factors. (It is worth noting that the converse of this situation is also expected, i.e., a gene that is not normally imprinted may become imprinted as a result of a mutation in a *cis* or *trans* element, resulting in predisposition to disease (Scrable et al. 1989; Feinberg 1995).

Whether the variation between individuals in the imprinting of *IGF2R* and *WT1* loci is caused by epigenetic changes or genetic changes that act in *cis* or in *trans* is unknown, but it should be possible to determine the answer to this question by standard genetic analysis. Although changes to both *cis* and *trans* elements involved in inactivation may result in polymorphism in the imprinting of any particular locus, the transmission of the failure-to-imprint phenotype will be different in each case. Mutations that affect the ability of the imprinted allele, itself, to be imprinted will segregate with that particular allele while mutations that affect the inactivation mechanism in *trans*, via a modifier gene, will segregate independently of the allele at the imprinted locus.

An additional consideration is that the process of imprinting need not give rise to monoallelic gene expression in all tissues or at all times during development. Imprinted genes may be imprinted only in certain tissues or during specific developmental time periods (reviewed in Latham 1995). The role of tissue and developmental stage-specific imprinting and the modifiers involved has yet to be determined. However, the failure of these modifications may be implicated in the genesis of several human diseases (Uyeno et al. 1996; Burger et al. 1997; Kim and Lee 1997; Nonomura et al. 1997; Okamoto, et al. 1997; Wu et al. 1997, Oda et al. 1998).

6
Modifiers of Imprinting: Possible Mode of Action

The number of biochemical pathways through which a modifier gene might act is potentially large. In addition to the many factors that might be involved directly in transcription, translation, chromatin structure and DNA modification, one must also entertain more classical notions of gene regulation. An increase or decrease in the level of any gene product in a metabolic pathway may result in feedback modulation of the levels of other gene products in the same metabolic pathway. Because modifier gene activity is defined at the level of phenotype, activity variants of one enzyme in a pathway may be identified as modifiers of the level of other gene products in the pathway. In most instances, there will be little reason, a priori, to favor one biochemical mechanism over another.

Although imprinted genes may be acted upon by the same types of modifiers as any other gene in the genome, we argue that the mechanism with the greatest potential for creating phenotypic variability due to the expression of imprinted genes is one that is implicit in the definition of an imprinted gene. Because imprinting is most often defined to result in the epigenetic silencing of one allele at a locus, dependent on parental origin, specific modifiers of imprinting must change this outcome. Logically, there are only two choices; to express the silenced allele, or to silence the expressed allele. These modifications may occur on a cell-by-cell, tissue-by-tissue, or individual-by-individual basis. At the extreme end of each of these possibilities, the imprint will have been modified to such an extent that it is no longer an imprint, i.e., all cells will express both alleles or no cell will express either allele; but between each of these extremes lies a continuum in which an individual may be a mosaic for cells that express the imprint and cells that do not.

It is worth noting that this type of modification would not, in general, be detected by methods that rely on analysing the presence or absence of a gene product by biochemical techniques that do not involve the analysis of individual cells. Only those assays that are both allele-specific and cell-specific may detect such modification.

The potential for somatic mosaicism in gene expression to produce a great range of phenotypic variability is demonstrated by the expression of X-linked recessive diseases in female humans (reviewed in Willard 1995), position-effect variegation in *Drosophila* (Locke et al. 1988; Sass and Henikoff 1998; see Pirotta, Chap. 10, this vol.) and mosaic expression of transgenes in mice (McGowan et al. 1989). These examples, together with observations on repression of the silent mating type information in yeast (reviewed in Holmes et al. 1996), predict a number of properties that may be relevant for the generation and modification of phenotype due to mosaic expression at imprinted loci:

1. Both *cis* and *trans* elements will be required for silencing the imprinted allele.
2. The *cis* elements, and several *trans* elements, are likely to be required for *initiation* of the imprint.
3. The *cis* elements are likely to be required for the inheritance of the imprint through mitosis.
4. The *trans* elements are likely to be required for the stable maintenance of the imprint in postmitotic cells but the *cis* elements may not be required.

At least some investigators within the genome imprinting field and almost everyone outside the field (see Sapienza 1995) assumes that the *cis* element involved in imprinting is a parental-origin specific DNA methylation mark (see Horsthemke et al., Chap. 5 and Razin and Shemer, Chap. 9, this vol.). Sites of parent-of-origin specific methylation have been found in the mouse *Igf2r* (Stöger et al. 1993), *Igf2* (Sasaki et al. 1992), *H19* (Bartolomei et al. 1991; Ferguson-Smith et al. 1993) and *Xist* genes (Zuccoti and Monk 1995; Ariel et al. 1995) and at sites within the human Prader-Willi and Angelman syndrome region (Dittrich et al. 1993; Horsthemke 1997). The idea that DNA methylation may be *the* gametic imprint is complicated somewhat by the fact that some loci show expression from both alleles (or, possibly, either allele) during preimplantation development (Latham et al. 1994) and that methylation appears to be associated with silencing in some cases (the paternal *H19* allele, and maternal *Xist* allele) and activation (or failure to silence) in other cases (the maternal *Igf2r* allele and maternal *Igf2* allele). If parental-origin-specific methylation is the *cis* element required as the imprint, then other factors must be required, not only to complete the silencing process by which an imprinted allele is recognized, but also to interpret whether the methylation event marks an allele to be silenced or an allele that is to remain active.

If the yeast and *Drosophila* silencing systems are to have any bearing on the silencing of alleles at imprinted loci, then many gene products involved in the formation of heterochromatin must be considered as candidates for *trans* factors involved in the initiation, somatic inheritance and maintenance of the imprint (Hendrich and Willard 1995; Holmes et al. 1996). Genetic and biochemical data gathered on such modifiers are consistent with the idea that they operate in a dosage-sensitive manner and give rise to stable, heritable chromatin structures in a cell and almost all of its descendants (see Pirotta Chap. 10, this vol.). Variability between individuals is achieved by changing the fraction of cells that do or do not express the affected allele, rather than changing the level of expression of an allele in all cells, i.e., expression in such cases is an all or nothing proposition for each cell. Variant alleles of any one of the factors that participate in the formation of the silenced structure may lead to a failure to silence, or an instability in silencing, such that somatic mosaicism in the silencing of one or more imprinted genes results.

The degree to which such mosaicism occurs in nature is unknown, in part because cell and allele specific assays in which such mosaicism could be detected have not been performed on many imprinted genes, but both *Mash2* and *H19* have been shown to exhibit such mosaic expression patterns (Guillemot et al. 1995; Adam et al. 1996), as have some imprinted transgenes (McGowan et al. 1989). A general prediction of this model is that many of the phenotypic differences due to "variable expressivity" of alleles at imprinted loci will be traced to differences in the fraction of cells in which the imprint survives.

Acknowledgments. We thank Richard Gordon and Keith Latham for helpful discussion. C.C.M. is supported by a Postdoctoral Fellowship from the Natural Sciences and Engineering Research Council (NSERC) of Canada.

References

Adam GIR, Cui H, Miller SJ, Flam F, Ohlsson R (1996) Allele-specific in situ hybridization (ASISH) analysis: a novel technique which resolves differential allelic usage of *H19* within the same cell lineage during human placental development. Development 122:839–847

Allen ND, Norris ML, Surani MA (1990) Epigenetic control of transgene expression and imprinting by genotype-specific modifiers. Cell 61:378–383

Ariel M, Robinson E, McCarrey JR, Cedar H (1995) Gamete-specific methylation correlates with imprinting of the murine *Xist* gene. Nat Genet 9:312–315

Baribault H, Penner J, Iozzo RV, Wilson-Heiner M (1994) Colorectal hyperplasia and inflammation in keratin 8-deficient FVB/N mice. Genes Dev 8:2964–2973

Bartolomei MS, Zemel S, Tilghman SM (1991) Parental imprinting of the mouse *H19* gene. Nature 351:153–155

Bonyadi M, Rusholme SA, Cousins FM, Su HC, Biron CA, Farrel M, Akhurst RJ (1997) Mapping of a major genetic modifier of embryonic lethality in *TGFβ1* knockout mice. Nat Genet 15:207–211

Burger J, Buiting K, Dittrich B, Gross S, Lich C, Sperling K, Horsthemke B, Reis A (1997) Different mechanisms and recurrence risks of imprinting defects in Angelman syndrome. Am J Hum Genet 61:88–93

Carvalho AB, Sampaio MC, Varandas FR, Klaczko LB (1998) An experimental demonstration of Fisher's principle: evolution of sexual proportion by natural selection. Genetics 148:719–731

Charlesworth B (1979) Evidence against Fisher's theory of dominance. Nature 287:848–849

Chastant S, Christians E, Campion E, Renard J-P (1996) Quantitative control of gene expression by nucleocytoplasmic interactions in early mouse embryos: consequence for reprogramation by nuclear transfer. Mol Reprod Dev 44:423–432

Comings DE, Comings BG (1986) Evidence for an X-linked modifier gene affecting the expression of Tourette syndrome and its relevance to the increased frequency of speech, cognitive, and behavioral disorders in males. Proc Natl Acad Sci USA 83:2551–2555

Corley-Smith GE, Lim CJ, Brandhorst BP (1996) Production of androgenetic zebrafish (*Danio rerio*). Genetics 142:1265–1276

Cormier RT, Hong KH, Halberg RB, Hawkins TL, Richardson P, Mulherkar R, Dove WF, Lander ES (1997) Secretory phospholipase *Pla2g2a* confers resistance to intestinal tumorigenesis. Nat Genet 17:88–91

Coussens LM, Hanahan D, Arbeit JM (1996) Genetic predisposition and parameters in K14-HPV16 transgenic mice. Am J Pathol 149(6):1899–1917

Devereux TR, Wiseman RW, Kaplan N, Garren S, Foley JF, White CM, Anna C, Watson MA, Patel A, Jarchow S, Maron RR, Anderson MW (1994) Assignment of a locus for mouse lung tumor susceptibility to proximal chromosome 19. Mamm Genome 5:749–755

Dietrich WF, Lander ES, Smith JS, Moser AR, Gould KA, Luongo C, Borenstein N, Dove W (1993) Genetic identification of *Mom-1*, a major modifier locus affecting *Min*-induced intestinal neoplasia in the mouse. Cell 75:631–639

Dittrich B, Buiting K, Gros S, Horsthemke B (1993) Characterization of a methylation imprint in the Prader-Willi syndrome region. Hum Mol Genet 2:1995–1999

Dobbie Z, Heinimann K, Bishop DT, Müller H, Scott RJ (1997) Identification of a modifier gene locus on chromosome 1p35-36 in familial adenomatous polyposis. Hum Genet 99:653–657

Easton DF, Ponder MA, Huson SM, Ponder BAJ (1993) An analysis of variation in expression of neurofibromatosis (NF) type I (*NFI*): evidence for modifying genes. Am J Hum Genet 53:305–313

Engler P, Naasch D, Pinkert CA, Doglio L, Glymour M, Brinster R, Storb U (1991) A strain-specific modifier on mouse chromosome 4 controls the methylation of independent transgene loci. Cell 65:939–947

Erickson RP (1996) Mouse models of human genetic disease: which mouse is more like a man? BioEssays 18:993–998

Feinberg AP (1995) A domain of abnormal imprinting in human cancer. In: Ohlsson R, Hall K, Ritzen M (eds) Genomic imprinting: causes and consequences. Cambridge University Press, Cambridge, pp 273–292

Ferguson-Smith AC, Sasaki H, Cattanach BM, Surani MA (1993) Parental-origin-specific epigenetic modifications of the mouse *H19* gene. Nature 362:751–755

Festing MF, Yang A, Malkinson AM (1994) At least four genes and sex are associated with susceptibility to urethane-induced pulmonary adenomas in mice. Genet Res 64:99–106

Fisher RA (1928a) The possible modification of the response of the wild type to recurrent mutations. Am Nat 62:115–126

Fisher RA (1928b) Two further notes on the origin of dominance. Am Nat 62:571–574

Forejt J, Gregorova S (1992) Genetic analysis of genomic imprinting: an *Imprintor-1* gene controls inactivation of the paternal copy of the mouse *Tme* locus. Cell 70:443–450

Forejt J, Gregorova S, Landikova M, Capkova J, Silver LM (1995) Genetic variation in parental imprinting on mouse chromosome 17. In: Ohlsson R, Hall K, Ritzen M (eds) Genomic imprinting: causes and consequences. Cambridge University Press, Cambridge, pp 29–45

Guillemot F, Caspary T, Tilghman SM, Copeland NG, Gilbert DJ, Jenkins NA, Anderson DJ, Joyner AL, Rossant J, Nagy A (1995) Genomic imprinting of *Mash-2*, a mouse gene required for trophoblast development. Nat Genet 9:235–241

Haldane JBS (1930) A note on Fisher's theory of the origin of dominance and a correlation between dominance and linkage. Am Nat 64:87–90

Hall J (1988) Review and hypothesis: somatic mosaicism: observations related to clinical genetics. Am J Hum Genet 43:355–363

Hendrich BD, Willard HF (1995) Epigenetic regulation of gene expression: the effects of altered chromatin structure from yeast to mammals. Hum Mol Genet 4:1765–1777

Holmes SG, Braunstein M, Broach JR (1996) Transcriptional silencing of the yeast mating-type genes. In: Epigenetic mechanisms of gene regulation. Cold Spring Harbor Laboratory Press, New York, pp 467–487

Horsthemke B (1997) Structure and function of the human chromosome 15 imprinting center. J Cell Physiol 173:237–241

Hummel KP (1958) The inheritance and expression of *disorganization*, an unusual mutation in the mouse. J Exp Zool 137:389–423

Hummel KP (1959) Developmental anomalies in mice resulting from action of the gene, *disorganization*, a semi-dominant lethal. Pediatrics 23:212–221

Iakoubova OA, Dushkin H, Beier DR (1995) Localization of a murine recessive polycystic kidney disease mutation and modifying loci that affect disease severity. Genomics 26:107–114

Jaenisch R (1997) DNA methylation and imprinting: why bother? Trends Genet 13:323–329

Jinno Y, Yun K, Nishiwaki K, Kubota T, Ogawa O, Reeve AE, Niikawa N (1994) Mosaic and polymorphic imprinting of the WT1 gene in humans. Nat Genet 6:305–309

Kim KS, Lee YI (1997) Biallelic expression of the *H19* and *Igf2* genes in hepatocellular carcinoma. Cancer Lett 119:143–148

Latham KE (1994) Strain-specific differences in mouse oocytes and their contributions to epigenetic inheritance. Development 120:3419–3426

Latham KE (1995) Stage-specific and cell type-specific aspects of genomic imprinting effects in mammals. Differentiation 59:269–282

Latham KE, Sapienza C (1998) Localization of genes encoding egg modifiers of paternal genome function to mouse chromosomes 1 and 2. Development 125:929–935

Latham KE, Solter D (1991) Effect of egg composition on the development capacity of androgenetic mouse embryos. Development 113:561–568

Latham KE, Doherty AS, Scott CD, Schultz RM (1994) *Igf2r* and *Igf2* gene expression in androgenetic, gynogenetic, and parthenogenetic preimplantation mouse embryos: absence of regulation by genomic imprinting. Genes Dev 8:290–302

Lee G-H, Drinkwater NR (1995) The *Hcr* (hepatocarcinogen resistance) loci of DBA/2J mice partially suppress phenotypic expression of the *Hcs* (hepatocarcinogen sensitivity) loci of C3H/HeJ mice. Carcinogenesis 16:1993–1996

Lee G-H, Bennett LM, Carabeo RA, Drinkwater NR (1995) Identification of hepatocarcinogen-resistance genes in DBA/2 mice. Genetics 139:387–395

Letts VA, Schork NJ, Copp AJ, Bernfield M, Frankel WN (1995) A curly-tail modifier locus, *mct1*, on mouse chromosome 17. Genomics 29:719–724

Locke J, Kotarski MA, Tartoff KD (1988) Dosage-dependent modifiers of position effect variegation in *Drosophila* and a mass-action model that explains their effect. Genetics 120:181–198

MacPhee M, Chepenik KP, Liddel RA, Nelson KK, Siracusa LD, Buchberg AM (1995) The secretory phospholipase A2 gene is a candidate for the *Mom1* locus, a major modifier of *Apc^{Min}* – induced intestinal neoplasia. Cell 81:957–966

Martin CC, McGowan R (1995a) Genotype-specific modifiers of transgene methylation and expression in the zebrafish, *Danio rerio*. Genet Res Camb 65:21–28

Martin CC, McGowan R (1995b) Parent-of-origin specific effects on the methylation of a transgene in the zebrafish, *Danio rerio*. Dev Genet 17(3):233–239

Mayo O, Bürger R (1997) The evolution of dominance: a theory whose time has passed? Biol Rev 72:97–110

McGowan R, Campbell R, Peterson A, Sapienza C (1989) Cellular mosiacism in the methylation and expression of hemizygous loci in the mouse. Genes Dev 3:1669–1676

McGrath J, Solter D (1984) Completion of mouse embryogenesis requires both maternal and paternal genomes. Cell 37:179–183

Moen CJA, Groot PC, Hart AAM, Snoek M, Demant P (1996) Fine mapping of colon tumor susceptibility (*Scc*) genes in the mouse, different from the genes known to be somatically mutated in colon cancer. Proc Natl Acad Sci USA 93:1082–1086

Mohlke KL, Nichols WC, Westrick RJ, Novak EK, Cooney KA, Swank RT, Ginsburg D (1996) A novel modifier gene for plasma von Willebrand factor level maps to distal chromosome 11. Proc Natl Acad Sci USA 93:15352–15357

Morgan TH, Bridges CB, Sturtevant AH (1925) The genetics of *Drosophila*. Bibl Genet 2:1–262

Moser AR, Dove WF, Roth KA, Gordon JI (1992) The *Min* (multiple intestinal neoplasia) mutation: its effects on gut epithelial cell differentiation and interaction with a modifier system. J Cell Biol 116:1517–1526

Neumann PE, Frankel WN, Letts VA, Coffin JM, Copp AJ, Bernfield M (1994) Multifactorial inheritance of neural tube defects: localization of the major gene and recognition of modifiers of *ct* mutant mice. Nat Genet 6:357–362

Nonomura N, Nishimura K, Miki T, Kanno N, Kojima Y, Yokoyama M, Okuyama A (1997) Loss of imprinting of the insulin-like growth factor II gene in renal cell carcinoma. Cancer Res 57:2575–2577

Obata M, Nishimori H, Ogawa K, Lee G-H (1996) Identification of the *Par2 (Pulmonary adenoma resistance)* locus on mouse chromosome 18, a major genetic determinant for lung carcinogen resistance in BALB/cByJ mice. Oncogene 13:1599–1604

Oda H, Kume H, Shimizu Y, Inoue T, Ishikawa T (1998) Loss of imprinting of *Igf2* in renal-cell carcinomas. Int J Cancer 75:343–346

Ogawa O, Eccles MR, Szeto J, McNoe LA, Yun K, Maw MA, Smith PJ, Reeve AE (1993) Relaxation of insulin-like growth factor II gene imprinting implicated in Wilms' tumour. Nature 362:749–751

Okamoto K, Morison IM, Taniguchi T, Reeve AE (1997) Epigenetic changes at the insulin-like growth factor II H19 locus in developing kidney is an early event in Wilms tumorigenesis. Proc Natl Acad Sci USA 94:5367–5371

Orr HA (1991) A test of Fischer's theory of dominance. Proc Natl Acad Sci USA 88:11413–11415

Prescott SM, White RL (1996) Self promotion? Intimate connections between *APC* and prostaglandin H synthase-2. Cell 87:783–786

Rainier S, Johnson LA, Dobry CJ, Ping AJ, Grundy PE, Feinburg AP (1993) Relaxation of imprinted genes in human cancer. Nature 362:747–749

Ridley RM, Farrer LA, Frith CD, Conneally PM (1992) A test of the hypothesis that age at onset in Huntington disease is controlled by an X-linked recessive modifier. Am J Hum Genet 50:536–543

Risch N, Merikangas K (1996) The future of genetic studies of complex human diseases. Science 273:1516–1517

Rozmahel R, Wilschanski M, Matin A, Plyte S, Oliver M, Auerbach W, Moore A, Forstner J, Durie P, Nadeau J, Bear C, Tsui LC (1996) Modulation of disease severity in cystic fibrosis transmembrane conductance regulator deficient mice by a secondary genetic factor. Nat Genet 12:280–287

Ruvinsky AO, Agulnik AI (1990) Gametic imprinting and the manifestation of the *Fused* gene in the house mouse. Dev Genet 11:263–269

Sapienza C (1995) Genome imprinting: an overview. Dev Genet 17:185–187

Sapienza C, Hall JG (1995) Genome imprinting in human disease. In: Scriver CR, Beaudet AL, Sly WS, Valle D (eds) The metabolic and molecular bases of inherited disease, 7th edn. McGraw-Hill, New York

Sapienza C, Paquette J, Tran TH, Peterson A (1989) Epigenetic and genetic factors affect transgene methylation imprinting. Development 107:165–168

Sasaki H, Jones PA, Chaillet RJ, Ferguson-Smith AC, Barton SC, Reik W, Surani MA (1992) Parental imprinting: potentially active chromatin of the repressed maternal allele of the mouse insulin-like growth factor II (*Igf2*) gene. Genes Dev 6:1843–1856

Sass GL, Henikoff S (1998) Comparative analysis of postion-effect variegation mutations in *Drosophila melanogaster* delineates the targets of modifiers. Genetics 148:733–741

Scheid OM, Afsar K, Paszkowski J (1998) Release of epigenetic silencing by *trans*-acting mutations in *Arabidopsis*. Proc Natl Acad Sci USA 95:632–637

Scrable H, Cavenee W, Ghavimi F, Lovell M, Morgan K, Sapienza C (1989) A model for embryonal rhabdomyosarcoma tumorigenesis that involves genomic imprinting. Proc Natl Acad Sci USA 86:7480–7484

Sibilia M, Wagner EF (1995) Strain-dependent epithelial defects in mice lacking the EGF receptor. Science 269:234–238

Stöger R, Kubicka P, Liu CG, Kafri T, Razin A, Cedar H, Barlow DP (1993) Maternal-specific methylation of the imprinted mouse *Igf2r* locus identifies the expressed locus as carrying the imprinting signal. Cell 73:61–71

Surani MA, Kothary R, Allen ND, Singh PB, Fundele R, Ferguson-Smith AC, Barton SC (1990) Genome imprinting and development in the mouse. Dev Suppl 89–98

Teicher LS, Caspari EW (1978) The genetics of *Blind* – a lethal factor in mice. J Hered 69:86–90

Thompson MW, McInnes RR, Willard HF (1991) Thompson & Thompson: Genetics in medicine, 5th edn. WB Saunders, Philadelphia

Threadgill DW, Dlugosz AA, Hansen LA, Tennenbaum T, Lichti U, Yee D, LaMantia C, Mourton T, Herrup K, Harris RC, Barnard JA, Yuspa SH, Coffey RJ, Magnuson T (1995) Targeted disruption of mouse EGF receptor: effect of genetic background on mutant phenotype. Science 269:230–233

Tomlinson IPM, Neale K, Talbot IC, Spigelman AD, Williams CB, Phillips RKS, Bodmer WF (1996) A modifying locus for familial adenomatous polyposis may be present on chromosome 1p35–p36. J Med Genet 33:268–273

Uyeno S, Aoki Y, Nata M, Sagisaka K, Kayama T, Yoshimoto T, Ono T (1996) *Igf2* but not *H19* shows loss of imprinting in human glioma. Cancer Res 56:5356–5359

Weichman K, Chaillet JR (1997) Phenotypic variation in a genetically identical population of mice. Mol Cell Biol 17(9):5269–5274

Wezel T van, Stassen APM, Moen CJA, Hart AM, van der Vald MA, Demant P (1996) Gene interaction and single tumor susceptibility in colon tumour susceptibility in mice. Nat Genet 14:468–470

Willard HF (1995) The sex chromosomes and X chromosome inactivation. In: Scriver CR, Beaudet AL, Sly WS, Valle D (eds) The metabolic and molecular bases of inherited disease (7th edn). McGraw Hill, New York, pp 719–737

Wright S (1929) Fisher's theory of dominance. Am Nat 63:274–279

Wu MS, Wang HP, Lin CC, Sheu JC, Shun. CT, Lee WJ, Lin JT (1997) Loss of imprinting and overexpression of Igf2 gene in gastric adenocarcinoma. Cancer Lett 120:9–14

Xu Y, Goodyer CG, Deal C, Polychronakos C (1993) Functional polymorphism in the parental imprinting of the human *IGF2R* gene. Biochem Biophys Res Commun 197:747–754

Zuccotti M, Monk M (1995) Methylation of the mouse *Xist* gene in sperm and eggs correlates with imprinted *Xist* expression and paternal X-inactivation. Nat Genet 9:316–320

Allelic *trans*-sensing and Imprinting

Andràs Pàldi and Yann Jouvenot

1
Introduction

Homologous chromosome pairing is known to be an essential point in the meiotic process, because it allows interchromosomal exchanges (crossovers) to occur in a balanced way. Interactions between homologous loci has been shown to occur in somatic cells of plants and animals. In general, somatic pairing is less extensive than in the germ cells. Although their functional significance remains unclear, recent observations show that homologous interactions are more widespread than previously thought. It has also become apparent that imprinted chromosomal regions can be involved in homologous pairing in mammalian somatic cells. This chapter will focus on the description of long-range homologous interactions and their implications in the conrext of imprinted chromosomal regions.

2
Homologous Interactions

Initial evidence for homologous interactions came from the observations that the expression of several genes of *Drosophila melanogaster* (Lewis 1954) and maize (Coe 1968) depended on some form of interaction between the homologous alleles. Since then, advances in the analysis of the nuclear architecture and the development of in situ hybridization techniques have given visual confirmation of these interactions.

INSERM U257; Institut Cochin de Génétique Moléculaire, 24, rue du Faubourg Saint Jacques; 75014 Paris, France

2.1
Functional Evidences for Homologue Interaction

2.1.1
Studies in D. melanogaster

The discovery that the expression of a gene is not only dependent on its chromosomal environment (effects in *cis*), but also on its allelic counterpart (effects in *trans*) was first made by Lewis in 1954 (Lewis 1954). While studying the complementation between two different mutant alleles of the *bithorax* complex (bx^{34e}/*Ubx*) in *Drosophila* (which gives rise to an attenuated mutant phenotype), he showed that this complementation was disrupted when one of the homologous chromosomes was rearranged. This resulted in an enhancement of the mutant phenotype. Naming this phenomenon transvection, he defined it as a Êposition effect that is revealed by modifying the *trans*-heterozygote by means of chromosomal rearrangements. It was the first demonstration that the alleles of a gene had a functional interaction, also alluding to the importance of homologous chromosome pairing in gene activity (the examples mentioned in this review are summarized in Table 1).

Several models for transvection have been suggested (reviewed by Wu 1993) and the term has been applied to many phenomena in which a role for homologue pairing is implied. In a review, Tartof and Henikoff (Tartof and Henikoff 1991) used the term *trans*-sensing effects to describe a general class of phenomena that share the common feature of one gene sensing the presence of its homologue in *trans*. This class obviously includes transvection, but also several other effects, such as dominant variegation and *zeste*-mediated repression.

Dominant variegation was described by Dreesen (Dreesen et al. 1991), who studied a variegating allele of the *brown* gene (see Table 1). *brown* is one of the genes determining the eye color in *Drosophila*. He showed that the dominant character of this mutation was disrupted if the variegating allele was not able to pair with the wild-type one (by chromosomal rearrangements).

Zeste-mediated repression was first described for the closely linked *white* gene on the X chromosome (Jack and Judd 1979). It appeared that when w^+ alleles are paired in females, the presence of a mutant *zeste* gene gives rise to a mutant phenotype (a yellow-colored eye). When the w^+ copies are unpaired, however, the *zeste* mutation no longer affects the phenotype and the wild-type eye color is restored (see Table 1). It is presently supposed that the *zeste* protein acts as a transcription factor, which instead of permitting communication between an enhancer and a promoter in *cis*, does so in *trans* (Jack and Judd 1979). *Zeste* mutations have also been shown to affect other genes (Wu and Goldberg 1989).

In these two phenomena, dominant variegation and *zeste*-mediated repression, we can see that the homologue pairing alters the normal gene expression,

Table 1. Examples of trans-sensing effects in *Drosophila*

Phenomenon	Model genotype	Phenotype (when homologous pairing is possible)	Phenotype (when homologous) pairing is not possible)
Transvection	bx^{34e}/Ubx	Attenuated haltere-to-wing transformation	Enhancement of the mutant phenotype
Dominant variegation	*brown*/+	Variegating eye color (brown and red	Red-coloured eye (wild-type)
Zeste-mediated repression	w^+/w^+ (and z^1/z^1 at the *zeste* locus)	Yellow-colouredeye	Wild-type eye (this rearrangement concerns the *w* locus)

which is restored by the disruption of close contact between the alleles. This is most likely mediated by long-range nucleotide sequence homology. Transvection represents the reciprocal situation where the interallelic contact is necessary for normal expression of the wild-type genes. In all instances, however, these data prove the existence of *trans*-allelic interactions in *Drosophila*. Other *trans*-sensing effects have been studied (for review, see Henikoff 1997) and they also lead to this conclusion, even if the mechanisms by which alleles communicate may not be the same.

2.1.2
Studies in Plants

Up to this point, we have referred only to functional evidence for homologous alleles interactions in *Drosophila*. Important work has also been carried out in the area of plant epigenetics, with the study of paramutations and other related phenomena involving interactions between homologous sequences. In 1956, Brink observed that when some alleles of the *r* gene in maize (where it encodes a transcription factor that regulates anthocyanin pigment production) were exposed to other specific alleles, their activity was reduced in an invariable and heritable way. He named this phenomenon paramutation, the alleles generating paramutations being paramutagenic and the alleles sensitive to their influence being paramutable. The term paramutation reflects the character of this event, because it produces a meiotically heritable modification of one allele's activity. The mechanism involved, i.e., the presence in the same genome of a paramutagenic allele that doesn't change the nucleotide sequence, justifies the use of the prefix para, relating this notion to the domain of epigenetics.

Several other endogenous plant genes were proved to exhibit paramutations, and this was also found for some plant transgenes (reviewed in Hollick et al. 1997). Their study showed that paramutated alleles can recover their original activity when they are in a hemizygous state or in the presence of neutral alleles (i.e., not affected by paramutation). This reversibility was not observed, however, in all the cases. It appears that paramutations can also be correlated with DNA methylation for some genes. At the moment, a common feature that can be drawn from the cloning of several paramutant genes is the presence of repetitive sequences, although their importance remains to be determined.

Although paramutation seems similar to dominant variegation in *Drososophila* (both events being *trans*-inactivations of one allele by the other), some distinctions can be made between the two phenomena. Firstly, in *Drosophila*, the inactive state is not transmitted through meiosis by the silenced wild-type allele. Secondly, the paramutation in the maize *b* and *r* genes was not affected by chromosomal rearrangements (Coe 1968). For this last point, however, we cannnot exclude the hypothesis of a transient pairing (Hollick et al. 1997), different from the homologous chromosome pairing that is supposed to be required in *Drosophila trans*-sensing mechanisms.

Another epigenetic phenomenon involving *trans*-inactivation in both plants and animals, is cosuppression (for review, see Bingham 1997). When additional copies of a gene are present in a genome, this phenomenon can result in a specific repression of some or all copies of that gene. In this case, the *trans*-inactivation does not occur between alleles of the same gene, but between supernumerary copies of a gene. We mention this event to highlight more evidence of functional interactions between homologues (even if they are non-allelic genes), though the mechanism of interaction is probably not the same as in the previous examples.

2.2
Morphological Evidence for Homologue Interaction

Other elements supporting the importance of homologue interactions come from visual observations of the pairing of homologous chromosomes or genes in somatic cells. The first clue was given by Metz in 1916, who showed that homologous chromosomes in metaphase neuroblast spreads are usually found close to each other (in *Drosophila*). The pairing theory had to wait until the recent development of in situ hybridization, however, which made it possible to see the position of genes in the interphase nuclei.

Several studies have shown the pairing of homologous regions in somatic cells. In 1989, Arnoldus et al. (Arnoldus et al. 1989) studied the pairing of chromosome 1 centromeres in human cerebellum interphase nuclei and showed that it was a cell-type-specific phenomenon. In 1993, Hiraoka et al. (Hiraoka et al. 1993) analyzed the histone gene cluster in *Drosophila* embryos and found that homologous pairing was temporally regulated. In this case the pairing occurs at the 14th nuclear cycle, in the developing embryo. They also demonstrated that homologous pairing was dependent on chromosomal position, since the frequency of association of the histone genes studied was strongly decreased in strains with chromosomal translocations.

Other studies have also confirmed homologous pairing in *Drosophila* (Kopczynski and Muskavitch 1992) and humans ((LaSalle and Lalande 1996, discussed below). By combining both functional and morphological data, therefore, we can be confident of the importance of homologue interactions. This should help us to approach other epigenetic mechanisms, such as parental imprinting, in which homologous interactions could be involved.

3
Homologue Interactions of Imprinted Chromosomal Domains

The idea that interactions between the homologous chromosomes could be involved in the imprinting process is not a new one. An allelic cross-talk model was proposed by Monk on the basis of theoretical speculations (Monk 1990). A transvection model of somatic transfer of genomic imprinting was suggested to explain the unexpected rescue from the lethal effect of the *T-associated maternal effect* (Tme) mutation by a duplicated paternal chromosome (Tsai and Silver 1991). Recently, several reports revealed that chromosomal domains containing imprinted genes are frequently involved in homologous interactions in both mouse and human cells. This conclusion was reached on the basis of both structural and functional observations.

3.1
Genomic Imprinting

Imprinting marks the parental origin of some chromosomal domains. The main distinguishing feature of imprinted genes is their capacity to be expressed exclusively from one parental allele. In fact, all current definitions of imprinting refer to this characteristic. Some imprinted genes are expressed monoallelically in almost all tissues, while others show monoallelic expression restricted to some tissues and/or to some stages of tissue differentiation. Using this criterion, therefore, the borderline between the group of imprinted and non-imprinted genes may not be clear.

A temporal difference has been observed in replication timing between the two parental homologues of imprinted regions during the S phase of the somatic cell cycle (Kitsberg et al. 1993; Knoll et al. 1994). These studies used fluorescent in situ hybridization to follow the replication of specific markers within large chromosomal domains. It was found that the asynchrony of replication is not restricted to imprinted genes but extended to all the genes residing in the same chromosomal region. This observation suggests that asynchronous replication of imprinted regions is related to the higher order structure of the whole domain, rather than to individual genes. In addition, there is no simple relationship between replication timing and the potential of an allele to be expressed. The *Igf2* and *H19* genes are both replicated earlier on the paternal chromosome, although the paternal *Igf2* is expressed and the paternal *H19* allele remains silent.

Imprinted regions in the human genome appear to recombine at different rates during male and female meiosis (Pàldi et al. 1995; Robinson and Lalande 1995). We proposed earlier that the difference in recombination frequency reflects a difference in chromatin structure in male and female meiotic cells that could contribute to the establishment of the initial gametic imprint by influencing the accessibility of the chromatin to epigenetic modifications (Pàldi et al. 1995).

There are two important points that have to be taken in consideration when examining genomic imprinting:

Firstly, none of the above listed features is exclusively limited to imprinted regions and genes. Whether the two parental alleles of non-imprinted genes are really always transcribed at the same rate and/or display similar methylation patterns remains largely unknown. Asynchronous replication also occurs in non-imprinted genomic regions (Chess et al. 1994). The same is true for the sex-dependent meiotic recombination frequency. There are many regions in the genome that recombine at much higher rate in one sex than in the other without any evidence for the presence of imprinted genes. The presence of parental imprint(s) in a chromosomal domain is correlated to the simultaneous occurrence of all the properties listed above and their concentration to defined genomic regions. The biological function of genomic imprinting remains unknown. Although the research is traditionally focused on monoallelic expression, it is difficult to determine what is the relative importance of the different properties for genomic imprint to fulfill its role.

Secondly, our present knowledge on imprinting is mainly based on the study of the two major regions where most of the known imprinted genes are clustered. These are: 1) the Prader-Willi/Angelman (PWS/AS) region on the human chromosome 15q11–13 (the mouse homologue is on the proximal chromosome 7) which contains at least seven imprinted genes; 2) the Beckwith-Wiedeman region on the human 11p15.5 chromosome (the mouse homologue is on the distal chromosome 7), also containing at least seven imprinted genes. One has to be very cautious, therefore, about the general value of conclusions drawn from the study of only two regions. The evidence for pairing and interaction between imprinted domains on homologous chromosomes also come from the studies of these two regions.

3.2
Morphological Observations

Three-dimensional fluorescence in situ hybridization studies of the PWS/AS region have shown a specific homologous association between the paternal and maternal chromosomes in human T lymphocytes (LaSalle and Lalande 1996). This association was transient and occurred only during the late S phase of the cell cycle. The analysis of the second cluster of imprinted genes using a probe specific for H19 produced similar results. The homologous association is specific for the imprinted domains, because no significant association was observed between the homologues of control chromosome 12 loci in any phase of the cell cycle. The authors analysed the chromosome 15q11–13 interhomologue association in the cells of Prader-Willi and Angelman syndrome patients. These patients carried two maternal (PWS) or two paternal (AS) copies of chromosome 15. There was no association between the two homologues in these cells. In addition, association was not observed in cells from patients with maternal or paternal deletion of 15q11–13. This is reminiscent of the observation in *zeste*-mediated repression of *white* gene in *Drosophila*. Alterations in the normal biparental contribution of 15q11–13 that affect the monoallelic expression of genes in the region and the parental chromosome specific methylation pattern, therefore, also disrupt homologous associations. This suggests that pairing of imprinted domains might be involved in the imprinting process. Alternatively, the parental imprint present in these domains may promote homologous pairing.

3.3
Functional Evidence

Physical association between the maternal and paternal 15q13 subregion have functional consequences on these domains. This is evidenced by the studies of the replication kinetics of the GABRB3/A5 intergenic region (LaSalle and Lalande 1995). A phage contig of this region was defined and these clones were used for replication studies, using DNA FISH analyses. This 50–60-kb-long domain of maternal early replication is surrounded by regions with earlier paternal replication in normal human cells (LaSalle and Lalande 1995). Parental origin is not, however, the only factor that determines the timing and kinetics of replication in this region. The differential replication timing of the intergenic region is altered in cells of patients with uniparental disomies (UPD). In cells with maternal UPD the replication of the two maternally inherited chromosomal domains is synchronous, but significantly delayed during the S phase, compared to the replication timing of the maternal chromosome in normal individuals. The opposite is true for the cells with paternal UPD. The two paternally inherited chromosomal regions are replicated earlier than the paternal homologue in normal cells. Both parental contributions are necessary, therefore, for the establishment of the replication domain organization in this region. We do not know if similar processes occur in other imprinted regions. As mentioned above, the other major cluster of imprinted genes on the human chromosome 11p15 also shows transient association, but the replication kinetics of this region have not been studied in detail.

If transient physical association between the paternal and maternal homologues is involved in the replication control of imprinted chromosomal domains, could it influence gene expression? Recent observations on the mouse *Ins2* gene suggest that it is indeed the case (Duvillié et al. 1998). *Ins2* is located on the distal end of the mouse chromosome 7, in the cluster homologous to the human 11p15 region. It is imprinted in a tissue-specific and developmental stage-dependent fashion (Deltour et al. 1995; Giddings et al. 1994). In a recent study (Duvillié et al. 1997) we generated two different null alleles of *Ins2* by homologous recombination (see also Fig. 1). In the first null allele (Neo) the promoter and the coding sequences were deleted and replaced by a Neo-cassette. In the second null allele (Zneo) the insulin gene coding sequences were replaced by the *LacZ* gene and the same Neo-cassette as in the first construct. The *LacZ* gene was under the control of the endogenous *Ins2* promoter and it was expected to have the same expression pattern as the endogenous gene. The neomycin gene has its own viral promoter and enhancer and is expected to be transcriptionally autonomous in both constructs. Comparing both single heterozygotes and compound heterozygotes made it possible to assess the imprinting pattern of the exogenous sequences introduced at the *Ins2* locus. In both mutants the maternally transmitted *Neo* allele is silent in all tissues at all developmental stages indicating that it fell under the influence of the imprinting regulatory elements. As anticipated, the maternally inherited *LacZ* is expressed in the embryo proper of (Zneo × wt) genotype, just like the endogenous *Ins2* gene. In (Zneo × Neo) compound heterozygotes, however, the maternally inherited *LacZ* was silenced (Fig. 1). The only difference between the (Zneo × wt) and the (Zneo × Neo) embryos is in the nature of the allele on the paternal chromosome. This phenomenon is reminiscent of dominant variegation in *Drosophila*. It suggests that the paternal allele at the *Ins2* locus can influence the expression of the maternal homologue in somatic cells. The ability to *trans*-inactivate the maternal homologue is a specific characteristic of the paternally transmitted Neo allele. It depends on the acquisition of the paternal specific imprint because it does not occur in the reciprocal genotype. The ability to respond to the effect of the Neo allele is also allele-dependent, because only the Zneo, but not the wild type *Ins2* allele is silenced. Although the mechanism is not understood, it might be important that the *Ins2* promoter is deleted in the Neo allele, but not in the Zneo allele.

An indirect confirmation of these observations came from human studies (Bennett et al. 1997). The *IDDM2* type 1 diabetes susceptibility locus was identified as an allelic variant of the (*INS*) gene. Among the distinct classes of alleles, Class I alleles predispose to type 1 diabetes in a recessive way, while class III alleles are dominantly protective. A particular class I allele does not predispose to disease when paternally transmitted. But this paternal effect is observed only when the father's other allele is class III. The paternal-specific effect at *IDDM2* depends, therefore, on the nature of the untransmitted pater-

Genotype   Chromosome structure

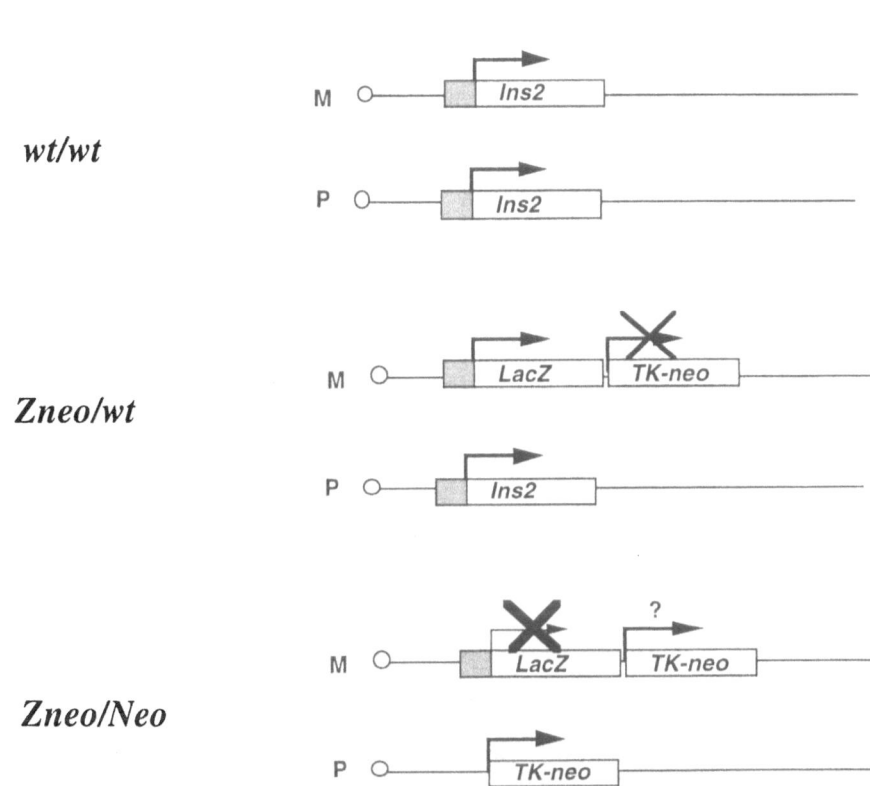

wt/wt

Zneo/wt

Zneo/Neo

Fig. 1. In wild-type embryos (*wt/wt*) both maternal (*M*) and paternal (*P*) alleles of *Ins2* are expressed, as indicated by the *arrows*. All the genes are shown as *open boxes*, the *filled box* represents the promoter of *Ins2*. In Zneo/wt embryos only the maternally inherited *LacZ* under the control of the endogenous promoter is expressed, but the *Neo* gene is not. However, when the *Ins2* gene with its promoter is replaced by the Neo cassette, the maternally transmitted *LacZ* remains silent. Differential expression of the maternal *Neo* cannot be tested

nal allele, suggesting the existence of heritable allelic interactions at the human *INS* locus. This is reminiscent of paramutations in plants.

Homologous interactions could also be extended to the neighbouring loci in the same region. A recent report suggests that the deletion of the maternal *H19* gene induces changes in *Igf2* methylation in both *cis* and *trans* (Forne et al. 1997). Functional homologous interactions may not occur at all stages, however; no role for maternal-paternal allelic interactions was found during preimplantation development (Szabo and Mann 1996).

Cosuppression-like phenomena were also observed in mice carrying extra copies of imprinted genes. In the first case, aberrant methylation of the endogenous *U2af1-rs1* gene caused by its own transgene has been observed (Hatada et al. 1997). In the second report, transactivation of the endogenous copy of the *Igf2* gene by its own transgene was detected (Sun et al. 1997). It is unclear at present, however, if these phenomena have any relevance for imprinting.

4
Discussion and Speculations

Altogether, the accumulated data provide good evidence that the homologues of the two best-characterized imprinted domains interact with each other in somatic cells in both mice and men. These interactions may influence all the characteristics commonly attributed to imprinted regions in somatic cells, such as monoallelic expression, asynchronous replication and differential methylation. Could the apparent association between the two phenomena result from a bias due to the intensive analysis of a limited number of examples? Should we add the allelic trans-sensing to the list of peculiar characteristics of imprinted regions? If the answer to the second question is yes, it is likely that the two phenomena are mechanistically related and the study of homologous interactions will help us to understand imprinting. In any case, the unusual interactions occurring at imprinted loci should warn us against simplistic interpretations of complex experimental results.

Eukaryotic cells carry a diploid chromosome complement. It appears that having two functional copies of the same chromosome (and the genes carried on it) is advantageous to the cell, as suggested by the evolutionary success of diploid organisms. Diploidy is a double-edged sword, however, since even partial loss or duplication of a chromosome usually has fatal consequences. To avoid accidents, diploid organisms had to develop defence mechanisms. This implies that the cell has to be aware of its diploid state, and the maintenance of the diploid state in the cell is likely to be an active process.

It would be teleologic to suppose that a special counting mechanism has evolved with the only function of supervising the correct chromosome number. It is more likely that the global surveillance of diploidy is achieved as a combined effect of several different mechanisms employed by the cell to coordinate various cellular functions. In addition to templating the ordered transcription of genes, each chromosome has to undergo a single round of replication during the somatic cell cycle. This has to be done in a way that allows the conservation, faithful replication and transmission to the daughter cells of all the epigenetic information necessary for the maintenance of the normal gene activity. The coordination of these functions is achieved with the help of a highly ordered, but dynamic and plastic organelle, the cell nucleus. The mechanisms in place include the forces responsible for the spatio-temporal organization of the chromosomes within the interphase nucleus.

According to a recent proposition, one type of local interactions that helps to build up large-scale global nuclear order is the long-range interaction between homologues (Marshall et al. 1997).

According to one hypothesis, genomic imprinting could play the role of a surveillance mechanism for chromosome loss (Thomas 1995). The postulated surveillance mechanism could make use of homologous interactions as a chromosome counting mechanism. In cells with normal diploid chromosome numbers, every chromosome can find its match. Transient homologous pairing of a small number of domains present on different chromosomes might, therefore, permit surveillance of normal chromosome number and facilitate counting in a similar way as in the *Xist*-driven X chromosome counting (Shemer et al. 1996).

Acknowledgments. We are grateful to Jacques Jami for stimulating discussions and for critical reading of the manuscript. We also thank Rajiv Joshi and Sukhvinder Sidhu for helpful comments.

References

Arnoldus E, Peters A, Bots G, Raap A, vander Ploeg M (1989) Somatic pairing of chromosome 1 centromeres in interphase nuclei of human cerebellum. Hum Genet 83:231–234

Bennett S, Wilson A, Esposito L, Bouzekri N, Undlien D, Cucca F, Nistico L, Buzzetti R, Group I, Bosi E, Pociot F, Nerup J, Chambon-Thomsen A, Pugliese A, Shield J, McKinney P, Bain S, Polychronakos C, Todd J (1997) Insulin VNTR allele-specific effect in type 1 diabetes depends on identity of untransmitted paternal allele. Nature Genetics 17:350–352

Bingham P (1997) Cosuppression comes to the animals. Cell 90: 385–387

Chess A, Simon I, Cedar H, Axel R (1994) Allelic inactivation regulates olfactory receptor gene expression. Cell 78:823–835

Coe E (1968) Heritable repression due to paramutation in maize. Science 162:925

Deltour L, Montagutelli X, Guenet J-L, Jami J, Pàldi A (1995) Tissue- and developmental stage-specific imprinting of the mouse proinsulin gene *Ins2*. Dev Biol 168:686–688

Dreesen T, Henikoff S, Loughney K (1991) A pairing-sensitive element that mediates *trans*-inactivation is associated with the *Drosophila brown* gene. Genes Devel 5:331–340

Duvillié B, Bucchini D, Tang T, Jami J, Pàldi A (1998) Imprinting at the mouse *Ins2* locus: Evidence for *cis*- and *trans*-allelic interactions. Genomics 47:52–57

Forne T, Oswald J, Dean W, Saam J, Bailleul B, Dandolo L, Tilghman S, Walter J, Reik W (1997) Loss of the maternal *H19* gene induces changes in *Igf2* methylation in both cis and trans. Proc Natl Acad Sci USA 94:10243–10248

Giddings S, King C, Harman K, Flood J, Carnaghi L (1994) Allele specific inactivation of insulin 1 and 2, in the mouse yolk sac, indicates imprinting. Nat Genet 6:310–313

Hatada I, Nabetani A, Arai Y, Ohishi S, Suzuki M, Miyabara S, Nishimune Y, Mukai T (1997) Aberrant methylation of an imprinted gene *U2af1-rs1(SP2)* caused by its own transgene. J Biol Chem 272:9120–9122

Henikoff S (1997) Nuclear organization and gene expression: homologous pairing and long-range interaction. Curr Opin Cell Biol 9:388–395

Hiraoka Y, Dernburg A, Parmelee S, Rykowski M, Agard D, Sedat J (1993) The onset of homologous chromosome pairing during *Drosophila melanogaster* embryogenesis. J Cell Biol 120:591–600

Hollick J, Dorweiler J, Chandler V (1997) Paramutation and related allelic interactions. Trends Genet 13:302–308

Jack JW, Judd BH (1979) Allelic pairing and gene regulation: a model for the zeste-white interaction in *Drosophyla melanogaster*. Proc Natl Acad Sci USA 76:1368–1372

Kitsberg D, Selig S, Brandeis M, Simon I, Keshet I, Driscoll D, Nicholls R, Cedar H (1993) Allele-specific replication timing of imprinted gene regions. Nature 364:459–463

Knoll J, Cheng S, Lalande M (1994) Allele specificity of DNA replication timing in the Angelman/Prader-Willi syndrome imprinted chromosomal region. Nat Genet 6:41–46

Kopczynski C, Muskavitch M (1992) Introns excised from the *Delta* primary transcript are localized near sites of *Delta* transcription. J Cell Sci 119:503–512

LaSalle J, Lalande M (1995) Domainorganization of allele-specific replication within the *GABRB3* gene cluster requires a biparental 15q11–13 contribution. Nat Genet 9:386–394

LaSalle JM, Lalande M (1996) Homologous association of oppositely imprinted chromosomal domains. Science 272:725–728

Lewis E (1954) The theory and application of a new method of detecting chromosomal rearrangements in *Drosophyla melanodaster*. Amer Nat 88:225–239

Marshall WF, Fung JC, and Sedat J (1997) Deconstructing the nucleus: global architecture from local interactions. Curr Opin Genet Devel 7:259–263

Monk M (1990) Variation in epigenetic inheritance. Trends Genet 6:110–114

Pàldi A, Gyapay G, and Jami J (1995) Imprinted chromosomal regions of the human genome display sex-specific meiotic recombination frequencies. Curr Biol 5:1030–1035

Robinson WP, and Lalande M (1995) Sex-specific meiotic recombination in the Prader-Willi/Angelman syndrome imprinted region. Human Mol Genet 4:801–806

Shemer R, Birger Y, Dean W, Reik W, Riggs A, and Razin A (1996) Dynapic methylation adjustment and counting as part of imprinting mechanisms. Proc Natl Acad Sci USA 93:6371–6373

Sun F, Dean W, Kelsey G, Allen N, and Reik W (1997) *Trans*-activation of *Igf2* in a mouse model of Beckwith-Wiedemann syndrome. Nature 389:809–815

Szabo P, Mann J (1996) Maternal and paternal genomes function independently in mouse ova in establishing expression of the imprinted genes *Snrpn* and *Igf2r*: no evidence for allelic trans-sensing and counting mechanisms. EMBO J 15:6018–6025

Tartof K, Henikoff S (1991) *Trans*-sensing effects from *Drosophila* to Humans. Cell 65:201–203

Thomas J (1995) Genomic imprinting proposed as a surveillance mechanism for chromosome loss. Proc Natl Acad Sci USA 92:480–482

Tsai J, Silver L (1991) Escape from genomic imprinting at the mouse *T-associated maternal* effect (*Tme*) locus. Genetics 129:1159–1166

Wu C (1993) Transvection, nuclear structure, and chromatin proteins. J Cell Biol 120:587–590

Wu CT, Goldberg ML (1989) The *Drosophila zeste* gene and transvection. Trends Genet 5:189–194

Nuclear Architecture

Wallace F. Marshall and John W. Sedat

Introduction: Control of Chromosome Interactions by Nuclear Architecture

Many important examples of epigenetic gene regulation involve interactions between loci on separate chromosomes. In this chapter we discuss the ways in which such interactions depend on the large scale structural context of the nucleus. Chromosomes are organized in a highly defined arrangement within the nucleus, as evidenced from a large number of structural studies in many different organisms (reviewed by Comings 1980; Marshall et al. 1997a). As a result of this organization, a particular genomic locus will tend to occupy a particular and reproducible spatial region within the nucleus. It has been demonstrated that interphase chromatin does not diffuse significantly over spatial scales larger than about 0.5 μm (Abney et al. 1997; Marshall et al. 1997b), so the only way that two loci could physically interact is if their positions within the nucleus are within 1 μm of each other. The relative proximity of any two loci, and hence their ability to interact with each other, is determined entirely by the nuclear architecture. Loci with similar positions in the nucleus will be able to interact. Indeed, their interactions will be facilitated because they will be maintained in close proximity throughout interphase, providing more opportunities to interact than if each was free to diffuse all throughout the nucleus. On the other hand, loci whose positions within the nucleus are far apart, will be prevented from interacting because they will never come into physical contact. Viewed in this way, the distance between two loci in the nucleus, as determined by the overall architectural arrangement of chromosomes, determines the relative concentration of the two loci. Thus, the large-scale three-dimensional structure of the nucleus is expected to exert a direct and powerful influence on which interactions between loci can occur.

We will begin with a review of nuclear architecture in general, followed by a more detailed discussion of the architecture of the Drosophila nucleus. With this structural context in mind, we will then describe two major types of

Department of Molecular, Cellular, and Developmental Biology, Yale University, New Haven, CT 06520
Dept. of Biochemistry and Biophysics, University of California, San Francisco, San Francisco, CA 94143

epigenetic phenomena in Drosophila, position effect variegation (PEV) and transvection, both of which involve interactions between distant chromosomal sites. Bearing in mind that nuclear architecture is expected to strongly influence such interactions, we will review recent evidence that directly supports a role for nuclear architecture in PEV and transvection. We will conclude by discussing how similar considerations may apply to imprinting and epigenetic regulation in humans.

Nuclear Architecture

The major result of decades of direct structural investigations using various types of microscopy is that a given locus appears to occupy a defined position in the nucleus. Because the nucleus is a complex three-dimensional object, the key to these studies has been the acquisition of three-dimensional images of chromosomes in situ. In this section we discuss the methods used to study nuclear architecture using three-dimensional light microscopy, and review the general features of nuclear architecture that these studies have brought to light.

The principle technical barrier to studying interphase nuclear organization is the fact that during interphase, chromosomes are decondensed and generally impossible to visualize clearly either by transmitted light methods (such as phase contrast or Nomarski) or by the use of general DNA stains. There are two primary ways to circumvent this difficulty. The first method is to image nuclei in which the interphase chromosomes are in fact condensed, either by inducing premature chromosome condensation (PCC) using cell fusion or anoxia (Sperling and Luedtke 1981; Foe and Alberts 1985) or by deliberately studying cell types whose chromosomes happen to remain condensed enough during interphase to enable them to be visualized (DuPraw 1965; Stack et al. 1977; Murray and Davies 1979; Agard and Sedat 1983). Unfortunately, the extent to which nuclear architecture in such cells represents organization found in more typical decondensed interphase nuclei is impossible to assess, and in the case of PCC there remains the unpleasant possibility that the observed organization may be to some extent artifactual. The second method for imaging interphase chromosomes is fluorescence in situ hybridization (FISH) which allows specific loci to be visualized by hybridizing fluorescently labeled DNA probes to individual chromosome sites. FISH has the advantage that it can be employed in any cell type, and positively identifies specific loci. A disadvantage with FISH, however, is that it does not generally allow the entire chromosome to be visualized simultaneously, and so a picture of overall nuclear organization must be built up by using a series of probes to different regions spanning the genome. Chromosome painting, a related technique using pools of probes spanning a large portion of a chromosome, circumvents this limitation to some degree. To date, a large number of studies have used FISH to visualize nuclear organization in situ (Arnoldus et al. 1989; Billia and

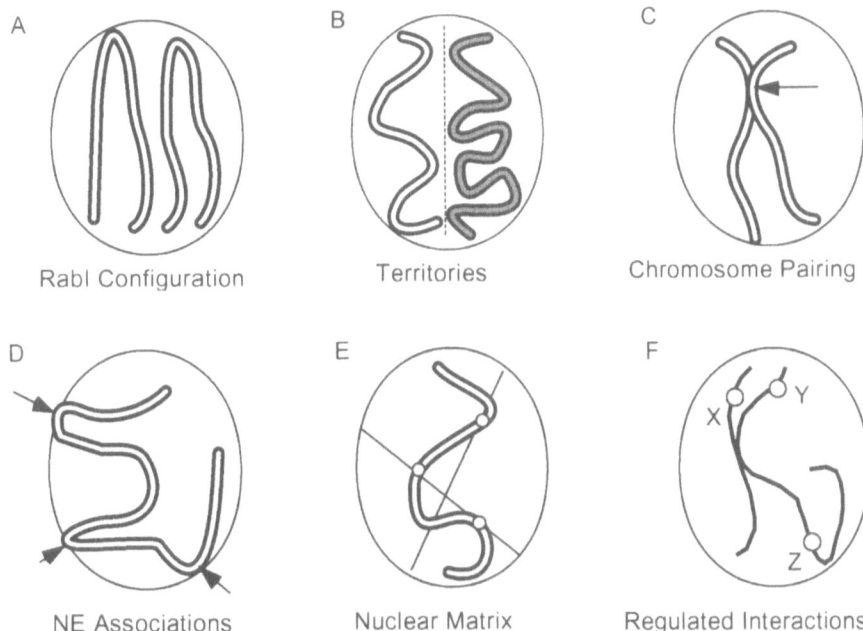

Fig. 1A–F. General features of nuclear architecture. A Rabl configuration. Anaphase chromosome arrangement is maintained in interphase, leading to a polarized nucleus with centromeres at one end and telomeres at the other. **B** Territories. Each chromosome occupies a limited region of the nucleus and adjacent chromosomes do not intertwine. Also thought to be a remnant of the anaphase chromosome arrangement. **C** Chromosome pairing. In some species and cell types, homologous chromosomes pair during interphase, either at a few discrete sites (*arrow*) or along their entire length. **D** Nuclear envelope associations. Specific chromosome regions interact with the nuclear envelope (*arrows*), and these interactions are likely to hold the chromosomes in place during interphase. **E** Nuclear matrix. Chromosomes are also thought to interact with an internal nuclear skeleton or matrix at defined chromosomal sites (*circles*). **F** The net result of this nuclear architecture is that interactions are regulated structurally. Despite being on different chromosomes, loci X and Y are spatially close together within the nucleus due to similar positions along the Rabl axis, and also due to a nearby chromosome pairing site, and thus are capable of interacting. Loci Z and Y, despite being on the same chromosome, are far apart due to the Rabl configuration, and therefore are unlikely to interact with each other

De Boni 1991; Chung et al. 1990; Cremer et al. 1993; Ferguson and Ward 1992; Funabiki et al. 1993; Gemkow et al. 1996; Guacci et al. 1994; Haaf and Ward 1995; Hiraoka et al. 1993; Hoefers et al. 1993; LaSalle and Lalande 1996; Manuelidis and Borden 1988; Marshall et al. 1996; Nagele et al. 1995; van Dekken et al. 1989; Vourc'h et al. 1993; Zalensky et al. 1995).

FISH studies have revealed, first and foremost, that the position of a given gene within the nucleus is not random, but is dictated by a global nuclear architecture. The nonrandom positioning of genes has been demonstrated most dramatically in Drosophila, in which over 40 individual loci were examined by FISH and each was found to reproducibly occupy a defined position in the nucleus (Marshall et al. 1996). This large-scale global architecture appears to arise from the interplay between a set of small-scale local interactions (reviewed by Marshall et al. 1997a) each of which provides a constraint on the possible chromosome configurations and, when taken together, force chromosomes into a globally defined arrangement. In the remainder of this section we review the most common organizational features of the nucleus.

Centromere-Telomere Polarization (Rabl Configuration)

The aspect of nuclear architecture that has been recognized for the longest time is the fact that in almost all nuclei, centromeres are generally located together at one end of the nucleus and telomeres at the other. This polarized chromosome arrangement, which is a remnant of the anaphase chromosome configuration, is named the Rabl configuration after its discoverer, Carl Rabl (reviewed in Comings 1980). A Rabl configuration has been demonstrated in many species including Drosophila (Ellison and Howard 1981; Foe and Alberts 1985; Hochstrasser et al. 1986; Marshall et al. 1996; Dernburg et al. 1996) and mammalian cells (Cremer et al. 1982; Haaf and Ward 1995). Because of the Rabl configuration, the spatial position of a locus within the nucleus is a strong function of the genomic position of the locus on the chromosome. By convention we will refer to the end of the nucleus containing the centromeres as the "top" and the end containing the telomeres as the "bottom" of the nucleus. In this frame of reference, loci whose genomic position is proximal to the centromeres will be found near the top of the nucleus, while more distal loci located further out on the arms will tend to be found near the bottom of the nucleus. This has an important and immediate implication for chromosome interactions: two loci on the same chromosome are increasingly less likely to interact as the genomic distance between them increases, because the spatial distance between them along the Rabl axis will be greater. Thus, when considering how nuclear architecture may influence interactions, it is important to keep in mind the positions of the two loci along the chromosome relative to each other and to the centromere. Another consequence of this Rabl configuration arises in species where the majority of heterochromatin is pericentric, such as in Drosophila. In this case, all the heterochromatin in the nucleus will be clustered in the upper portion of the nucleus, with the euchromatin extending out below it.

Territories

A very large amount of chromatin is packed into a very tiny nucleus, and one might expect that the nuclear interior would be highly entangled. Interestingly, this is not the case. When large regions or entire chromosomes are visualized, it is seen that each chromosome occupies a limited subregion or territory within the nucleus (Schardin et al. 1985; Hochstrasser et al. 1986; Cremer et al. 1993), and moreover, the territories of different chromosome do not overlap to any significant extent, even when they are adjacent to each other in the nucleus. This territorial organization has also been recognized on the basis of the pattern of gamma radiation induced chromosome interchanged in Drosophila embryos, which suggests that each chromosome arm occupies a limited sub-region of the nucleus (Hilliker 1986). One ramification of this territorial organization is that different chromosomes can only interact across the boundary between adjacent territories, so that only a limited subset of loci on a chromosome, those located at this boundary, are capable of interchromosomal interactions. Indeed, chromatin deep within a territory may be relatively inaccessible, as evidenced by the demonstration that transcription and splicing appear to take place mainly at the outer boundary of chromosome territories (Zirbel et al. 1993). Consistent with this idea, the majority of genes appear to be positioned at these territory boundaries (Kurz et al. 1996). The demonstration that the territories occupied by the active and inactive X chromosomes in mammalian cells have dramatically different shapes (Eils et al. 1996) may also suggest a link between territorial organization and gene silencing.

How are these territories established and maintained? One possibility is that after mitosis, chromosomes decondense and are able to undergo some diffusional motion within the nucleus, but then at some point the chromatin becomes anchored and diffusion is constrained (Marshall et al. 1997b) so that further spreading or tangling is blocked. In this model, mixing between chromosomes is limited because they start out as spatially separate condensed chromosomes with no overlap, and only a limited period of time is available for mixing, during which time the chromosomes are still somewhat condensed and hence less likely to become entangled. Therefore, establishment of territories does not necessarily require any special forces to act during interphase. This model also implies that if there is any organization to the condensed mitotic chromosome, such order will be reflected within the interphase territory. Specifically, if particular loci are near the central core of the mitotic chromosome (Saitoh and Laemmli 1994; Bickmore and Oghene 1996) they should end up deep within the interphase territory, while loci on the outer loops of the condensed chromosome (Bickmore and Oghene 1996) are more likely to end up on the boundary of the territory. Chromosome territories clearly represent an important area of nuclear architecture that warrants further investigation, and dramatically illustrate the degree to which nuclear architecture can influence chromosome interactions.

Homolog Association

Diploid eukaryotic nuclei contain two homologous copies of each chromosome. When specific loci are visualized in such nuclei using FISH, one frequently observes a single spot of fluorescence, indicating that the two homologous loci are in close proximity. Certainly homologs are well known to pair during meiosis, but homologous pairing is also a feature of somatic interphase nuclei, although this is species and cell-type specific. For example, somatic homolog pairing has been observed in interphase cells of Drosophila (Lifschytz and Hareven 1982; Hiraoka et al. 1993) and in humans at some loci (Arnoldus et al. 1989; LaSalle and Lalande 1996), though some loci are clearly unpaired in human and mouse cells (Kitsberg et al. 1993). Homologs appear not to be paired in interphase cells of S. cerevisiae (Guacci et al. 1994) or human cells during metaphase (Nagele et al. 1995). In addition, homologous pairing appears to be developmentally regulated. In Drosophila embryos, homologous chromosomes are initially unpaired during the first 12 syncytial division cycles, but then the frequency of homolog pairing appears to increase markedly (Hiraoka et al. 1993). Moreover, at a given developmental stage, only a subset of loci may be homologously paired, that is, pairing may not be uniform along the entire length of the chromosome (Fung et al. 1998). Thus, while we must bear in mind the potential for homolog pairing, whether or not homologs actually are paired at a given site in a given cell type, must be separately determined for each specific case.

NE Association

The maintenance of a defined nuclear architecture suggests that chromatin is anchored to some sort of immobile superstructure. The most visibly obvious structure to which chromatin could be anchored is the nuclear envelope (NE), and interactions between chromatin and the NE are thought to play an important role in setting up and maintaining nuclear organization. FISH experiments have demonstrated that specific loci are reproducibly associated with the NE during interphase (Hochstrasser et al. 1986; Manuelidis and Borden 1988; van Dekken et al. 1990; Chung et al. 1990; Vourc'h et al. 1993; Marshall et al. 1996).

What is the molecular basis of this site-specific chromatin-NE association? A large number of in vitro binding studies have revealed that binding interactions can occur between nuclear lamins and naked DNA (Luderus et al. 1994; Baricheva et al. 1996), between lamins and nucleosomes (Yuan et al. 1991), between lamins and core histones (Taniura et al. 1995), between lamins and chromosomes (Glass et al. 1993), between a lamin associated protein and chromosomes (Foisner and Gerace 1993), and between the lamin B receptor and the heterochromatin protein HP-1 (Ye and Worman 1996). In addition, a DNA binding motif has been recognized in a component of the nuclear pore complex (Sukegawa and Blobel 1993). While the lamin-DNA interactions may be sequence specific, at least to some extent, it is unclear whether the other interactions, for example between lamins and core histones, could actually be the basis for site-specific NE association, although it is conceivable that site-specific differences in nucleosome folding, supercoiling, etc., could lead to the type of specific NE associations seen in interphase. The extent to which these binding interactions, detected in vitro using purified components at high concentration, actually take place in vivo also remains to be determined. Certainly, it seems unlikely that all these potential interactions are equally important biologically, and it will be interesting to see, in future studies, which if any plays a dominant role.

Internal Nuclear Skeleton

The nuclear envelope is certainly the most visually obvious large-scale structure to which chromatin could be anchored, but it may not be the only one. A variety of biochemical preparations have suggested the presence of a protein-based nuclear skeleton, sometimes referred to as the nuclear scaffold or nuclear matrix (Berezney and Coffey 1974). In such preparations, histones and other proteins are extracted from the nucleus leaving behind an insoluble residue which presumably represents a network of interacting proteins. Such a network is not typically observed in electron micrographs of nuclei, but does appear when nuclei are mounted using a special resinless embedding method (Capco et al. 1982). Importantly, specific DNA sequences have been found to preferentially associate with the nuclear matrix (Gasser and Laemmli 1986). These regions, called MARs (matrix attachment regions) or SARs (scaffold attachment regions) may play a major role in maintaining nuclear architecture by anchoring chromatin to the matrix, and may explain the fact that diffusion of interphase chromatin appears highly constrained (Abney et al. 1997; Marshall et al. 1997b).

Despite a great many studies on the composition of these nuclear skeleton preparations, there remains a great deal of controversy regarding the extent to which they represent structures actually present inside living cells, as opposed to artifacts produced by the matrix preparation procedure. Obviously, the only convincing way to find out if living cells have a nuclear skeleton is to attempt to detect it inside living cells, but surprisingly few studies of this type have been reported. In one recent report, it was shown that a protein constituent of the nuclear matrix in Drosophila embryos, CP60, localizes to the nucleus during interphase and appears to form a fibrous network (Oegema et al. 1997). Importantly, when the nuclear envelope breaks down in prophase, and chromosomes are condensed, the CP60 network remains intact in the shape of the nucleus, and this is maintained until anaphase. Therefore, inside a living cell, CP60 is in an insoluble non-diffusing form. The most straightforward interpretation is that CP60 is part of, or bound to, a large-scale nuclear skeleton of some sort. Similar experiments localizing other putative nuclear skeleton components in living cells in other species will help to finally resolve this long standing question.

Functions of Nuclear Architecture

Why, teleologically, do cells have a defined nuclear architecture? Nuclear architecture has been proposed to play a role in meiosis by facilitating homologous pairing of telomeres. During meiosis, it has been observed that in many organisms homolog pairing begins at the telomeres which are in contact with the nuclear envelope. It has been proposed that NE anchorage of telomeres may facilitate their interaction by allowing homologous telomeres to search for each other in a two-dimensional space, rather than the three dimensional space of the nuclear interior (reviewed in Loidl 1990). Computer simulations have confirmed the intuitive view that such confinement could dramatically increase homology search efficiency (Loidl and Langer 1993).

Another proposed function for a defined nuclear organization is represented by the gene gating hypothesis (Blobel 1985) in which it is proposed that active genes associate with the nuclear envelope in order to facilitate transcript export. In cases of targeted mRNA localization, it has been further proposed that by localizing a gene to a particular region on the nuclear surface, transcript can be specifically exported through that region and thus be directed out of the nucleus in a specified direction. Thus, targeting of active genes to the NE is proposed to play a role in rapid transcript export as well as transcript localization. The concept of gene gating received some support when active regions of the genome were visualized by a clever scheme that detected DNaseI cleavable sites in situ. Such experiments (Hutchison and Weintraub 1985) showed that the majority of DNaseI cleavable sites, which presumably reflect the distribution of active chromatin (Weintraub and Groudine 1976), were localized near the nuclear envelope. The gene-gating model has been particularly compelling in the case of the apical localization of Drosophila pair-rule genes, whose localization seems fixed relative to the position of the nucleus (Davis et al. 1993). In this case, the known strong Rabl configuration in Drosophila (see below) causes the nucleus to be highly polarized. Any genes localized to the NE near the apical end of the nucleus would be predicted to export their transcript into the apical cytoplasm thereby giving the observed transcript localization. Experimental evidence tends to argue against this hypothesis in the case of Drosophila, however. First of all, pair-rule genes are distributed throughout the genome, and are not clustered near the centromeres, so based on the Rabl configuration there is no reason to think that they will be apically localized in the nucleus. When these pair-rule genes are inserted into different places in the genome, their transcripts still localize correctly, and if two different genes whose transcripts localize differently are inserted in tandem to the same genomic region, their individual transcripts are still localized correctly (Davis et al. 1993), which is inconsistent with the position of the locus playing a determinative role in transcript localization. Detection of pair rule gene nascent transcripts in the nucleus using immunohistochemical in situ hybridization (Davis et al. 1993), as well as localization of pair rule genes themselves by DNA FISH (Marshall and Sedat unpublished data) has revealed that pair rule genes are not localized apically and are also not associated with the nuclear envelope. These experiments strongly argue against gene gating in the case of pair-rule transcript localization. With respect to gene gating to facilitate transcript export apart from a role in localization, there also appears to be no correlation: the highly transcribed ubx and histone genes in Drosophila are not NE associated (Gemkow et al. 1996; Marshall et al. 1996). Therefore, it appears rather unlikely that NE association or the Rabl configuration play a direct role in transcript export or localization, at least in the case of Drosophila where this issue has been most extensively investigated.

Another possible function of a defined nuclear architecture is to regulate or control interactions between loci. As discussed above, the fact that chromatin diffusion is highly constrained implies that the relative position of any two loci in the nucleus will, to a large extent, determine their ability to interact. Certainly, local sequence information at the two loci, recognized by specific binding proteins which bring about an interaction, can play a major role in determining the interactions between loci, by determining the affinity constant for the interaction. But nuclear architecture can increase or decrease the tendency of the loci to interact just as effectively, not by altering the intrinsic binding affinity, but simply by increasing or decreasing the relative concentrations of two loci. For example, the pattern of radiation induced chromosome interchanges appears to be strongly influenced by the territorial organization of chromosomes in the nucleus (Hilliker 1986). But nuclear architecture also appears to play a role in a variety of other chromosome interactions. In particular, there are a variety of epigenetic phenomena that appear to involve interactions between loci that are not adjacent to each other in the genome. In such cases, nuclear architecture is likely to play an important functional role. In the following sections we will further consider this possibility using specific examples from Drosophila. We begin with an overview of Drosophila nuclear architecture.

Architecture of the Drosophila Nucleus

The preeminence of Drosophila melanogaster as a model system for genetic studies of chromosome behavior have made it an organism of choice for studies of nuclear architecture. Recent studies have begun to relate chromosome behavior, as determined from genetic studies, to chromosome arrangement determined by structural studies. Before discussing these experiments, we will first review what is known about the architecture of the Drosophila nucleus.

The polytene chromosomes in nuclei of most Drosophila larval tissues, particularly the salivary glands, consist of thousands of identical chromatids lined up side by side, giving rise to large, easily visualized chromosomes with a recognizable banded appearance when viewed with transmitted light microscopy. These bands can also be recognized when these chromosomes are stained with the DNA-binding fluorescent dyes. In one of the first applications of three-dimensional fluorescence microscopy (Agard and Sedat 1983), Sedat and coworkers acquired high resolution three-dimensional images from polytene nuclei. Interactive computer modeling techniques were then used to trace the path of each chromosome through the nucleus, and the banding pattern using fluorescent staining was related to the banding pattern seen in standard polytene squash preparations allowing the position of each locus, as defined by the unique pattern of bands and interbands, to be determined within the nucleus in three dimensions. Then, several quantitative measures of position and chromosome organization were measured at each site (Hochstrasser et al. 1986). This allowed a comparison of the position and conformation at a given site between a large collection of nuclei, to look for patterns in common. This analysis reached several important conclusions. First of all, these chromosomes were arranged in a very pronounced Rabl configuration, with all centromeres clustered at one end of the nucleus, and all telomeres at the other. In between, the chromosomes ran down through the nucleus. The chromosomes themselves were not perfectly straight but were coiled in a right-handed helix. A second dramatic conclusion was that the chromosomes were not intertwined at all, but rather, each chromosome occupied a separate territory or sector within the nucleus. Another dramatic observation was the fact that a set of sites were identified that were always in close contact with the nuclear envelope. This, together with the strong Rabl configuration, suggested that there may be some predetermined positioning for different loci within the nucleus. A fundamental assumption in this work was that the position of a locus in the polytene nucleus is likely to reflect the position of that locus in the diploid nucleus at the onset of polyteny, but this has never been explicitly proven. Therefore, these results provided the impetus to study nuclear organization within diploid nuclei.

The picture of Drosophila nuclear architecture obtained by studies of polytene nuclei has subsequently been verified and extended by similar studies in diploid interphase nuclei of the Drosophila embryo. Because diploid interphase chromosomes cannot normally be visualized, these studies have, for the most part, relied on fluorescence in situ hybridization to identify the position of specific loci. The main result of these studies is that for all genomic loci examined by FISH in Drosophila (well over 40 loci have been tested), each was found to reproducibly occupy a defined region in the nucleus (Marshall et al. 1996). This nonrandom positioning appears to involve several architectural features outlined below.

Perhaps the most dramatic feature of Drosophila nuclear architecture is the high level of somatic homolog pairing in diploid cells. In almost all cells, at almost all developmental stages, chromosomes seem to be paired along their entire length. This was initially recognized for mitotic chromosomes (Metz 1916) but has subsequently been confirmed in interphase nuclei by FISH (Lifschytz and Hareven 1982; Hiraoka et al. 1993; Fung et al. 1998). This homolog pairing is known to influence gene expression under some circumstances (see below).

Prior to the application of FISH technology, it had already been known for some time that Drosophila embryo nuclei were polarized, based on staining with AT-specific DNA stains (Ellison and Howard 1981) which showed a bright focus of staining at one end of each nucleus, the end nearest the surface of the embryo. Examination of nuclei in which interphase chromosomes were induced to undergo premature condensation due to anoxia (Foe and Alberts 1985), and of recently condensed prophase chromosomes in embryo nuclei (Hiraoka et al. 1990a), also supported the presence of a Rabl configuration. Subsequent studies using FISH have confirmed that this represents a Rabl configuration present in the interphase nucleus, such that centromeric regions are clustered at one end of the nucleus nearest the embryo surface while telomeres are mainly found at the other end of the nucleus, pointing towards the embryo interior (Hiraoka et al. 1990b, Gunawardena et al. 1995; Marshall et al. 1996; Dernburg et al. 1996).

Studies of Drosophila embryo nuclei using FISH have also confirmed another major result of the polytene studies, namely that a set of discrete loci is nonrandomly associated with the nuclear envelope (Marshall et al. 1996). Of those polytene NE association sites tested by FISH, all were found to be NE associated in the embryo, but additional sites were seen to be NE associated that were not NE associated in polytene nuclei. The reasons for this discrepancy are unclear, but there is some evidence that chromosome-NE associations are gradually lost during the course of polytenization (Quick 1980). The molecular basis of this interaction is currently unknown, but preliminary experiments using a combination of FISH and immunofluorescence indicate that NE associated sites appear to colocalize with nuclear lamin fibers rather than nuclear pore complexes (O. Beske, W. F. Marshall, and J. W. Sedat, unpubl. results). While the chromosomal determinants of NE association have not yet been identified, chromosome rearrangements have indicated that the NE association of a locus is determined locally, by chromosomal elements in cis. Specifically, if an NE associated site is moved to another region of the genome using chromosome rearrangements, it can cause regions near the site of insertion that are not normally NE associated to become targeted to the nuclear envelope (Dernburg et al. 1996).

Other aspects of nuclear architecture measured in polytene nuclei, such as helical coiling of the chromosome, separation of chromosomes into discrete territories, and nonrandom interchromosomal and intrachromosomal interactions, are more difficult to assess by FISH as they require the simultaneous use of several FISH probes, but such studies are currently underway.

Epigenetic Phenomena in Drosophila – Position Effect Variegation & Transvection

In order to explore the role of nuclear architecture in epigenetic processes, we need a model system which is genetically manipulable and in which the nuclear organization is well understood. Drosophila melanogaster fits the bill on both accounts. Drosophila has long played a central role in studies of chromosome behavior. In particular, epigenetic processes in which gene expression is affected by large scale genomic context, were discovered in Drosophila early on and have been the subject of genetic experimentation ever since. Moreover, nuclear architecture in Drosophila has been perhaps more thoroughly investigated and analyzed than in any other organism. This is, of course, not a coincidence. Drosophila was deliberately selected as the subject of investigations of chromosome organization precisely because of its popularity as a genetic system for studies of chromosome function. In this section we review two epigenetic processes in Drosophila, transvection and position effect variegation (PEV), which are thought to involve long range interactions at the nuclear architectural level.

Transvection

In Drosophila, a number of mutations have been discovered in which the expression of a gene depends on whether or not the two homologous chromosomes are physically paired in the vicinity of the gene (Tartof and Henikoff 1991; Wu 1993). One example is the zeste-white interaction (reviewed in Wu and Goldberg 1989). zeste encodes a transcription factor that binds upstream of the white gene, and in *zeste* mutants, the white gene becomes inactivated but only when the two homologous copies of white are paired (Jack and Judd 1979). Chromosome rearrangements that disrupt homolog pairing at white restore white expression. A similar phenomenon is transvection, a type of intragenic complementation that requires homolog pairing. As a specific example, in flies carrying two mutant alleles of the *yellow* gene, one of which disrupts the upstream enhancer/promoter region but not the coding region and the other of which disrupts the coding region but not the enhancers, the *yellow* gene is expressed, but only if the two homologs are able to pair (Geyer et al. 1990). Presumably, the enhancer elements on one homolog can act in trans on the promoter of the other homolog, and this trans-activation requires physical association of the homologs. Goldsborough and Kornberg (1996) have recently provided evidence that homologous pairing has a general enhancing effect on transcription at the Ubx locus. These pairing-dependent phenomena provide a clear-cut example in which large-scale nuclear architecture (in this case, physical association between homologs) exerts a strong epigenetic influence on gene expression. Moreover, the ability of homologs to pair is itself influenced by other aspects of nuclear architecture. It has recently been shown that somatic homolog pairing in Drosophila is disrupted during anaphase, and is re-established during interphase, a slow process that appears to involve a random-walk search of the homologs within the nucleus (Fung et al. 1998). If pairing at a given locus has not yet been established by the time the gene's product is required, an effect on transvection may be observed. The distance between homologs prior to the onset of pairing, which is determined by the overall nuclear architecture, determines the time required to establish pairing, and can therefore determine whether or not transvection occurs. The ability of chromosome rearrangements to disrupt transvection is probably due to the fact that loci generally occupy defined regions within the nucleus, so that homologous loci start out closer together than non-homologs. In rearranged chromosomes, the homologous loci are no longer in the same region of the nucleus, due to the Rabl orientation, and thus require more time to pair over this greater distance. Consistent with this interpretation, it has been shown that somatic homolog pairing in Drosophila embryos occurs less frequently for a rearranged chromosome (Hiraoka et al. 1993) and requires more for complete pairing (Fung et al. 1998). It has also been shown that mutations that slow the rate of cell division, thus providing more time for homologs to locate each other and pair, are able to restore a transvection effect that has been disrupted by chromosome rearrangements (Golic and Golic 1996).

Position Effect Variegation

In Drosophila, heterochromatin is clustered around the centromeres on the autosomes and the X chromosome. When chromosome rearrangements move a portion of heterochromatin into a normally euchromatic region, expression of euchromatic genes near the breakpoint becomes variegated, that is, gene expression becomes repressed to a variable extent from cell to cell. The degree to which a given gene is repressed depends on how close it is to the breakpoint, genes nearest the heterochromatin insertion are repressed the most strongly. This phenomenon is known as position effect variegation or PEV (reviewed by Weiler and Wakimoto 1995 and Henikoff 1990). PEV is most easily observed when genes involved in generating eye pigment become variegated. For example, when the *white* gene becomes variegated, white patches of variable size are seen in the otherwise red eye. The fact that variegated expression often occurs in such patches implies that the variegated state is clonally inherited over many cell divisions. PEV thus represents a clear case of epigenetic regulation or somatic imprinting. A number of mutations have been identified that interfere with PEV, thus revealing genes whose products are involved in establishment or maintenance of the heterochromatic state. One such gene, known as Su(var)205, suggests a tenuous link between nuclear envelope association and PEV: the product of Su(var)205 is a heterochromatin-associated protein called HP-1 (James and Elgin 1986) and it has been reported that a mammalian

HP-1 homolog is capable of binding to the lamin B receptor, a component of the nuclear envelope (Ye and Worman 1996). However, the relevance of this result for PEV in Drosophila is currently unclear. Known HP-1 binding sites in Drosophila do not appear to correlate with nuclear envelope association sites (Marshall et al. 1996). Moreover, FISH analysis of NE association sites in Su(var)205 null mutant embryos did not reveal any effect of the mutation on NE association (W. Marshall, J. W. Sedat, and R. Kellum, unpublished data). Thus, the role of NE interactions in PEV, if any, is not obvious.

One form of PEV, known as dominant position effect variegation, is particularly interesting in that it combines the phenomenon of position effect variegation with the trans-sensing of transvection. Specifically, the brown (*bw*) eye color gene has an allele known as *brown*^{Dominant} (*bw*^{D}) that is caused by an insertion of a large block of heterochromatin into the *bw* coding region (Slatis 1955) so that expression of the *bw* gene is completely eliminated in this allele. Interestingly, in *bw*^{D}/*bw*$^{+}$ heterozygotes, the expression of the wild-type *bw*+ allele is repressed, even though there is no heterochromatin in cis. Somehow, the heterochromatin on the rearranged chromosome is able to cause variegation of the wild-type gene on the homologous chromosome. The ability of *bw*^{D} to silence its partner depends on the ability of the *bw*^{D} chromosome to pair with its homologous partner (Dreesen et al. 1991). In this respect, dominant PEV resembles transvection. But the behavior of *bw*^{D} also depends on the distance of the *bw*^{D} locus from the centromere (Talbert et al. 1994; Henikoff et al. 1995), a hallmark feature of PEV.

A model for dominant PEV is based on the strong Rabl configuration of Drosophila (see above) which means that, because heterochromatin is generally clustered around the centromeres, all the heterochromatin in the nucleus will be clustered at the top of the nucleus in a special heterochromatic domain. A nuclear architecture-based model for PEV (Henikoff 1994) proposes that the cluster of heterochromatin defines a special region of the nucleus, and any euchromatic genes brought into this region will become heterochromatic and transcriptionally repressed. In this model, silencing by PEV would require movement of the silenced gene up to the top of the nucleus where the majority of the heterochromatin is located. Presumably, this movement would take place by random Brownian motion of the chromosomes (Marshall et al. 1997b). During this random motion, loci closer to the centromere would start out closer to, and hence be more likely to make contact with, centromeric heterochromatin, allowing them to be silenced more effectively. Thus, this model can explain the distance effect of PEV. In the specific case of *bw*^{D}, it is proposed that the *bw*^{D} allele pairs with its homologous copy and then recruits it to the heterochromatic domain of the nucleus in a stochastic manner, where it becomes repressed. Recently, FISH has been used to directly verify this prediction.

When the *bw* locus was visualized by FISH in wild-type cells, it was found that, as expected from its distal location on the chromosome arm and from the strong Rabl orientation in Drosophila, the gene was found at the bottom of the nucleus near the telomeres (Dernburg et al. 1996). In contrast, in bw^D larvae, it was found that in a large fraction of cells, the bw^D locus was localized near the top of the nucleus along with the rest of the heterochromatin (Csink and Henikoff 1996; Dernburg et al. 1996). This repositioning of bw^D to the heterochromatic domain is less frequent when the locus is moved farther away from the centromere or in Su(var) mutants that reduce PEV (Csink and Henikoff 1996), thus, the same factors that affect variegation have an equivalent effect on localization to the heterochromatic domain, consist with the model. Thus, dominant PEV appears to provide a very dramatic example of the importance of nuclear architecture on epigenetic gene regulation.

Role of Nuclear Architecture in Mammalian Imprinting

While it is possible that the various epigenetic phenomena of Drosophila are specific to this organism, recent studies indicate that in fact, phenomena like transvection and PEV may be more widespread than was previously thought. Chromosome pairing-dependent phenomena have been demonstrated in plants (Coen and Carpenter 1988) and fungi (Aramayo and Metzenberg 1996). There also appears to be a link in humans between somatic chromosome pairing and imprinting. The Prader-Willi syndrome / Angelman syndrome locus is oppositely imprinted in males and females. Using FISH, LaSalle and Lalande (1996) demonstrated that both a paternally imprinted and a maternally imprinted homolog are required for somatic homolog pairing at this locus. Thus, for this locus at least, imprinting is necessary for homolog pairing. Whether the converse will also be true, that homolog pairing plays an essential role in imprinting, has yet to be determined. Another study has revealed a potential similarly between gene silencing in humans and PEV in Drosophila. Just as Drosophila variegating loci are physically recruited to a special distinct heterochromatic compartment in the nucleus (see above), Brown et al. (1997) have demonstrated that repressed genes in B lymphocytes become physically associated with clusters of centromeric heterochromatin in the nucleus.

Thus, the relationship between nuclear architecture and epigenetic gene regulation is likely to be a general one. Certainly, a great deal more work must be done in Drosophila, exploiting its advantages as a model system. But at the same time, further characterization of nuclear architecture in mammalian cells is clearly of a high priority. It is to be expected that the combined genetic and structural approach that has proved so fruitful in Drosophila, will also yield insights into the connection between nuclear architecture and gene regulation in mammals as well.

References

Abney JR, Cutler B, Filbach ML, Axelrod D, Scalettar BA (1997) Chromatin dynamics in interphase nuclei and its implications for nuclear structure. J Cell Biol 137:1459–1468

Agard DA, Sedat JW (1983) Three-dimensional architecture of a polytene nucleus. Nature 302:676–681

Aramayo R, Metzenberg RL (1996) Meiotic transvection in fungi. Cell 86:103–113

Arnoldus EP, Peters AC, Bots GT, Raap AK, van der Ploeg M (1989) Somatic pairing of chromosome I centromeres in interphase nuclei of human cerebellum. Hum Genet 83:231–234

Baricheva EA, Berrios M, Bogachev SS, Borisevich IV, Lapik ER, Sharakhov IV, Stuurman N, Fisher PA (1996) DNA from Drosophila melanogaster beta-heterochromatin binds specifically to nuclear lamins in vitro and the nuclear envelope in situ. Gene 171:171–176

Berezney R, Coffey D (1974) Identification of a nuclear protein matrix. Biochem Biophys Res Commun 60:1410–1419

Bickmore WA, Oghene K (1996) Visualizing the spatial relationships between defined DNA sequences and the axial region of extracted metaphase chromosomes. Cell 84:95–104

Billia F, De Boni U (1991) Localization of centromeric satellite and telomeric DNA sequences in dorsal root ganglion neurons in vitro. J Cell Sci 100:219–226

Blobel G (1985) Gene gating: a hypothesis. Proc Natl Acad Sci USA 82:8527–8529

Brown KE, Guest SS, Smale ST, Hahm K, Merkenschlager M, Fisher AG (1997) Association of transcriptionally silent genes with ikaros complexes at centromeric heterochromatin. Cell 91:845–854

Capco DG, Wan KM, Penman S (1982) The nuclear matrix: three-dimensional architecture and protein composition. Cell 29:847–858

Chung HM, Shea C, Fields S, Taub RN, van der Ploeg LHT, Tse DB (1990) Architectural organization in the interphase nucleus of the protozoan Trypanosoma brucei: location of telomeres and mini-chromosomes. EMBO J 9:2611–2619

Coen ES, Carpenter R (1988) A semi-dominant allele, niv-525, acts in trans to inhibit expression of its wild-type homolog in Antirrhinum majus. EMBO J 7:877–883

Comings DE (1980) Arrangement of chromatin in the nucleus. Hum Genet 53:131–143

Cremer T, Cremer C, Baumann H, Luedtke EK, Sperling K, Teuber V, Zorn C (1982) Rabl's model of the interphase chromosome arrangement tested in chinese hamster cells by premature chromosome condensation and laser-UV-microbeam experiments. Hum Genet 60:46–56

Cremer T, Kurz A, Zirbel R, Dietzel S, Rinke B, Schroeck E, Speicher MR, Mathieu U, Jauch A, Emmerich P, Scherthan H, Ried T, Cremer C, Lichter P (1993) Role of chromosome territories in the functional compartmentalization of the cell nucleus. Cold Spring Harbor Symp Quant Biol 58:777–792

Csink A, Henikoff S (1996) Genetic modification of heterochromatic association and nuclear organization in Drosophila. Nature 381:529–531

Davis I, Francis-Lang H, Ish-Horowicz D (1993) Mechanisms of intracellular transcript localization and export in early Drosophila embryos. Cold Spring Harbor Symp Quant Biol 58:793–798

Dernburg AF, Broman KW, Fung JC, Marshall WF, Philips J, Agard DA, Sedat JW (1996) Perturbation of nuclear architecture by long-distance chromosome interactions. Cell 85:745–759

Dreesen TD, Henikoff S, Loughney K (1991) A pairing-sensitive element that mediates trans-inactivation is associated with the Drosophila brown gene. Gen Dev 5:331–340

Dupraw EJ (1965) The organization of nuclei and chromosomes in honey bee embryonic cells. Proc Natl Acad Sci USA 53:161–168

Eils R, Dietzel S, Bertin E, Schroeck E, Speicher MR, Ried T, Robert-Nicoud M, Cremer C, Cremer T (1996) Three-dimensional reconstruction of painted human interphase chromosomes: active and inactive X chromosome territories have similar volumes but differ in shape and surface structure. J Cell Biol 135:1427–1440

Ellison JR, Howard GC (1981) Non-random position of the A-T rich DNA sequences in early embryos of Drosophila virilis. Chromosoma 83:555–561

Ferguson M, Ward DC (1992) Cell cycle dependent chromosomal movement in pre-mitotic human T-lymphocyte nuclei. Chromosoma 101:96–106

Foe VE, Alberts BM (1985) Reversible chromosome condensation induced in Drosophila embryos by anoxia: visualization of interphase nuclear organization. J Cell Biol 100:1623–1636

Foisner R, Gerace L (1993) Integral membrane proteins of the nuclear envelope interact with lamins and chromosomes, and binding is modulated by mitotic phosphorylation. Cell 46:521–530

Funabiki H, Haga I, Uzawa S, Yanagida M (1993) Cell cycle-dependent specific positioning and clustering of centromeres and telomeres in fission yeast. J Cell Biol 121:961–976

Fung JC, Marshall WF, Dernburg AF, Agard DA, Sedat JW (1998) Homologous chromosome pairing in Drosophila melanogaster proceeds through multiple, independent initiations. J Cell Biol 141:5–20

Gasser SM, Laemmli UK (1986) Cohabitation of scaffold binding regions with upstream/enhancer elements of three developmentally regulated genes of D. melanogaster. Cell 46:521–530

Gemkow MJ, Buchenau P, Arndt-Jovin DJ (1996) FISH in whole-mount Drosophila embryos. RNA: activation of a transcriptional locus, DNA: gene architecture and expression. Bioimaging 4:107–120

Geyer PK, Green MM, Corces VG (1990) Tissue-specific transcriptional enhancers may act in trans on the gene located in the homologous chromosome: the molecular basis of transvection in Drosophila. EMBO J 9:2247–2256

Glass CA, Glass JR, Taniura H, Hasel KW, Blevitt JM, Gerace L (1993) The alpha-helical rod domain of human lamins A and C contains a chromatin binding site. EMBO J 12:4413–4424

Goldsborough AS, Kornberg TB (1996) Reduction of transcription by homologue asynapsis in Drosophila imaginal discs. Nature 381:807–810

Golic MM, Golic KG (1996) A quantitative measure of the mitotic pairing of alleles in Drosophila melanogaster and the influence of structural heterozygosity. Genetics 143:385–400

Guacci V, Hogan E, Koshland D (1994) Chromosome condensation and sister chromatid pairing in budding yeast. J Cell Biol 125:517–530

Gunawardena S, Heddle E, Rykowski M (1995) 'Chromosomal puffing' in diploid nuclei of Drosophila melanogaster. J Cell Sci 108:1863–1872

Haaf T, Ward DC (1995) Rabl orientation of CENP-B box sequences in Tupaia belangeri fibroblasts. Cyt Cell Genet 70:258–262

Henikoff S (1990) Position-effect variegation after 60 years. Trends Genet 6:422–426

Henikoff S (1994) A reconsideration of the mechanism of position effect. Genetics 138:1–5

Henikoff S, Jackson JM, Talbert PB (1995) Distance and pairing effects on the brown[Dominant] heterochromatic element in Drosophila. Genetics 140:1007–1017

Hilliker AJ (1986) Assaying chromosome arrangement in embryonic interphase nuclei of Drosophila melanogaster by radiation induced interchanges. Genet Res 47:13–18

Hiraoka Y, Agard DA, Sedat JW (1990a) Temporal and spatial coordination of chromosome movement, spindle formation, and nuclear envelope breakdown during prometaphase in Drosophila melanogaster embryos. J Cell Biol 111:2815–2828

Hiraoka Y, Rykowski MC, Lefstin JA, Agard DA, Sedat JW (1990b) Three-dimensional organization of chromosomes studied by in situ hybridization and optical sectioning microscopy. Proc Society of Photo-optical Instrumentation Engineers 1205:11–19

Hiraoka Y, Dernburg AF, Parmelee SJ, Rykowski MC, Agard DA, Sedat JW (1993) The onset of homologous chromosome pairing during Drosophila melanogaster embryogenesis. J Cell Biol 120:591–600

Hoefers C, Baumann P, Hummer G, Jovin TM, Arndt-Jovin DJ (1993) The localization of chromosome domains in human interphase nuclei. Three-dimensional distance determinations of fluorescence in situ hybridization signals from confocal laser scanning microscopy. Bioimaging 1:96–106

Hochstrasser M, Mathog D, Gruenbaum Y, Saumweber H, Sedat JW (1986) Spatial organization of chromosomes in the salivary gland nuclei of Drosophila melanogaster. J Cell Biol 102:112–123

Hutchison N, Weintraub H (1985) Localization of DNAase I-sensitive sequences to specific regions of interphase nuclei. Cell 43:471–482

Jack JW, Judd BH (1979) Allelic pairing and gene regulation: a model for the zeste-white interaction in Drosophila melanogaster. Proc Natl Acad Sci USA 76:1368–1372

James TC, Elgin SCR (1986) Identification of a nonhistone chromosomal protein associated with heterochromatin in Drosophila melanogaster and its gene. Mol Cell Biol 6:3862–3872

Kitsberg D, Selig S, Brandeis M, Simon I, Keshet I, Driscoll DJ, Nicholls RD, Cedar H (1993) Allele-specific replication timing of imprinted gene regions. Nature 364:459–463

Kurz A, Lampel S, Nickolenko JE, Bradl J, Brenner A, Zirbel RM, Cremer T, Lichter P (1996) Active and inactive genes localize preferentially in the periphery of chromosome territories. J Cell Biol 135:1195–1205

LaSalle JM, Lalande M (1996) Homologous association of oppositely imprinted chromosomal domains. Science 272:725–728

Lifschytz E, Hareven D (1982) Heterochromatin markers: arrangement of obligatory heterochromatin, histone genes and multisite gene families in the interphase nucleus of D. melanogaster. Chromosoma 86:443–455

Loidl J (1990) The initiation of meiotic chromosome pairing: the cytological view. Genome 33:759–770

Loidl J, Langer H (1993) Evaluation of models of homologue search with respect to their efficiency on meiotic pairing. Heredity 71:342–351

Luderus ME, den Blaauwen JL, de Smit OJ, Compton DA, van Driel R (1994) Binding of matrix attachment regions to lamin polymers involves single-stranded regions and the minor groove. Mol Cell Biol 14:6297–6305

Manuelidis L, Borden J (1988) Reproducible compartmentalization of individual chromosome domains in human CNS cells revealed by in situ hybridization and three-dimensional reconstruction. Chromosoma 96:396–410

Marshall WF, Dernburg AF, Harmon B, Agard DA, Sedat JW (1996) Interactions of chromatin with the nuclear envelope: positional determination within the nucleus in Drosophila melanogaster. Mol Biol Cell 7:825–842

Marshall WF, Fung JC, Sedat JW (1997a) Deconstructing the nucleus: global architecture from local interactions. Curr Op Gen Dev 7:259–263

Marshall WF, Straight AF, Marko JF, Dernburg AF, Swedlow J, Belmont A, Murray AW, Agard DA, Sedat JW (1997b) Interphase chromosomes undergo constrained diffusional motion in living cells. Curr Biol 7:930–939

Metz CW (1916) Chromosome studies on the Diptera. II. The paired association of chromosomes in the Diptera, and its significance. J Exp Zool 21:213–279

Murray AB, Davies HG (1979) Three-dimensional reconstruction of the chromatin bodies in the nuclei of mature erythrocytes from the newt Triturus cristatus: the number of nuclear envelope-attachment sites. J Cell Sci 35:59–66

Nagele R, Freeman T, McMorrow L, Lee HY (1995) Precise spatial positioning of chromosomes during prometaphase: evidence for chromosomal order. Science 270:1831–1835

Oegema K, Marshall WF, Sedat JW, Alberts BM (1997) Two proteins that cycle asynchronously between centrosomes and nuclear structures: Drosophila CP60 and CP190. J Cell Sci 110:1573–1583

Quick P (1980) Junctions of polytene chromosomes and the inner nuclear membrane. Experientia 36:456–457

Saitoh Y, Laemmli UK (1994) Metaphase chromosome structure: bands arise from a differential folding path of the highly AT-rich scaffold. Cell 76:609–622

Schardin M, Cremer T, Hager HD, Lang M (1985) Specific staining of human chromosomes in Chinese hamster x man hybrid cell lines demonstrates interphase chromosome territories. Hum Genet 71:281-287

Slatis HM (1955) A reconsideration of the *brown-dominant* position effect. Genetics 40:246-251

Stack SM, Brown DB, Dewey WC (1977) Visualization of interphase chromosomes. J Cell Sci 26:281-299

Sperling K, Luedtke EK (1981) Arrangement of prematurely condensed chromosomes in cultured cells and lymphocytes of the Indian muntjac. Chromosoma 83:541-553

Sukegawa J, Blobel G (1993) A nuclear pore complex protein that contains zinc finger motifs, binds DNA, and faces the nucleoplasm. Cell 72:29-38

Talbert PB, LeCiel CD, Henikoff S (1994) Modification of the Drosophila heterochromatic mutation brown[Dominant] by linkage alterations. Genetics 136:559-571

Taniura H, Glass C, Gerace L (1995) A chromatin binding site in the tail domain of nuclear lamins that interacts with core histones. J Cell Biol 131:33-44

Tartof KD, Henikoff S (1991) Trans-sensing effects from Drosophila to humans. Cell 65:201-203

van Dekken H, van Rotterdam A, Jonker R, van der Voort HTM, Brakenhoff GJ, Baumann JGJ (1990) Confocal microscopy as a tool for the study of the intranuclear topography of chromosomes. J Microsc 158:207-214

Vourc'h C, Taruscio D, Boyle AL, Ward DC (1993) Cell cycle-dependent distribution of telomeres, centromeres, and chromosome-specific subsatellite domains in the interphase nucleus of mouse lymphocytes. Exp Cell Res 205:142-151

Weiler KS, Wakimoto BT (1995) Heterochromatin and gene expression in Drosophila. Annu Rev Genet 29:577-605

Weintraub H, Groudine M (1976) Chromosomal subunits in active genes have an altered conformation. Science 193:848-856

Wu CT, Goldberg ML (1989) The Drosophila zeste gene and transvection. Trends Genet 5:189-194

Wu CT (1993) Transvection, nuclear structure, and chromatin proteins. J Cell Biol 120:587-590

Ye Q, Worman HJ (1996) Interactions between an integral protein of the nuclear envelope inner membrane and human chromodomain proteins homologous to Drosophila HP1. J Biol Chem 271:14653-14656

Yuan J, Simos G, Blobel G, Georgatos SD (1991) Binding of lamin A to polynucleosomes. J Biol Chem 266:9211-9215

Zalensky AO, Allen MJ, Kobayashi A, Zalenskaya IA, Balhron R, Bradbury EM (1995) Well-defined genome architecture in the human sperm nucleus. Chromosoma 103:577-590

Zirbel RM, Mathieu UR, Kurz A, Cremer T, Lichter P (1993) Evidence for a nuclear compartment of transcription and splicing located at chromosome domain boundaries. Chromosome Res 1:92-106

Appendix: Imprinted Genes and Regions in Mouse and Human

Colin V. Beechey

1
The Mouse Imprinting Map and Human Homologous Regions

1.1
Introduction

The imprinting maps (Figs. 1–7) illustrate regions of the mouse genome that have been screened for developmental anomalies that could be attributable to imprinting (imprinting effects, reviewed in Cattanach and Beechey 1997). The maps also include chromosomes 9,14,18 and 19 that contain imprinted genes but which have not shown clear imprinting effects in genetic tests. The map is a modified version of an earlier imprinting map (ref. 7: Cattanach and Beechey 1997), that was updated annually in *Mouse Genome* and is now available on the WWW (http://www.mgu.har.mrc.ac.uk). It is based on mouse genetic studies (Cattanach and Beechey 1997) in which Robertsonian (Rb) and reciprocal (T) translocations have been used to generate mice with uniparental disomies and uniparental duplications (partial disomies) of whole or selected chromosome regions respectively (see Sect. 1.2 Methodology). These mice have been screened for abnormal phenotypes (imprinting effects) and autosomal chromosome regions where such effects have been found are illustrated on both the G-banded and genetic maps. The imprinting effects discovered so far are diverse (Cattanach and Beechey 1997). They include early embryonic lethalities, late foetal lethalities, neonatal abnormalities often with inviability, growth effects and more subtle effects upon postnatal development such as those shown by the mouse model of Angelman syndrome (Cattanach et al. 1997). The chromosomal location of the 28 currently known autosomal imprinted genes are shown in bold and the repressed parental allele indicated. The majority of imprinted genes locate to regions associated with imprinting phenotypes, the exceptions being *Rasgrf1* on Chr 9, *Htr2a* on Chr 14, *Ins1* on Chr 19, and possibly *Impact* on Chr 18. Fifteen imprinted genes are located in two domains in central and distal Chr 7. Recent evidence suggests that other

Medical Research Council, Mammalian Genetics Unit, Harwell, Didcot, Oxfordshire OX11 0RD, UK

Key to mouse imprinting maps

MatDp = Maternal duplication
PatDp = Paternal duplication

established imprinting regions with phenotype

regions with no discernible imprinting phenotype

untested or not investigated for post-natal effects

◄ *Igf2 etc* = imprinted gene, maternally repressed

Mash2 etc ► = imprinted gene, paternally repressed

p etc = positions in cM of phenotypic marker loci and imprinted genes

T26H, = chromosome anomalies that define imprinting regions (see text)
Rb9Lub,
Del56H,
Is1Ct etc

Fig. 1. Key to mouse imprinting map

domains may be located on proximal Chr 11 (Cattanach et al. 1998) and distal Chr 2 (Kikyo et al. 1997; Williamson et al. 1998). The breakpoints of chromosome anomalies that currently define each imprinting region are illustrated, and their positions shown, on both the genetic and G-banded maps. The locations of imprinted genes and marker loci on the genetic map are taken from The mouse genome database (MGD 1998). Other chromosome anomalies such as deletions, (Del) and insertions (Is), that have been used to define or investigate imprinting regions are also identified, as are human homologues (Lyon et al. 1997) for mouse chromosome regions with imprinting effects or imprinted genes.

1.2
Methodology

1.2.1
Uniparental Disomies

Mice with both copies of a specific chromosome inherited from the same parent, or uniparental disomies, can be generated by intercrossing heterozygotes for Robertsonian translocations (nonhomologous chromo-

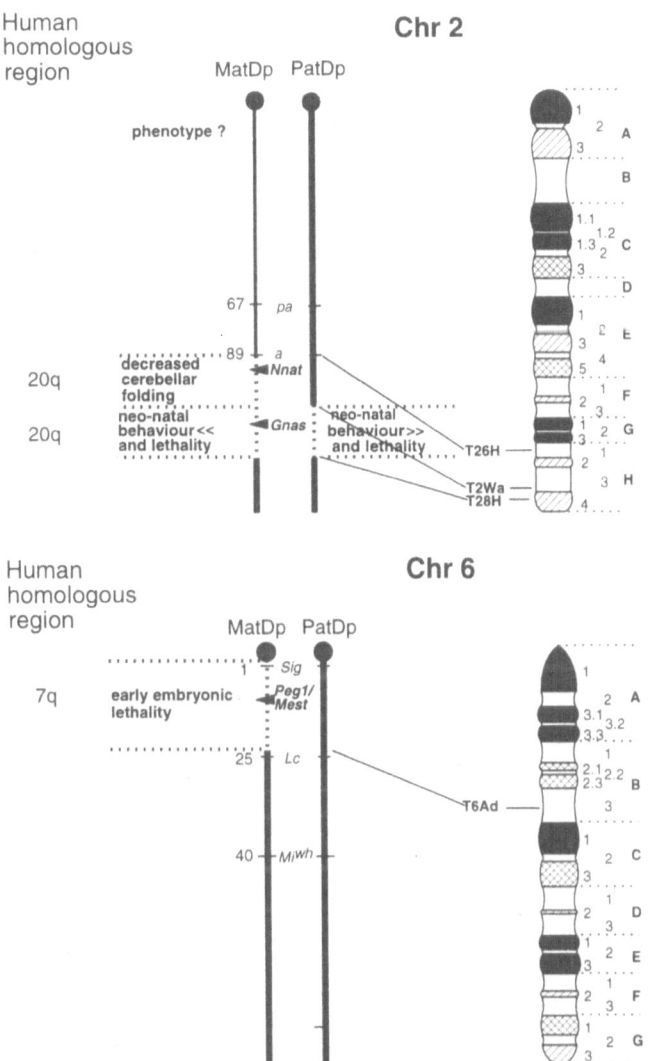

Fig. 2-7. Linkage and G-banded maps of mouse chromosomes 2, 6, 7, 9, 11, 12, 14, 17, 18, and 19. Regions that show developmental abnormalities (imprinting effects) with maternal or paternal duplication (MapDp or PatDp) are shown on Chrs 2, 6, 7, 11, 12, 17, 18, and 19. Chromosomes 9, 14 and 19 have not shown clear imprinting effects in genetic tests but contain imprinted genes. The breakpoints of chromosome anomalies that define imprinting regions are shown on both the linkage and G-banded maps. Human homologies to mouse imprinted regions are shown on the *left*

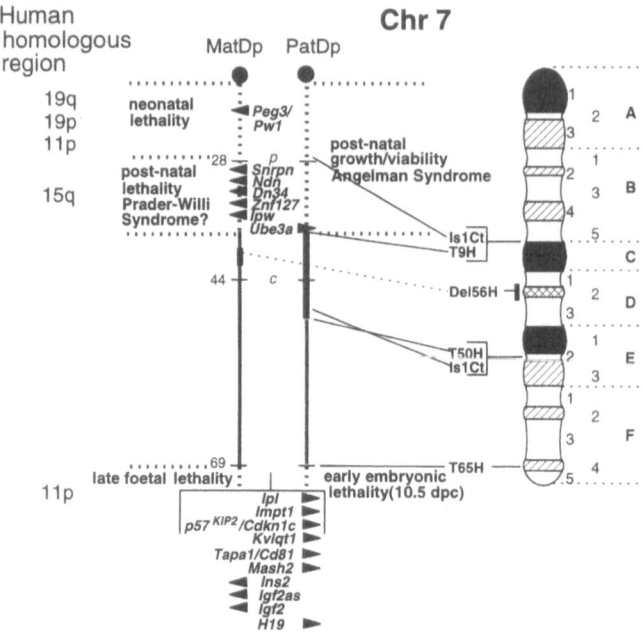

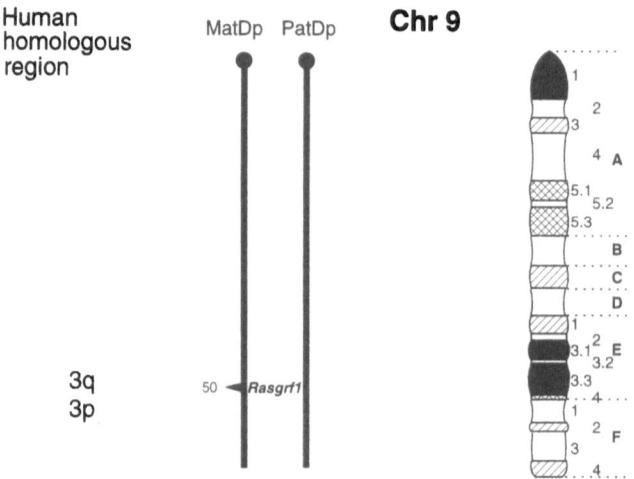

Fig. 2-7. *Continued*

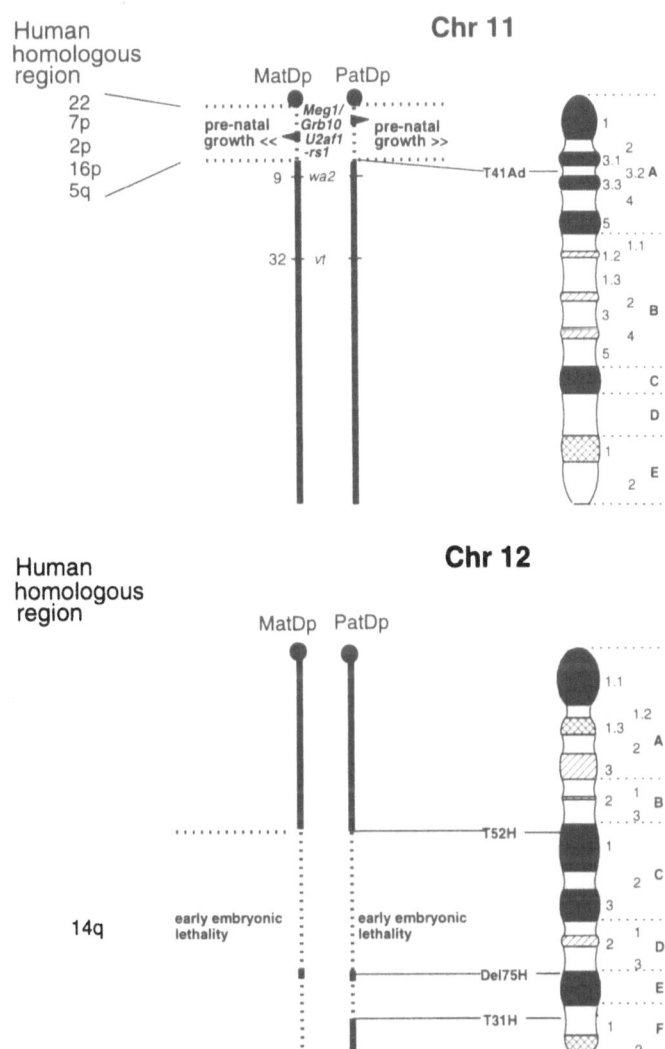

Fig. 2-7. *Continued*

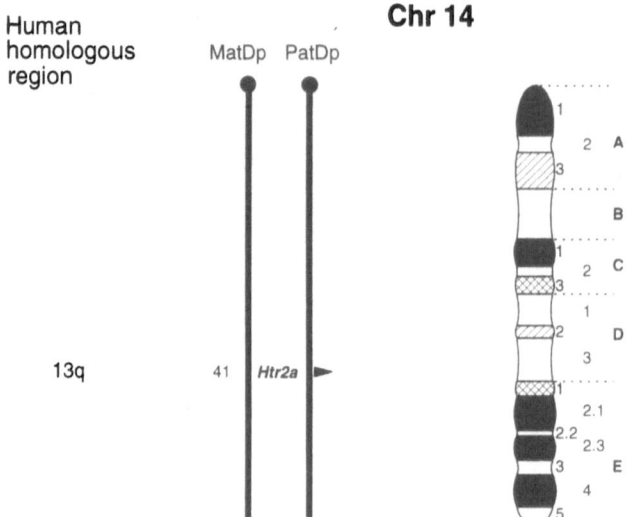

Fig. 2-7. *Continued*

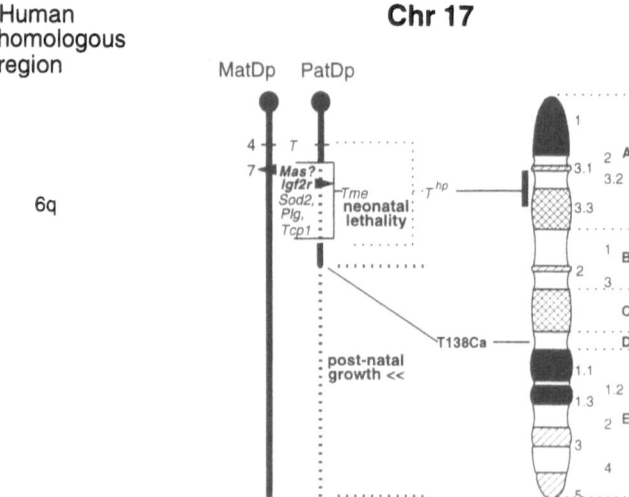

Fig. 2-7. *Continued*

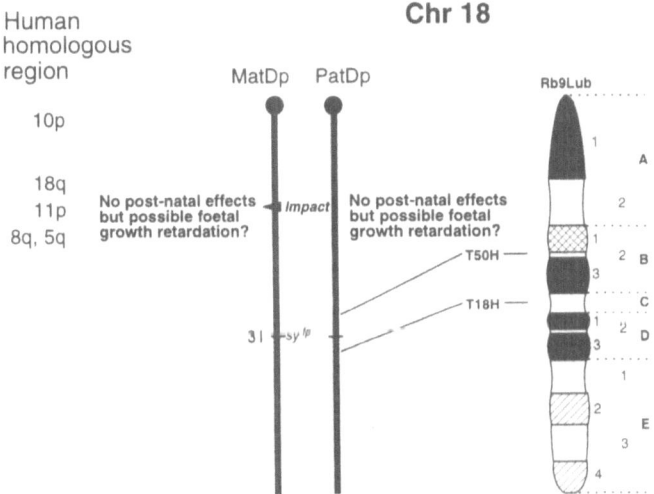

Fig. 2-7. *Continued*

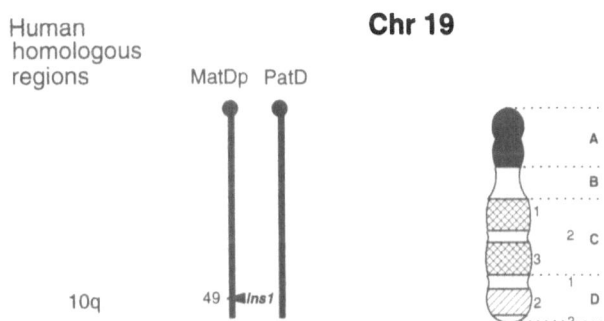

Fig. 2-7. *Continued*

somes fused at the centromere). The simplest system utilizes intercrosses between heterozygotes for a single Robertsonian translocation, preferably with a high rate of nondisjunction, one parent in the cross being homozygous for a suitable marker gene on one of the translocation chromosomes (Fig. 8). Nondisjunction during gamete formation in both parents leads to gametes that are disomic or nullisomic for one or other of the chromosomes involved in the translocation. These unbalanced gametes can usually complement each other to form chromosomally balanced viable zygotes with maternal (MatDi) or paternal disomy (PatDi). These can be recognized according to which parent was homozygous for the marker gene, and their phenotype studied for any imprinting effects. The frequency of each type of disomy among the total

Production of uniparental disomies

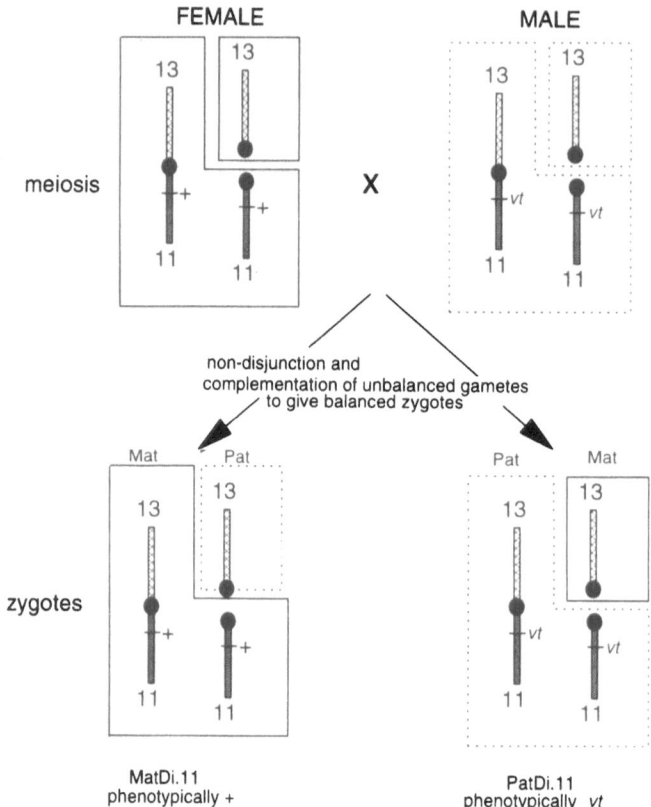

Fig. 8. Diagram of single Robertsonian system for generating uniparental disomies. The example presented illustrates how, as a result of meiotic nondisjunction in both parents heterozygous for a single (11:13) Robertsonian translocation, offspring with paternal disomy (PatDi) and maternal disomy (MatDi) for Chr 11 can be generated. If one parent, in this example the male, is homozygous for the visible Chr 11 marker gene *vt* and the female parent is homozygous normal, then PatDi offspring will show the *vt* phenotype. Offspring with maternal disomy (MatDi) are phenotypically wild type (+) like their normal sibs (not shown) but they can be detected in the reciprocal cross in which the female parent is homozygous for *vt*. Chr 13 uniparental disomies (not shown) are also generated but without Chr 13 marker genes these cannot be identified. Offspring with unbalanced chromosome complements, e.g., trisomy or monosomy for Chrs 11 and 13 are also generated (not shown) but die prenatally

progeny can reach about 5% but is commonly lower. Chromosomally unbalanced zygotes, such as monosomies and trisomies, are also generated but these usually die before birth.

A more efficient system (not shown) uses intercrosses between double heterozygotes for two Robertsonian translocations sharing a common arm (monobrachial homology) with one parent homozygous for a marker gene on the shared chromosome. In such crosses higher frequencies of nondisjunction are found than with single Robertsonian heterozygotes, allowing up to 10% of uniparental disomies to be recovered. Litter sizes can be severely reduced, however, due to the high frequencies of chromosomally unbalanced zygotes produced.

1.2.2
Uniparental Duplications (Partial Disomies)

Mice with uniparental inheritance of only parts of specific chromosomes, rather than whole chromosomes, can be generated by intercrossing heterozygotes for reciprocal translocations (reciprocal exchange of segments between nonhomologous chromosomes; see Figs. 9 and 10). Unbalanced gametes with duplications and/or deficiencies of chromosome regions are produced and, as with the Robertsonian system, these can usually complement each other to form chromosomally balanced zygotes. Marker genes on one or both of the chromosomes concerned allows mice with paternal (PatDp) or maternal duplication (MatDp) to be recognized, and their phenotype studied for imprinting effects.

This system for the production of PatDp and MatDp mice has two complicating factors, however. The first is that regions proximal and distal to the translocation breakpoints have to be considered separately. Recovery of mice with duplication of distal regions (Fig. 9) is dependent upon the common adjacent-1 meiotic segregation of gametes, giving a recovery rate of PatDp and MatDp offspring of around 16%. Recovery of mice with duplications of regions proximal to translocation breakpoints (Fig. 10) is dependent upon the much rarer adjacent-2 segregation. The frequency of this varies with each translocation and MatDp and PatDp offspring are recovered with a frequency of around only 5% or less.

The second complication is that with reciprocal translocations the other chromosome segments involved in the exchange must be considered. Thus, zygotes with MatDp for one chromosome involved in the translocation also have PatDp for the other. Consequently, whenever an imprinting effect is found with a reciprocal translocation it could be attributable to either chromosome involved. Which chromosome is responsible can only be determined by conducting further tests with other translocations involving the same regions of one or other chromosome. However, one useful feature is that because two chromosomes are involved, the marker genes used can be on either chromo-

Production of uniparental duplications for chromosome regions distal to translocation breakpoint

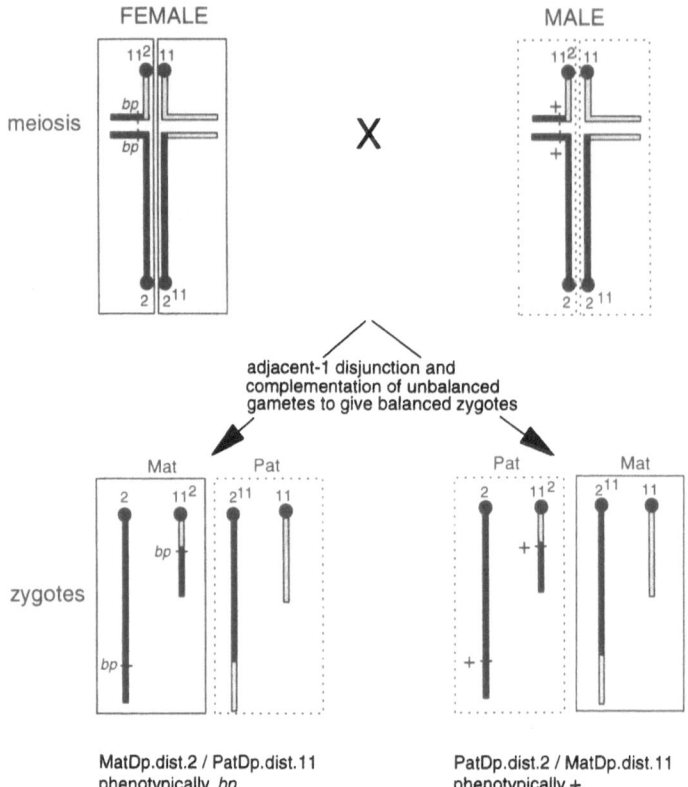

Fig. 9. Diagram of system for generating uniparental duplications (partial disomies) for chromosome regions distal to reciprocal translocation breakpoints. The reciprocal translocation shown in the example is between Chrs 2 and 11 and the region distal to the breakpoint is marked in the female parent with the visible Chr 2 marker gene *bp*, while the male parent is homozygous normal. Adjacent-1 disjunction at meiosis in both parents and complementation of unbalanced gametes can generate offspring with maternal and paternal duplication (MatDp and PatDp) for distal regions of Chrs 2 and 11. The class with MatDp.dist2 (and therefore also PatDp.dist11) are identifiable by their *bp* phenotype. The reciprocal cross in which the male parent is homozygous for *bp* similarly identifies PatDp.dist2 (and therefore also MatDp.dist11). The other balanced and unbalanced zygotes produced in these crosses are not shown

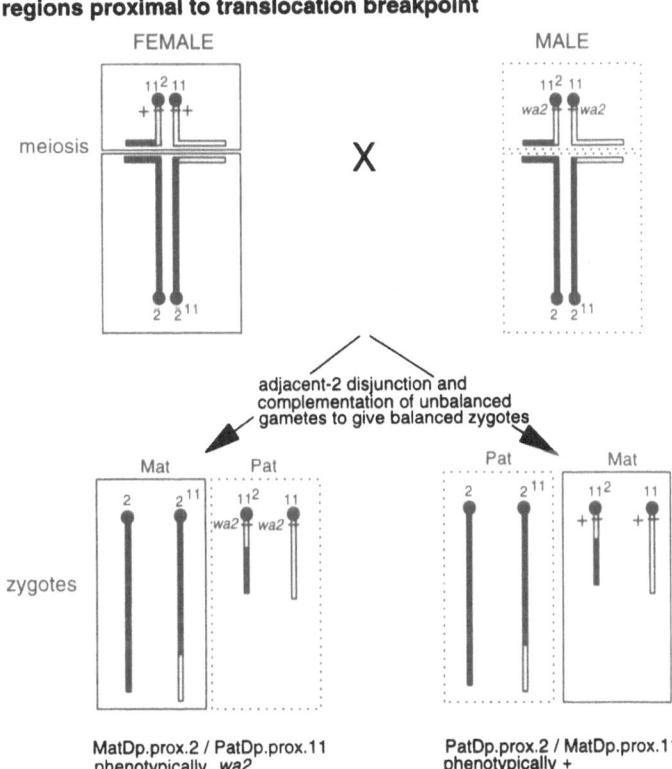

Production of uniparental duplications for chromosome regions proximal to translocation breakpoint

Fig. 10. Diagram of system for generating uniparental duplications (partial disomies) for chromosome regions proximal to reciprocal translocation breakpoints. The region proximal to the breakpoint of the (2;11) reciprocal translocation is marked in the male parent by the visible Chr 11 marker gene *wa2*, while the female parent is homozygous normal. Rare adjacent-2 disjunction at meiosis in both parents and complementation of unbalanced gametes can generate offspring with MatDp and PatDp for the proximal regions of Chrs 2 and 11. Progeny with PatDp.prox11 (and therefore also MatDp.prox2) are identifiable by their *wa2* phenotype. The reciprocal cross similarly identifies MatDp.prox11 (and therefore also PatDp.prox2). The other balanced and unbalanced zygotes produced in these crosses are not shown

some, not just on the chromosome of interest. More importantly, through the use of several translocations involving a chromosome with an imprinting effect, the extent and location of the region can be refined.

2
Imprinted Genes Identified in Mouse and Human

Table 1 lists the G-band positions of known mouse imprinting regions and describes the imprinting phenotypes resulting from maternal and paternal

Table 1. Mouse autosomal regions with imprinting phenotypes

Imprinting region	Boundaries on G-band map		Parental origin	Imprinting phenotype	Reference	Imprinted loci within region (not in order)
	Distal	Proximal				
Prox Chr 2?	Centromere (A1)	T26H (H1)	MatDp	?	6	–
Dist Chr 2	T26H (H1)	T2Wa (H3)	MatDp	Decreased cerebellar folding	43,91,92	Nnat
Dist Chr 2	T2Wa (H3)	T28H (H4)	MatDp	Hypokinetic behavior/lethality,	6	Gnas
Dist Chr 2	T28H (H4)	T2Wa (H3)	PatDp	Hyperkinetic behavior/lethality,	6	Gnas
Prox Chr 6	Centromere (A1)	T6Ad (B3)	MatDp	Early embryonic lethality	6	Peg1/Mest
Prox Chr 7	Centromere (A1)	Is1Ct/T9H (B5/C)	MatDp	Neonatal lethality	6	Peg3/Pw1
Prox Chr 7	Centromere (A1)	T9H (B5/C)	PatDp	Angelman syndrome homologue, postnatal growth/viability	6,10	Peg3/Pw1, Ube3a, Ipw, Znf127, Dn34, Ndn, Snrpn
Cent Chr 7	Is1Ct (B5/C)	T9H (B5/C)	MatDp	Prader-Willi syndrome homologue, postnatal lethality	6,8	Ube3a, Ipw, Znf127, Dn34, Ndn, Snrpn
Dist Chr 7	T65H (F4/5)	Telomere (F5)	MatDp	Late fetal lethality	6	H19, Igf2, Igf2as, Ins2, Mash2, P57^{KIP2}/Cdkn1c, Ipl/Tdag51, Impt1, Kvlqt1, Tapa1/Cd81
Dist Chr 7	T65H (F4/5)	Telomere (F5)	PatDp	Early embryonic lethality (10.5 dpc)	6,58	H19, Igf2, Igf2as, Ins2, Mash2, P57^{KIP2}/Cdkn1c, Ipl/Tdag51, Impt1, Kvlqt1, Tapa1/Cd81
Prox Chr 11	Centromere (A1)	T41Ad (A3.2)	MatDp	Reduced prenatal growth	6,9,11	U2af1-rs1, Meg1/Grb10
Prox Chr 11	Centromere (A1)	T41Ad (A3.2)	PatDp	Enhanced prenatal growth	6,9,11	U2af1-rs1, Meg1/Grb10
Dist Chr 12	T52H (C1)	Telomere (F2) but excluding prox E (De175H)	MatDp	Early embryonic lethality	6	–
Dist Chr 12	T52H (C1)	Telomere (F2) but excluding prox E (De175H) and F1/2 (T31H)	PatDp	Early embryonic lethality	6	–
Prox Chr 17	T^{hp} (A3)	T^{hp} (A3)	PatDp	Neonatal lethality (Tme)	6,78	Igf2r, (Mas possibly not imprinted)
Dist Chr 17	T138Ca (D)	Telomere (E5)	PatDp	Reduced postnatal growth	6	–
Chr 18	?	?	MatDp	No postnatal effects but possible fetal growth retardation?	6,7,68	Impact
Chr 18	?	?	PatDp	No postnatal effects but possible fetal growth retardation?	6,7,68	Impact

Table 2. Autosomal imprinted genes in mouse and human

| MOUSE | | | | | | HUMAN | | | |
Mouse loci	Chr	Repressed allele	Location cM	Location G-band	Reference	Human loci	Location	Repressed allele	Reference
Nnat	2	Mat	89	H1-H3	38,62,92	GNAS1	20q13	–	14,20,93
Gnas	2	Mat	104	H1/H3	43,62,90	MEST	7q32	Pat	46,67
Peg1/Mest	6	Mat	–	B1	40,46,79	PEG3	19q13.4	–	44
Peg3/Pwl	7	Mat	–	A5/B1	40,44,47,76	SNRPN	15q11-13	Mat	24,74
Snrpn	7	Mat	29	B5/C	8,51,62	NDN	15q11-13	Mat	34,55
Ndn	7	Mat	–	B5/C	55,88	DN34/FNZ127	15q11-13	Mat	7,16,25,66
Dn34	7	Mat	–	B5/C	37	ZNF127	15q11-13	Mat	7,16,66
Znf127/Zfp127	7	Mat	–	B5/C	37	IPW	15q11-13	Mat	73
Ipw	7	Mat	–	B5/C	89	UBE3A	15q11-13	Pat	64,45,57,82
Ube3a	7	Pat	28.6	B5/C	1,62	PAR1	15q11-13	Mat	81
–						PAR5	15q11-13	Mat	81
Ipl/Tdag51	7	Pat	–	F4/5	71	IPL/TDAG51	11p15.5	Pat	71
Impt1	7	Pat	–	F4/5	13	IMPT1	11p15.5	Pat	13
–	–	–				2G3.8	11p15.5	?	Tycko Chap. 7, this vol.
$P57^{KIP2}$/Cdkn1c	7	Pat	69.0	F4/5	5,30,62,95,96	CDKN1C	11p15.5	Pat	31,56,84
Kvlqt1	7	Pat	–	F4/5	26,75	KVLQT1	11p15.5	Pat	49,65
Tapa1/Cd81	7	?	69.2	F4/5	5,62	TAPA1/CD81	11p15.5	Pat	2,Tyto Chap. 10, this vol.
Mash2	7	Pat	69.3	F4/5	5,27,28,58,62	ASCL2/HASH2	11p15.5	–	62
Ins2	7	Mat	69.2	F4/5	5,22,62	INS2	11p15	–	23
Igf2as	7	Mat	69.0	F4/5	61	IGF2AS?	11p15	Mat	60
Igf2	7	Mat	69.0	F4/5	5,15,18,48,62,85	IGF2	11p15	Mat	21
H19	7	Pat	69	F4/5	4,5,52,62,85,86	H19	11p15	Pat	72,97
Rasgrf1	9*	Mat	50	–	19,62,70	–	–	–	–
Megl/Grb10	11	Pat	9.0	A1-A4	60,62	GRB10	7p11.2-12	–	35
U2af1-rs1	11	Mat	13	A1/3.2	11,17,32,33,62,63,83	U2AFBPL	5q23-31	N.I	39,69
Htr2a	14*	Pat	41	D2/E2	42,53,62	HTR2A	13q14	Pat	42
Mas	17	Pat?	7.4	–	62,80,87	Mas	6q	N.I	59,77
Igf2r	17	Pat	7.3	A3	3,62,80	IGF2R	6q25-q27	Pat	48,94
Impact	18	Mat	–	A2/B2	29	–	–	–	–
Ins1	19*	Mat	49.0	–	22,62	WT1	11p13	Pat	36
–	–	–	–			–	–	–	–

* Chromosomes with no discernible imprinting phenotype (see imprinting map).
Mat = maternal, Pat = tpaternal, N.I = not imprinted.

Physical Map of Human 11p15.5/Mouse 7F4/5

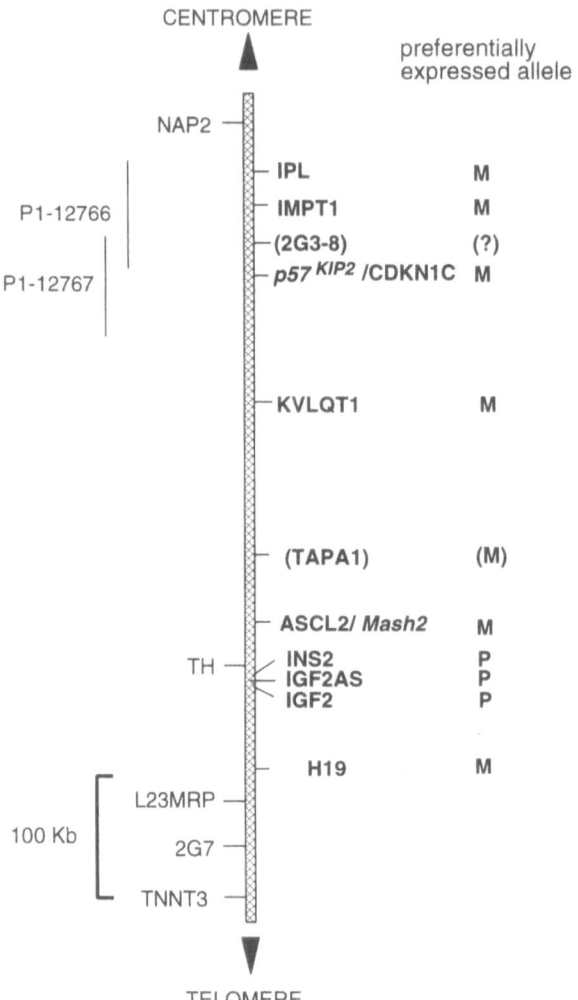

Fig. 11. Physical map of the human 11p15/mouse 7F4/5 imprinted domain, drawn approximately to scale from information supplied by B. Tycko (see Tycko, Chap. 7, this vol.) and Wolf Reik. Imprinted genes are shown in bold on the right and the parental origin (M maternal; P paternal) of the expressed allele is indicated. The orientation (telomere – centromere) of the homologous mouse imprinted domain has not yet been established

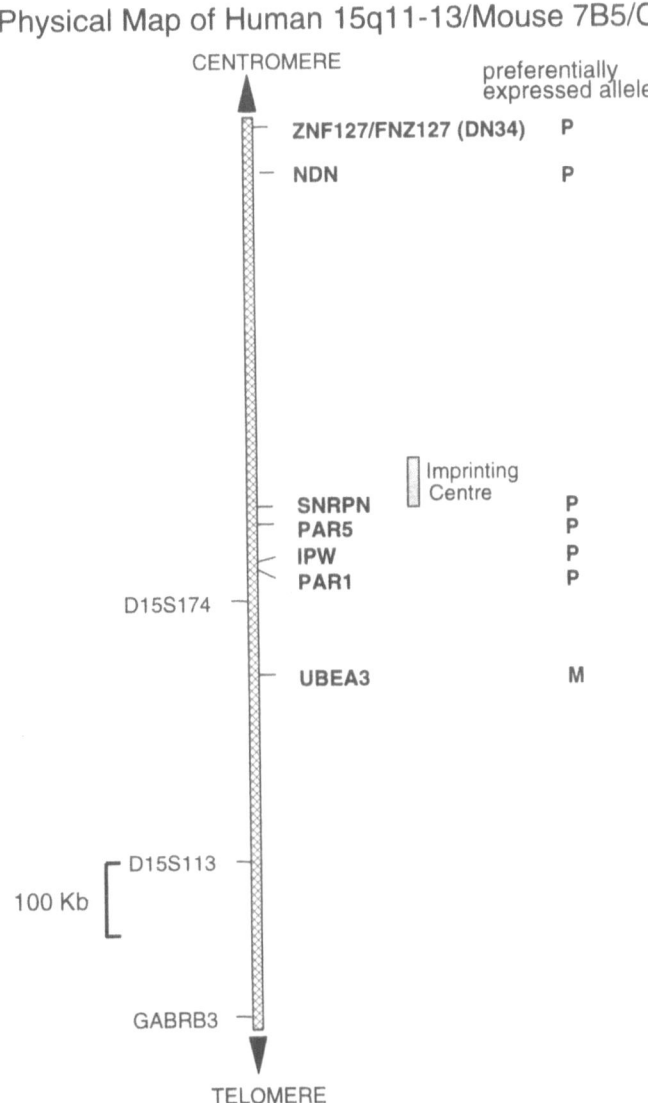

Physical Map of Human 15q11-13/Mouse 7B5/C

Fig. 12. Physical map of the human 15q11–13/Mouse 7B5/C imprinted domain, drawn approximately to scale from information supplied by B. Horsthemke and references [24, 34, 64, 81, and 25]. Imprinted genes are shown in bold on the right and the paternal origin of the expressed allele is indicated. The orientation (telomere-centromere) of the homologous mouse imprinted domain is probably the same

duplication (Mat and PatDp) for these regions, with references. Imprinted genes locating to these regions are also shown.

Table 2 lists known imprinted genes in mouse and human and includes chromosomal location, repressed allele and key references.

Acknowledgments. I would like to thank Drs Bruce Cattanach, Ted Evans and Jo Peters for helpful comments and Theresa Kent for extensive help in the preparation of this appendix.

References

1. Albrecht U, Sutcliffe JS, Cattanach BM, Beechey CV, Armstrong D, Eichele G, Beaudet AL (1997) Imprinted expression of the murine Angelman syndrome gene, *Ube3a*, in hippocampal and Purkinje neurons. Nat Genet 17:75–78
2. Andria ML, Hsieh CL, Oren R, Francke U, Levy S (1991) Genomic organisation and chromosomal localisation of the TAPA-1 gene. J Immunol 147:1030–1036
3. Barlow DP, Stoger R, Herrmann BG, Saito K, Schweifer N (1991) The mouse insulin-like growth factor type-2 receptor is imprinted and closely linked to the *Tme* locus. Nature 349:84–87
4. Bartolomei MS, Zemel S, Tilghman SM (1991) Parental imprinting of the mouse *H19* gene. Nature 351:153–55
5. Beechey CV, Ball ST, Townsend KMS, Jones J (1997) The mouse chromosome 7 distal imprinting domain maps to G-bands F4/F5. Mamm Genome 8:236–40
6. Cattanach BM, Beechey CV (1997) Genomic imprinting in the mouse: possible final analysis. In: Reik W, Surani A (eds) Genomic imprinting: frontiers in molecular biology. IRL Press, Oxford University Press, Oxford, 18:118–145
7. Cattanach BM, Beechey CV (1998) Unpublished results
8. Cattanach BM, Barr JA, Evans EP, Burtenshaw M, Beechey CV, Leff SE, Brannan CI, Copeland NG, Jenkins NA, Jones J (1992) A candidate mouse model for Prader-Willi syndrome which shows an absence of *Snrpn* expression. Nat Genet 2:270–274
9. Cattanach BM, Beechey CV, Rasberry C, Jones J, Papworth D (1996) Time of initiation and site of action of the mouse chromosome 11 imprinting effects. Genet Res Camb 68:35–44
10. Cattanach BM, Barr JA, Beechey CV, Martin J, Noebels J, Jones J (1997) A candidate model for Angelman syndrome in the mouse. Mamm Genome 7:472–478
11. Cattanach BM, Shibata H, Hayashizaki Y, Townsend KMS, Ball S, Beechey CV (1998) Association of a redefined proximal mouse chromosome 11 imprinting region and *U2afbp-rs/U2af1-rs1* expression. Cytogenet Cell Genet 80:41–47
12. Clayton-Smith J, Webb T, Cheng XJ, Pembrey ME, Malcolm S (1993) Duplication of chromosome 15 in the region 15q11-13 in a patient with developmental delay and ataxia with similarities to Angelman syndrome. J Med Genet 30:529–531
13. Dao D, Frank D, Qian N, O'Keefe D, Vosatka RJ, Walsh CP, Tycko B (1998) *Impt1*, an imprinted gene similar to polyspecific transporter and multiple-drug resistance genes. Human Mol Gen 7:597–608
14. Davies SJ, Hughes HE (1993) Imprinting in Albright's hereditary osteodystrophy. J Med Genet 30:101–103
15. DeChiara TM, Robertson EJ, Efstratiadis A (1991) Parental imprinting of the mouse insulin-like growth factor II gene. Cell 64:849–859

16. Driscoll DJ, Waters MF, Williams CA, Zori RT, Glenn CC, Avidano KM, Nicholls RD (1992) A DNA methylation imprint, determined by the sex of the parent, distinguishes the Angelman and Prader-Willi Syndromes. Genomics 13:917–924

17. Feil R, Boyano MD, Allen ND, Kelsey G (1997) Parental chromosome-specific chromatin confirmation in the imprinted U2af1-rs1 gene in the mouse. J Biol Chem 272:20893–20900

18. Ferguson-Smith AC, Cattanach BM, Barton SC, Beechey CV, Surani MA (1991) Embryological and molecular investigations of parental imprinting on mouse chromosome 7. Nature 351:667–670

19. Gariboldi M, Sturani E, Canzian F, Degregorio L, Manenti G, Dragani TA, Pierotti MA (1994) Genetic-mapping of the mouse cdc25mm gene, a ras-specific guanine nucleotide-releasing factor, to chromosome-9. Genomics 21:451–453

20. Gejman PV, Weinstein LS, Martinez M, Spiegel AM, Cao Q, Hsieh WT, Hoehe MR, Gershon ES (1991) Genetic-mapping of the gs-alpha subunit gene (GNAS1) to the distal long arm of chromosome 20 using a polymorphism detected by denaturing gradient gel-electrophoresis. Genomics 9:782–783

21. Giannoukakis N, Deal C, Paquette J, Goodyer CG, Polychronakos C (1993) Parental genomic imprinting of the human IGF2 gene. Nat Genet 4:98–101

22. Giddings SJ, King CD, Harman KW, Flood JF, Carnaghi LR (1994) Allele specific inactivation of insulin 1 and 2, in the mouse yolk sac, indicates imprinting. Nat Genet 6:310–313

23. Glaser T, Gerhard D, Jones C, Albritton L, Lalley P, Housman D (1985) A fine structure deletion map of chromosome 11p. (Abstract) Cytogenet Cell Genet 40:643

24. Glenn CC, Porter KA, Jong MTC, Nicholls RD, Driscoll DJ (1993) Functional imprinting and epigenetic modification of the human SNRPN gene. Hum Mol Genet 2:2001–2005

25. Glenn CC, Driscoll DJ, Yang TP, Nicholls RD (1997) Genomic imprinting: potential function and mechanisms revealed by the Prader-Willi and Angelman syndromes. Mol Hum Reprod 3:321–332

26. Gould TD, Pfeifer K (1998) Imprinting of mouse Kvlqt1 is developmentally regulated. Hum Mol Gen 7:483–487

27. Guillemot F, Nagy A, Auerbach A, Rossant J, Joyner AL (1994) Essential role of Mash2 in extraembryonic development. Nature 371:333–336

28. Guillemot F, Caspary T, Tilghman SM, Copeland NG, Gilbert DJ, Jenkins NA, Anderson DJ, Joyner AL, Rossant J, Nagy A (1995) Genomic imprinting of Mash2, a mouse gene required for trophoblast development. Nat Genet 9:235–242

29. Hagiwara Y, Hirai M, Nishiyama K, Kanazawa I, Ueda T, Sakaki Y, Ito T (1997) Screening for imprinted genes by allelic message display: identification of a paternally expressed gene Impact on mouse chromosome 18. Proc Natl Acad Sci USA 94:9249–9254

30. Hatada I, Mukai T (1995) Genomic imprinting of p57^{KIP2}, a cyclin-dependent kinase inhibitor, in mouse. Nat Genet 11:204–206

32. Hatada I, Sugama T, Mukai T (1993) A new imprinted gene cloned by a methylation-sensitive genome scanning method. Nucleic Acids Res 21:5577–5582

31. Hatada I, Inazawa J, Abe T, Nakayama M, Kaneko Y, Jinno Y, Niikawa N, Ohashi H, Fukushima Y, Iida K, Yutani C, Takahashi S, Chiba Y, Ohishi S, Mukai T (1996) Genomic imprinting of human p57^{KIP2} and its reduced expression in Wilms' tumors. Hum Mol Genet 5:783–788

33. Hayashizaki Y, Shibata H, Hirotsune S, Sugino H, Okazaki Y, Sasaki N, Hirose K., Imoto H, Okuizumi H, Muramatsu M, Komatsubara H, Shiroishi T, Moriwaki K, Katsuki M, Hatano N, Sasaki H, Ueda T, Mise N, Takagi N, Pass C, Chapman VM (1994) Identification of an imprinted U2af binding protein related sequence on mouse chromosome 11 using the RLGS method. Nat Genet 6:33–40.

34. Jay P, Rougeulle C, Massacrier A, Moncla A, Mattei MG, Malzac P, Roeckel N, Taviaux S, Lefranc JLB, Cau P, Berta P, Lalande M, Muscatelli F (1997) The human necdin gene, NDN, is maternally imprinted and located in the Prader-Willi Syndrome chromosomal region. Nat Genet 17:357–361

35. Jerome CA, Scherer SW, Tsui LC, Gietz RD, Triggsraine B (1997) Assignment of growth factor receptor-bound protein 10 (GRB10) to human chromosome 7p11.2-P12. Genomics 40:215–216

36. Jinno Y, Yun K, Nishiwaki K, Kubota T, Ogawa O, Reeve AE, Niikawa N (1994) Mosaic and polymorphic imprinting of the WT1 gene in humans. Nat Genet 6:305–309

37. Jones J, Cattanach BM (1998) Unpublished results

38. Kagitani F, Kuroiwa Y, Wakana S, Shiroishi T, Miyoshi N, Kobayashi S, Nishida M, Kohda T, Kaneko-Ishino T, Ishino F (1997) *Peg5/Neuronatin* is an imprinted gene located on sub-distal chromosome 2 in the mouse. Nucleic Acids Res 25:3428–3432

39. Kalcheva I, Plass C, Sait S, Eddy R, Shows T, Watkins-Chow D, Camper S, Shibata H, Ueda T, Takagi N, Hayashizaki Y, Chapman V (1995) Comparative mapping of the imprinted *U2afbpL* gene on mouse chromosome 11 and human chromosome 5. Cytogenet Cell Genet 68:19–24

40. Kaneko-Ishino T, Kuroiwa Y, Miyoshi N, Kohda T, Suzuki R, Yokoyama M, Viville S, Barton SC, Ishino F, Surani MA (1995) *Peg1/Mest* imprinted gene on chromosome 6 identified by cDNA subtraction hybridization. Nat Genet 11:52–59

41. Kato MV, Shimizu T, Nagayoshi M, Kaneko A, Sasaki MS, Ikawa Y (1996) Genomic imprinting of the human serotonin-receptor (HTR2) gene involved in development of retinoblastoma. Am J Hum Genet 59:1084–1090

42. Kato MV, Ikawa Y, Hayashizaki Y, Shibata H (1998) Paternal imprinting of mouse serotonin receptor 2A gene *Htr2* in embryonic eye: a conserved imprinting regulation on the *RB/Rb* locus. Genomics 47:146–148

43. Kikyo N, Williamson CM, John RM, Barton SC, Beechey CV, Ball ST, Cattanach BM, Surani MA, Peters J (1997) Genetic and functional analysis of neuronatin in mice with maternal or paternal duplication of distal Chr 2. Dev Biol 190:66–77

44. Kim J, Ashworth L, Branscomb E, Stubbs L (1997) The human homolog of a mouse-imprinted gene, *Peg3*, maps to a zinc finger gene-rich region of human chromosome 19q13.4. Genome Res 7:532–540

45. Kishino T, Lalande M, Wagstaff J (1997) UBE3A/E6-AP mutations cause Angelman Syndrome. Nat Genet 15:70–73

46. Kobayashi S, Kohda T, Miyoshi N, Kuroiwa Y, Aisaka K, Tsutsumi O, Kaneko-Ishino T, Ishino F (1997) Human PEG1/MEST, an imprinted gene on chromosome 7. Hum Mol Genet 6:781–786

47. Kuroiwa Y, Kaneko-Ishino T, Kagitani F, Kohda T, Li-Lan L, Tada M, Suzuki R, Yokoyama M, Shiroishi T, Wakana S, Barton SC, Ishino F, Surani MA (1996) *Peg3* imprinted gene on proximal chromosome 7 encodes for a zinc finger protein. Nat Genet 12:186–190

48. Laureys G, Barton DE, Ullrich A, Francke U (1988) Chromosomal mapping of the gene for the type-1 insulin-like growth- factor receptor cation-independent mannose 6-phosphate receptor in man and mouse. Genomics 3:224–229

49. Lee MP, Hu R-J, Johnson LA, Feinberg AP (1997) Human KVLQT1 gene shows tissue-specific imprinting and encompasses Beckwith-Wiedemann Syndrome chromosomal rearrangements. Nat Genet 15:181–185

50. Lefebvre L, Viville S, Barton SC, Ishino F, Surani MA (1997) Genomic structure and parent-of-origin-specific methylation of PEG1. Hum Mol Genet 6:1907–1915

51. Leff SE, Brannan CI, Reed ML, Ozcelik T, Francke U, Copeland NG, Jenkins NA (1992) Maternal imprinting of the mouse *Snrpn* gene and conserved linkage homology with Prader-Willi syndrome region of humans. Nat Genet 2:259–264

52. Leighton PA, Saam JR, Ingram RS, Stewart CL, Tilghman SM (1995) An enhancer deletion affects both *H19* and *Igf2* expression. Genes Dev 9:2080–2089

53. Liu J, Chen Y, Kozak CA, Yu L (1991) The 5-HT$_2$ serotonin receptor gene *htr-2* is tightly linked to *es-10* on mouse chromosome 14. Genomics 11:231–234

54. Lyon MF, Cocking Y, Gao X (1997) Mouse chromosome atlas. Mouse Genome 95:731–788

55. Macdonald HR, Wevrick R (1997) The necdin gene is deleted in Prader-Willi Syndrome and is imprinted in human and mouse. Hum Mol Genet 6:1873–1878

56. Matsuoka S, Thompson JS, Edwards MC, Barletta JM, Grundy P, Kalikin LM, Harper JW, Elledge SJ, Feinberg AP (1996) Imprinting of the gene encoding a human cyclin-dependent kinase inhibitor, p57^{KIP2}, on chromosome 11p15. Proc Natl Acad Sci USA 93:3026–3030

57. Matsuura T, Sutcliffe JS, Fang P, Galjaard RJ, Jiang YH, Benton CS, Rommens JM, Beaudet AL (1997) De novo truncating mutations in E6-Ap ubiquitin-protein ligase gene (UBE3A) in Angelman Syndrome. Nat Genet 15:74–77

58. McLaughlin KJ, Szabo P, Haegel H, Mann JR (1966) Mouse embryos with paternal duplication of an imprinted chromosome 7 region die at midgestation and lack placental spongiotrophoblast. Development 122:265–270

59. Miller N, McCann AH, O'Connell D, Pedersen IS, Spiers V, Gorey T, Dervan PA (1997) The MAS proto-oncogene is imprinted in human breast tissue. Genomics 46:509–512

60. Miyoshi N, Kuroiwa Y, Kohda T, Shitara H, Yonekawa H, Kawabe T, Hasegawa H, Barton SC, Surani MA, Kaneko-Ishino T, Ishino F (1998) Identification of the Meg1/Grb10 imprinted gene on mouse proximal chromosome 11, a candidate for the Silver-Russell Syndrome gene. Proc Natl Acad Sci USA 95:1102–1107

61. Moore T, Constancia M, Zubair M, Bailleul B, Feil R, Sasaki H, Reik W (1997) Multiple imprinted sense and antisense transcripts, differential methylation and tandem repeats in a putative imprinting control region upstream of mouse Igf2. Proc Natl Acad Sci USA 94:12509–12514

62. Mouse genome database (MGD) (1998) Mouse genome Informatics, Jackson Laboratory, Bar Harbor, Maine. World Wide Web (URL:http://www.informatics.jax.org) January 1998

63. Nabetani A, Hatada I, Morisaki H, Oshimura M, Mukai T (1997) Mouse U2af1-rs1 is a neomorphic imprinted gene. Mol Cell Biol 17:789–798

64. Nakao M, Sutcliffe JS, Durtschi B, Mutirangura A, Ledbetter DH, Beaudet AL (1994) Imprinting analysis of three genes in the Prader-Willi/Angelman region: SNRPN, E6-associated protein, and PAR-2 (D15s225e). Hum Mol Genet 3:309–315

65. Neyroud N, Tesson F, Denjoy I, Leibovici M, Donger C, Barhanin J, Faure S, Gary F, Coumel P, Petit C, Schwartz K, Guicheney P (1997) A novel mutation in the potassium channel gene KVLQT1 causes the Jervell and Lange-Nielsen cardioauditory syndrome. Nat Genet 15:186–189

66. Nicholls RD (1994) New insights reveal complex mechanisms involved in genomic imprinting. Am J Hum Genet 54:733–740

67. Nishita Y, Yoshida I, Sado T, Takagi N (1996) Genomic imprinting and chromosomal localization of the human MEST gene. Genomics 36:539–542

68. Oakey RJ, Matteson PG, Litwin S, Tilghman SM, Nussbaum RL (1995) Nondisjunction rates and abnormal embryonic development in a mouse cross between heterozygotes carrying a (7, 18) Robertsonian translocation chromosome. Genetics 141:667–674

69. Pearsall RS, Shibata H, Brozowska A, Yoshino K, Okuda K, Dejong PJ, Plass C, Chapman VM, Hayashizaki Y, Held WA (1996) Absence of imprinting in U2AFBPL, a human homolog of the imprinted mouse gene U2afbp-rs. Biochem Biophys Res Commun 222:171–177

70. Plass C, Shibata H, Kalcheva I, Mullins L, Kotelevtseva N, Mullins J, Kato R, Sasaki H, Hirotsune S, Okazaki Y, Held WA, Hayashizaki Y, Chapman VM (1996) Identification of Grf1 on mouse chromosome 9 as an imprinted gene by RLGS-M. Nat Genet 14:106–109

71. Qian N, Frank D, Okeefe D, Dao D, Zhao L, Yuan L, Wang Q, Keating M, Walsh C, Tycko B (1997) The IPL gene on chromosome 11p15.5 is imprinted in humans and mice and is similar to TDAG51, implicated in fas expression and apoptosis. Hum Mol Genet 6:2021–2029

72. Rachmilewitz J, Goshen R, Ariel I, Schneider T, de Groot N, Hochberg A (1992) Parental imprinting of the human H19 gene. FEBS Lett 309:25–28

73. Rachmilewitz J, Elkin M, Looijenga LHJ, Verkerk AJMH, Gonik B, Lustig O, Werner D, de Groot N, Hochberg A (1996) Characterization of the imprinted IPW gene: allelic expression in normal and tumorigenic human tissues. Oncogene 13:1687–1692

74. Reed ML, Leff SE (1994) Maternal imprinting of human SNRPN, a gene deleted in Prader-Willi syndrome. Nat Genet 6:163–167
75. Reik W, Paulsen M, Davies K, Franck O, Villar AJ, Walter J (1998) Pers Comm
76. Relais F,Wei X-J, Wu X, Sassoon DA (1998) *Peg3/Pwl* is an imprinted gene involved in the TNF-NFKB signal transduction pathway. Nat Genet 18:287–290
77. Riesewijk AM, Schepens MT, Mariman EM, Ropers HH, Kalscheuer VM (1996) The MAS protooncogene is not imprinted in humans. Genomics 35:380–382
78. Rogers I, Okano K, Varmuza S (1997) Paternal transmission of the mouse T^{hp} mutation is lethal in some genetic backgrounds. Dev Genet 20:23–28
79. Sado T, Nakajima N, Tada M, Takagi N (1993) A novel mesoderm-specific cdna isolated from a mouse embryonal carcinoma cell-line. Dev Growth Differ 35:551–560
80. Schweifer N, Valk PJM, Delwel R, Cox R, Francis F, Meier-Ewert S, Lehrach H, Barlow DP (1997) Characterization of the C3 YAC contig from proximal mouse chromosome 17 and analysis of allelic expression of genes flanking the imprinted *Igf2r* gene. Genomics 43:285–297
81. Sutcliffe JS, Nakao M, Christian S, Orstavik H, Tommerup N, Ledbetter DH, Beaudet AL (1994) Deletions of a differentially methylated CpG island at the SNRPN gene define a putative imprinting control region. Nat Genet 8:52–58.
82. Sutcliffe JS, Jiang Y-H, Galjaard R-J, Matsuura T, Fang P, Kubota T, Christian SL, Bressler J, Cattanach B, Ledbetter DH, Beaudet AL (1997) The E6-AP ubiquitin-protein ligase (UBE3A) gene is localized within a narrowed Angelman Syndrome critical region. Genome Res 7:368–377
83. Tada M, Tada T, Takagi N, Hayashizaki Y, Shibata H, Hirotsune S, Okazaki Y, Muramatsu M, Sasaki H, Ueda T, Chapman VM (1994) Localization of mouse imprinted gene *U2af1-rs1* to A3.2–4 band of chromosome 11 by FISH. Mamm Genome 5:655–657
84. Taniguchi T, Okamoto K, Reeve AE (1997) Human p57 (KIP2) defines a new imprinted domain on chromosome 11p but is not a tumour suppressor gene in Wilms tumour. Oncogene 14:1201–1206
85. Tilghman SM, Bartolomei MS, Webber AL, Brunkow ME, Saam J, Leighton PA, Pfeifer K (1995) In: Ohlsson R, Hall H, Ritzen M (eds) Genomic imprinting: causes and consequences. Cambridge University Press, Cambridge pp 170–181
86. Tremblay KD, Saam JR, Ingram RS, Tilghman SM, Bartolomei MS (1995) A paternal-specific methylation imprint marks the alleles of the mouse *H19* gene. Nat Genet 9:407–413
87. Villar AJ, Pedersen RA (1994) Parental imprinting of the *Mas* protooncogene in mouse. Nat Genet 8:373–379
88. Watrin F, Roeckel N, Lacroix L, Mignon C, Mattei MG, Disteche C, Muscatelli F (1997) The mouse necdin gene is expressed from the paternal allele only and lies in the 7c region of the mouse chromosome 7, a region of conserved synteny to the human Prader-Willi Syndrome region. Eur J Hum Genet 5:324–332
89. Wevrick R, Francke U (1997) An imprinted mouse transcript homologous to the human imprinted in Prader-Willi syndrome (IPW) gene. Hum Mol Genet 6:325–332
90. Williamson CM, Schofield J, Dutton ER, Seymour A, Beechey CV, Edwards YH, Peters J (1996) Glomerular-Specific imprinting of the mouse *Gs-Alpha* gene: how does this relate to hormone resistance in Albright hereditary osteodystrophy. Genomics 36:280–287
91. Williamson CM, Beechey CV, Ball ST, Dutton ER, Tease C, Peters J (1997) A new imprinting region in Band H1 of mouse chromosome 2. Mouse Genome 95:687–688
92. Williamson CM, Beechey CV, Ball ST, Dutton ER, Cattanach BM, Tease C, Ishino F, Peters J (1998) Localisation of the imprinted gene neuronatin, *Nnat*, to band H1 on mouse chromosome 2 identifies a new imprinting region. Cytogenet Cell Genet 81:73–78
93. Wilson LC, Ooude-Luttikhuis MEM, Clayton PT, Fraser WD, Trembath RC (1994) Parental origin of Gsalpha gene mutations in Albright's hereditary osteodystrophy. J Med Genet 31:835–839

94. Xu YQ, Goodyer CG, Deal C, Polychronakos C (1993) Functional polymorphism in the parental imprinting of the human IGF2R gene. Biochem Biophys Res Commun 197:747–754

95. Yan YM, Lee MH, Massague J, Barbacid M (1997) Ablation of the Cdk inhibitor *P57(Kip2)* results in increased apoptosis and delayed differentiation during mouse development. Genes Dev 11:973–983

96. Zhang P, Liegeois NJ, Wong C, Finegold M, Hou H, Thompson JC, Silverman A, Harper JW, DePinho RA, Elledge SJ (1997) Altered cell differentiation and proliferation in mice lacking *p57^{KIP2}* indicates a role in Beckwith-Wiedemann syndrome. Nature 387:151–158

97. Zhang YH, Tycko B (1992) Monoallelic expression of the human H19 gene. Nat Genet 1:40–44.

Index